T3-BPW-973

IEE CONTROL ENGINEERING SERIES 49

POLYNOMIAL METHODS IN OPTIMAL CONTROL AND FILTERING

Other volumes in this series:

Volume 1 **Multivariable control theory** J. M. Layton
Volume 2 **Elevator traffic analysis, design and control** G. C. Barney and S. M. dos Santos
Volume 3 **Transducers in digital systems** G. A. Woolvet
Volume 4 **Supervisory remote control systems** R. E. Young
Volume 5 **Structure of interconnected systems** H. Nicholson
Volume 6 **Power system control** M. J. H. Sterling
Volume 7 **Feedback and multivariable systems** D. H. Owens
Volume 8 **A history of control engineering, 1800-1930** S. Bennett
Volume 9 **Modern approaches to control system design** N. Munro (Editor)
Volume 10 **Control of time delay systems** J. E. Marshall
Volume 11 **Biological systems, modelling and control** D. A. Linkens
Volume 12 **Modelling of dynamical systems—1** H. Nicholson (Editor)
Volume 13 **Modelling of dynamical systems—2** H. Nicholson (Editor)
Volume 14 **Optimal relay and saturating control system synthesis** E. P. Ryan
Volume 15 **Self-tuning and adaptive control: theory and application** C. J. Harris and S. A. Billings (Editors)
Volume 16 **Systems modelling and optimisation** P. Nash
Volume 17 **Control in hazardous environments** R. E. Young
Volume 18 **Applied control theory** J. R. Leigh
Volume 19 **Stepping motors: a guide to modern theory and practice** P. P. Acarnley
Volume 20 **Design of modern control systems** D. J. Bell, P. A. Cook and N. Munro (Editors)
Volume 21 **Computer control of industrial processes** S. Bennett and D. A. Linkens (Editors)
Volume 22 **Digital signal processing** N. B. Jones (Editor)
Volume 23 **Robotic technology** A. Pugh (Editor)
Volume 24 **Real-time computer control** S. Bennett and D. A. Linkens (Editors)
Volume 25 **Nonlinear system design** S. A. Billings, J. O. Gray and D. H. Owens (Editors)
Volume 26 **Measurement and instrumentation for control** M. G. Mylroi and G. Calvert (Editors)
Volume 27 **Process dynamics estimation and control** A. Johnson
Volume 28 **Robots and automated manufacture** J. Billingsley (Editor)
Volume 29 **Industrial digital control systems** K. Warwick and D. Rees (Editors)
Volume 30 **Electromagnetic suspension—dynamics and control** P. K. Sinha
Volume 31 **Modelling and control of fermentation processes** J. R. Leigh (Editor)
Volume 32 **Multivariable control for industrial applications** J. O'Reilly (Editor)
Volume 33 **Temperature measurement and control** J. R. Leigh
Volume 34 **Singular perturbation methodology in control systems** D. S. Naidu
Volume 35 **Implementation of self-tuning controllers** K. Warwick (Editor)
Volume 36 **Robot control** K. Warwick and A. Pugh (Editors)
Volume 37 **Industrial digital control systems (revised edition)** K. Warwick and D. Rees (Editors)
Volume 38 **Parallel processing in control** P. J. Fleming (Editor)
Volume 39 **Continuous time controller design** R. Balasubramanian
Volume 40 **Deterministic control of uncertain systems** A. S. I. Zinober (Editor)
Volume 41 **Computer control of real-time processes** S. Bennett and G. S. Virk (Editors)
Volume 42 **Digital signal processing: principles, devices and applications** N. B. Jones and J. D. McK. Watson (Editors)
Volume 43 **Trends in information technology** D. A. Linkens and R. I. Nicolson (Editors)
Volume 44 **Knowledge-based systems for industrial control** J. McGhee, M. J. Grimble and A. Mowforth (Editors)
Volume 45 **Control theory—a guided tour** J. R. Leigh
Volume 46 **Neural networks for control and systems** K. Warwick, G. W. Irwin and K. J. Hunt (Editors)
Volume 47 **A history of control engineering, 1930-1956** S. Bennett
Volume 48 **MATLAB toolboxes and applications** A. J. Chipperfield and P. J. Fleming (Editors)

POLYNOMIAL METHODS IN OPTIMAL CONTROL AND FILTERING

Edited by
K. J. HUNT

Peter Peregrinus Ltd., on behalf of the Institution of Electrical Engineers

Published by: Peter Peregrinus Ltd., London, United Kingdom

Peter Peregrinus Ltd.,
Michael Faraday House,
Six Hills Way, Stevenage,
Herts. SG1 2AY, United Kingdom

British Library Cataloguing in Publication Data

A CIP catalogue record for this book
is available from the British Library

ISBN 0 86341 295 5

Printed in England by Short Run Press Ltd., Exeter

Contents

List of Authors

Anders Ahlén
Automatic Control and Systems Analysis Group
Department of Technology
Uppsala University
Box 27
S-751 03 Uppsala
Sweden
aa@gimli.teknikum.uu.se

Alessandro Casavola
Dipartimento di Sistemi e Informatica
Universita di Firenze
Via di S. Marta
3 - 50139 Firenze
Italy
udini@ifiidg.bitnet

Michael Grimble
Industrial Control Centre
Department of Electronic and Electrical Engineering
University of Strathclyde
Glasgow G1 1QE
Scotland
grimble@uk.ac.strath.icu

Kenneth Hunt
Control Group
Department of Mechanical Engineering
University of Glasgow
Glasgow G12 8QQ
Scotland
ken@uk.ac.gla.eng.ctrl

Jan Ježek
Institute of Information Theory and Automation
Czechoslovak Academy of Sciences
Czechoslovakia
jezek@cspgas11.bitnet

Vladimir Kučera
Institute of Information Theory and Automation
Czechoslovak Academy of Sciences
182 08 Prague 8
Czechoslovakia
kucera@cspgas11.bitnet

Jean Jacques Loiseau
Laboratoire d'Automatique de Nantes
URA CNRS 823
Ecole Centrale de Nantes
44072 Nantes Cedex 03
France
loiseau@lan01.ensm-nantes.fr

Edoardo Mosca
Dipartimento di Sistemi e Informatica
Universita di Firenze
Via di S. Marta
3 - 50139 Firenze
Italy
udini@ifiidg.bitnet

Michael Šebek
Institute of Information Theory and Automation
Czechoslovak Academy of Sciences
182 08 Prague 8
Czechoslovakia
sebek@cspgas11.bitnet

Mikael Sternad
Automatic Control and Systems Analysis Group
Department of Technology
Uppsala University
Box 27
S-751 03 Uppsala
Sweden
ms@frej.teknikum.uu.se

Petr Zagalak
Institute of Information Theory and Automation
Czechoslovak Academy of Sciences
182 08 Prague 8
Czechoslovakia
zagalak@cspgas11.bitnet

Preface

This book provides a broad coverage of the *Polynomial Equation Approach* in a range of linear control and filtering problems. Kučera's book of 1979, *Discrete Linear Control: the polynomial equation approach*, is the landmark work in this area. Although polynomials had been used much earlier in control synthesis problems, *Discrete Linear Control* was the first text to rigorously formalise the approach; it incorporated Kučera's earlier work on parameterising all stabilising feedback controllers and, directly building on this most fundamental result, derived the design equations for some key multivariable regulation problems. The development, resting entirely on polynomial matrix algebra, was completed by the inclusion of numerical algorithms. It is on these foundations that the range of subsequent work described in this volume lies; we describe the extension of Kučera's ideas into a very wide spectrum of control and filtering problems.

The principal feature of the approach is the description of systems in *fractional form* using transfer-functions. This representation leads quite naturally and directly to the parameterisation of all 'acceptable' feedback controllers for a given problem in the form of a *Diophantine equation* over polynomials. In the Polynomial Equation Approach this direct parameterisation is explicitly carried through to the synthesis of controllers and filters, and further, to the computer implementation of numerical algorithms. A detailed discussion of these motivating factors is given in chapter 1 which, fittingly, has been written by Vladimir Kučera.

Direct polynomial methods have received comparatively little attention in relation to the alternative Wiener-Hopf transfer-function method, and to state-space methods which rely on Riccati equations. It is the aim of this book to demonstrate the power and breadth of polynomial methods in control and filtering. Although our coverage is broad, we do restrict our attention to linear time-invariant systems - a brief discussion of the extension to non-linear and time-varying systems, together with the appropriate references, is given in the introductory chapter by Kučera.

The book is likely to be of interest to students, researchers and engineers with some control and systems theory, or signal processing, background. It could be used as the basis of a graduate level course in optimal control and filtering. The book proceeds from the necessary background material presented at a tutorial level, through recent theoretical and practical developments, to a detailed presentation of numerical algorithms. The structure of the book reflects this development.

Following the introductory chapter the book is structured into five principal parts which deal with the following topics:

1. H_2 (LQ) control optimisation and self-tuning - chapters 2 and 3,
2. H_∞ control optimisation - chapter 4,
3. H_2 and H_∞ filtering problems - chapters 5 and 6,
4. system structure - chapters 7 and 8,
5. numerical algorithms - chapters 9 and 10.

In chapter 2 Hunt, Šebek and Kučera discuss the polynomial approach to H_2 (LQ) control optimisation. They first review results for the explicit problems of regulation, tracking and feedforward. Then, the solution to the very general H_2 standard problem is presented. The focus of chapter 3, by Sternad and Ahlén, is on the application of LQ control in the self-tuning context. The integration of recursive identification algorithms to achieve this is discussed. They also describe a new variational procedure for the derivation of LQ controllers for regulation, tracking and feedforward problems. A concise discussion of practical issues relating to the implementation of polynomial-based controllers is given.

In chapter 4, Casavola and Mosca move to H_∞ control optimisation. Their use of an H_∞ criterion in the feedback part of the controller is motivated by the desire for robustness in the face of unstructured plant uncertainty. The controller is enhanced with an H_2 feedforward controller for tracking stochastic reference signals; thus, these authors tackle a mixed H_2/H_∞ problem.

Ahlén and Sternad apply the polynomial approach to a range of optimal filtering problems in chapter 5. Their derivations are based on the variational procedure outlined previously in chapter 3. They begin by formulating a general problem within which a range of specific filtering problems lie. Filtering problems considered include: filtering, prediction and smoothing; multivariable deconvolution and equalisation; numerical differentiation and state estimation; and decision feedback equalisation. In chapter 6, Grimble extends the range of filtering problems covered by the polynomial approach by using an H_∞ criterion. Again, Grimble's development is based upon a general standard formulation which covers many practical cases of interest.

Chapters 7 and 8 utilise the polynomial approach to examine some deeper structural issues in systems theory. In chapter 7, Šebek describes the extension of polynomial techniques to n-D systems. This work has particular relevance for applications such as image processing, seismic data processing and control of time delay systems. Chapter 8, by Zagalak, Kučera and Loiseau, deals with the assignment of invariant factors via state feedback in linear descriptor systems. The chapter first reviews some theoretical concepts relating to descriptor systems. It then develops, using the polynomial approach, some fundamental theorems using particular forms of state feedback.

The final part of the book considers numerical algorithms for a variety of polynomial operations underlying the design methods covered in the preceding chapters. Ježek, in chapter 9, explains the process of abstraction from systems and control theory to algebra. He then describes the fundamental notions of polynomial algebra, and the various forms of equation arising from control and filtering problems. Algorithms for their solution are described. Finally, in chapter 10, Šebek describes algorithms for J-spectral factorisation. Šebek motivates his presentation by considering the polynomial approach to H_∞ optimal regulation, and in particular the standard problem; J-spectral factorisation is a fundamental operation which underlies the solution of the standard H_∞ problem.

The book provides a broad coverage of the use of polynomial methods in control and filtering. It is hoped that the reader can follow-up the references given in each chapter to pursue his or her own special interests.

Editorial Note

This book was printed by the publisher using camera-ready copy supplied by the editor. Nine of the ten chapters were typeset initially by the authors using LaTeX and sent electronically to the editor. The editor typeset and edited the final copy for the book after including those chapters within the LaTeX `book` documentstyle.

Kenneth Hunt
Glasgow: September 22, 1992

Chapter 1

The Algebraic Approach to Control System Design

V. Kučera

1.1 Introduction

One of the most elegant and powerful approaches to the design of control systems is based on the properties of rings. The theory of rings is a cornerstone of abstract algebra, and it made its way to control theory during the last three decades.

We shall first explain the basic ideas of this approach and then trace back their origins in the control literature. This will hopefully enhance the tutorial value of this contribution and help to show many results, obtained in isolation, to be closely related.

A typical control problem can be stated as follows. Given a plant, together with a set of performance specifications, design a controller such that the resulting system meets the performance requirements. These requirements are of two types: (i) the system is "desirable" in some sense and (ii) the system is "optimal" in some sense.

In the first category, one is given a plant, together with a specification of a class of desirable systems, and any controller is deemed to be acceptable so long as it results in a system that lies within the desirable class. A typical example is the problem of stabilizing an unstable plant.

In the second category, one is given a plant, a class of desirable systems, as well as a performance measure, and the objective is to choose, among all the controllers that result in a desirable system, one that optimizes the performance measure. Typical examples include optimal regulation, tracking, and disturbance rejection.

The first step of the algebraic approach is to describe the systems under consideration by *transfer functions.* The class of desirable systems is then specified by a set of desirable transfer functions. If the sums and products of desirable transfer functions (or parallel and series connections of the underlying systems) are once again desirable, then the set is a *ring.*

The second step is to fix the control system configuration. For many reasons we shall concentrate on *feedback systems.* The transfer function of a feedback system is a ratio that involves the two component transfer functions. This in turn specifies the class of transfer functions in which the plant and the controller transfer functions may lie in order for the feedback system to be desirable: it consists of all *ratios* of desirable transfer functions.

The third step is to find an acceptable controller for the given plant. Provided the plant transfer function belongs to the *field of fractions* associated with the ring of desirable transfer functions, the desirability of the resulting feedback system corresponds to a divisibility condition in the ring. This condition leads to a linear equation in the ring and its solution defines an acceptable controller. Such equations are known as *diophantine equations.*

By now we have solved the problem of type (i). In fact we have got much more: not only does the diophantine equation provide one acceptable controller but the solution set of this equation determines the set of all acceptable controllers for the given plant. And since the solution set can be *parametrized* in a simple manner so too can the set of all acceptable controllers.

We are now in a position to solve the problems of type (ii) in a systematic manner. The starting point is the parametrized set of acceptable controllers. From this we deduce a parametrization of the performance measure. The value of the parameter that optimizes the performance measure will then determine the optimal controller. The point is that the measure of performance depends in a much simpler manner on the parameter than on the controller itself.

The objective of this contribution is to explain the fundamentals and the rationale of the algebraic approach and guide the reader through various examples of control system design. The emphasis is on the ideas, not on the technicalities. This introductory exposition will hopefully motivate the reader to pursue the details in the references and help him or her to appreciate the power and elegance of the algebraic methods in control system design.

1.2 Historical Notes

Quite naturally, these ideas originated in the realm of linear time-invariant and finite-dimensional systems that evolve in *discrete time.* The correspondence of time-domain and transfer-function quantities is particularly simple and transparent in this case, and can be captured by polynomial rings [20] . The first attempts to use *polynomial equations* in the design of control systems date back to [49], [43], [1] and [40]. The algebraic background of these polynomial manipulations was explained later in [24], [25], [28] and the polynomial equation approach to the design of discrete-time linear systems was fully developed in [29].

The counterpart results for *continuous-time* systems followed in due course [42], [14], [39], [2] and [31]. A particular impetus was provided by the descriptor (or semi-state, or implicit, or singular) systems, which can easily be studied by means of transfer functions and which have drawn attention to the impulsive behaviour of linear systems [46], [36] and [33], [32]. The algebraic approach has reached a state of maturity in [48]; this work shows the relevance of the ring of *stable rational functions* rather than that of polynomials for the design of continuous-time systems.

The landmark contributions along this way, however, are [52] and [7]. The former work provides an explicit parametrization of all controllers that stabilize a given plant and makes use of it to optimize the system performance. The latter work reduces the parametrization problem to its algebraic essentials and shows how to set up the problem in an appropriate ring.

Another step forward in extending the algebraic approach was the case of *time-varying* linear systems. The field of coefficients is replaced by a differential ring of functions over an interval of time [22], [18]. The resulting rings are no longer commutative, for multiplication and differentiation do not commute. Nevertheless, the same master ideas can be used to reduce the analysis and synthesis of control systems to the analysis and solution of linear diophantine equations [18]. Some control strategies, when applied over a *finite horizon*, lead to time-varying control laws no matter whether the properties of the underlying system vary in time or not. So these control problems naturally fall within this framework.

A more delicate extension of this algebraic approach is needed to study *infinite-dimensional* linear systems. The main obstacle is that desirable transfer functions of these systems form more general rings for which coprime fractions cannot be assumed a priori to exist [21], [3], [47]. If they do exist, however, one can apply the algebraic approach to advantage [48]. New control strategies for *distributed* systems [4], [17] and for *time-delay* systems [23], [44] were formulated and solved in this way. Another class of systems that can be studied by introducing more general rings are *n-D systems*, or systems whose behaviour evolves in a vector-valued time set [19]. The diophantine equations over the ring of polynomials in n indeterminates were studied in [45].

Perhaps the most surprising feature of the fractional approach is that it is not restricted to linear systems. In fact, suitable fractions for the input-output maps of *non-linear* systems were defined in [13] and used to solve some control problems of practical importance, including that of stabilization [12], [38].

Tracing back the origins of the algebraic approach to control system synthesis, we have encountered a large body of mathematical and control literature. Some results, obtained in isolation, are now seen to be related. The points of view presented here first appeared in [35] and reflect the author's own experience in this area.

1.3 Algebraic Preliminaries

We explain briefly some notions from algebra which will be used throughout the chapter. For a rigorous and comprehensive treatment the reader is referred to [16] or [53] . The appendices in [48] make a good introduction to the subject.

A *ring* is a set equipped with two operations, addition and multiplication, that forms a commutative group with respect to addition, a semigroup with respect to multiplication and satisfies distributive laws connecting the two operations. If the multiplication is commutative, the ring is called commutative. If the ring has an identity, then its elements having a multiplicative inverse are called *units* of the ring. A ring in which every non-zero element is a unit is called a *field.*

Examples of fields: real numbers R, rational functions $R(q)$ in the indeterminate q over R.

Examples of rings: integers Z, rational functions from $R(q)$ that are analytic in a subset of the complex plane, such as

proper rational functions $R_p(q)$, analytic at $q = \infty$;

proper and Hurwitz-stable rational functions $R_H(q)$, analytic in $Re\ q \geq 0$ including $q = \infty$;

proper and Schur-stable rational functions $R_S(q)$, analytic in $|q| \geq 1$;

proper and finite-expansion rational functions $R_f(q)$, analytic for every $q \neq 0$ (polynomials in q^{-1});

polynomials $R[q]$ in q, analytic for every $q \neq \infty$.

The units of $R_p(q)$ are rational functions of relative degree zero, those of $R_H(q)$ and $R_S(q)$ are minimum-phase rational functions and those of $R_f(q)$ and $R[q]$ are non-zero constants.

An element A of $R_p(q)$ is said to be *strictly proper* if $A(\infty) = 0$.

A ring is said to be a *domain* if the product of every pair of non-zero elements is non-zero.

A subring of a ring is said to be an *ideal* if it is closed with respect to multiplication by elements of the ring. An ideal is *principal* if it is generated by multiples of a single element. A domain in which every ideal is principal is a principal ideal domain; one in which every finitely generated ideal is principal is called a Bezout domain. Note that our example rings are all principal ideal domains.

The linear equation $AX = B$ in a ring has a solution if and only if A divides B in that ring. An equation of the type

$$AX + BY = C$$

is called *diophantine*, and it is solvable in a Bezout domain if and only if every common divisor of A and B divides C. If X', Y' is a solution pair of this equation, then

$$X = X' + BW, \quad Y = Y' - AW$$

is also a solution pair for an arbitrary element W of the ring [29].

Two elements of a ring are *coprime* if their only common divisors are units of the ring. For A and B coprime, the above formula for X and Y generates *all* solution pairs of the diophantine equation.

1.4 Systems and Stability

The fundamentals of the algebraic approach will be explained for causal *linear* systems with *rational* transfer functions and whose input u and output y are scalar quantities. We suppose that u and y lie in a space of functions mapping a time set into a value set. The time set is a subset of real numbers bounded on the left, say R_+ (the non-negative reals) in the case of continuous-time systems and Z_+ (the non-negative integers) for discrete-time systems. The value set is taken to be the field of real numbers R; later we shall generalize.

The transfer function of a continuous-time system is the Laplace transform of its impulse response $g(t)$,

$$G(s) = \int_0^\infty g(t)\mathrm{e}^{-st}\mathrm{d}t.$$

For discrete-time systems, the transfer function is defined as the z-transform of its unit pulse response $(h_0, h_1, \ldots)$,

$$H(z) = \sum_{i=0}^{\infty} h_i z^{-i},$$

and it is always proper.

Systems which have the desirable property that a "bounded input" produces a "bounded output" are usually called stable. More precisely, let the input and output spaces of a continuous-time system be the spaces of locally (Lebesgue) integrable functions f from R_+ into R and define a norm

$$\|f\|_\infty = \operatorname*{ess\,sup}_{t \geq 0} |f(t)| \cdot$$

The corresponding normed space is denoted by L^∞. Then the system is said to be *bounded-input bounded-output* (BIBO) *stable* if any input $u \in L^\infty$ produces an output $y \in L^\infty$. It is a well-known fact [8] that a system with a rational transfer function $G(s)$ is BIBO-stable if and only if $G(s)$ is proper and Hurwitz-stable, i.e. belongs to $R_H(s)$.

In the study of discrete-time systems, we let the input and output spaces be the spaces of infinite sequences $f = (f_0, f_1, \ldots)$ in R and define a norm

$$\|f\|_\infty = \sup_{i=0,1,\ldots} |f_i|.$$

The corresponding normed space is denoted by l^∞. Then a system is said to be BIBO-stable if any input $u \in l^\infty$ gives rise to an output $y \in l^\infty$. As shown in [8], a system with a proper rational transfer function $H(z)$ is BIBO-stable if and only if $H(z)$ is Schur-stable, i.e. belongs to $R_S(z)$.

1.5 Fractional Description

We define the set of desirable transfer functions as those corresponding to BIBO-stable systems, namely $R_H(s)$ in continuous time and $R_S(z)$ in discrete time, and observe that these sets are rings, in particular principal ideal domains. Their field of fractions is in both cases the field of rational functions, $R(s)$ or $R(z)$. And since the systems under consideration are just those having a rational transfer function, say G, we can write

$$G = \frac{B}{A}$$

where A and B are elements of either $R_H(s)$ or $R_S(z)$. When A and B are coprime, the fractional representation of G is unique up to the units of the ring.

The field of rational functions is the quotient field of many other rings, including all the example rings. These are useful in achieving other "desirable" properties than BIBO-stability. For example, $R_p(s)$ is instrumental in the study of impulse-free behaviour of continuous-time systems while $R_f(z)$ can be used to advantage in the design of finite impulse response systems. The polynomial rings $R[s]$ or $R[z]$

are well matched with the ultimate goal of achieving desired modal properties (pole placement).

For example, neither the integrator with transfer function

$$I(s) = \frac{1}{s}$$

nor the differentiator with transfer function

$$D(s) = s$$

are BIBO-stable systems: we can write them in terms of $R_H(s)$ as follows:

$$I(s) = \frac{\frac{1}{s+a}}{\frac{s}{s+a}}, \quad D(s) = \frac{b}{\frac{b}{s}}$$

where $a > 0$ and b are real numbers.

A specific feature of discrete-time systems results from the inclusion

$$R_f(z) \subset R_S(z).$$

We observe that the ring $R_f(z)$ of rational functions analytic outside $z = 0$ is isomorphic with the ring of polynomials in the delay operator z^{-1}, denoted by $R[z^{-1}]$. In fact, polynomials in z^{-1} have been used in the analysis and design of discrete-time systems since the early days [49], [43], [29], for instance

$$\frac{b}{z+a} = \frac{\frac{b}{z}}{\frac{z+a}{z}} = \frac{bz^{-1}}{1+az^{-1}}.$$

Now we better understand why: the polynomials in z^{-1} are in fact special proper and Schur-stable rational functions in z and hence suitable for the study of finite as well as stable and causal processes [34].

1.6 Feedback Systems

To control a system means to alter its dynamics so that a desired behaviour may be obtained. This can be done through feedback. A typical feedback system consists of two subsystems, S_1 and S_2, connected as shown in the figure.

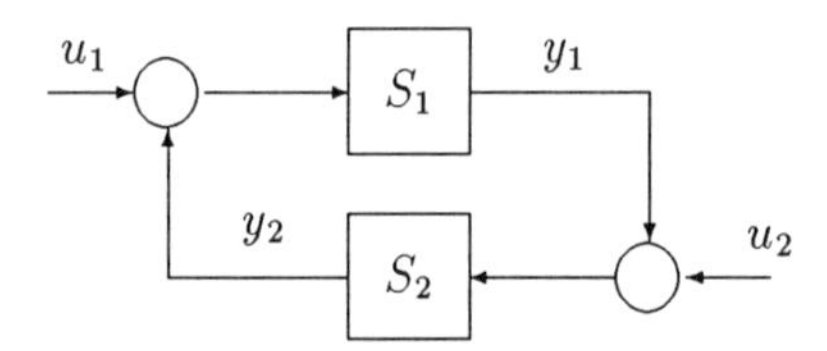

In many applications, it is desirable that the feedback system be *internally* BIBO-stable in the sense that whenever the exogenous inputs u_1 and u_2 are bounded so too are the internal signals y_1 and y_2.

In order to study this property, we express the transfer functions of S_1 and S_2 in the form

$$S_1 = \frac{B}{A}, \quad S_2 = -\frac{Y}{X}$$

where A, B and X, Y are two couples of coprime elements of $R_H(s)$ or $R_S(z)$, according to the nature of the system.

By definition, the feedback system will be internally BIBO-stable if and only if the four transfer functions from u_1, u_2 to y_1, y_2

$$y_1 = \frac{BX}{AX + BY} u_1, \qquad y_1 = \frac{-BY}{AX + BY} u_2$$

$$y_2 = \frac{-BY}{AX + BY} u_1, \quad y_2 = \frac{-AY}{AX + BY} u_2$$

(of which two are the same) reside in $R_H(s)$ or $R_S(z)$, respectively. This is the case if and only if [5], [7], [48] the common denominator $AX + BY$ is a unit of $R_H(s)$ or $R_S(z)$.

We illustrate on the example where S_1 is a differentiator and S_2 is an invertor. Since

$$S_1 = s, \quad S_2 = -1$$

we can take

$$A = \frac{1}{s + a}, \quad B = \frac{s}{s + a}, \quad X = 1, \quad Y = 1$$

for any real $a > 0$. Then

$$AX + BY = \frac{s + 1}{s + a}$$

is a unit of $R_H(s)$. Hence the resulting feedback system is internally BIBO-stable.

Any loop closed around a *discrete-time* system involves some information delay, however small [35]. Indeed, a control action applied to S_1 cannot affect the measurement from which it was calculated in S_2. Therefore, either S_1 or S_2 must include a one-step *delay*; we shall always think of it as having been included in S_1.

1.7 Parametrization of Stabilizing Controllers

The design of feedback control systems consists of the following: given one subsystem, say S_1, we are to determine the other subsystem, S_2, so that the resulting feedback system has a desirable property. We call S_1 the *plant* and S_2 the *controller*. Our focus is on achieving the desirable property of internal BIBO-stability. Any controller S_2 that BIBO-stabilizes the plant S_1 will be called a *stabilizing* controller for the plant.

Suppose S_1 is a continuous-time plant that gives rise to the transfer function

$$S_1 = \frac{B}{A}$$

for some coprime elements A and B of $R_H(s)$. It follows from the foregoing analysis that a stabilizing controller exists and that all controllers that stabilize the given plant are generated by all solution pairs X, Y with $X \neq 0$ of the equation

$$AX + BY = 1$$

in $R_H(s)$.

This is a result of fundamental importance [26], [27], [51], [52], [7] which can be considered to have launched this entire area of research. It provides a *parametrization* of the family of all stabilizing controllers for S_1 in a particularly simple form,

$$S_2 = -\frac{Y' - AW}{X' + BW}$$

where X', Y' are any elements of $R_H(s)$ such that $AX' + BY' = 1$ and W is a parameter that varies over $R_H(s)$ while satisfying $X' + BW \neq 0$.

In order to determine the set of all controllers S_2 that stabilize S_1, one needs to do two things: (i) express S_1 as a ratio of two coprime elements from $R_H(s)$ and (ii) find a particular solution in $R_H(s)$ of a diophantine equation. The first step is very easy for rational systems. The second step is equivalent to finding *one* controller that stabilizes S_1. Once these two steps are completed, the formula above provides a parametrization of the set of all stabilizing controllers for S_1. The condition $X' + BW \neq 0$ is not very restrictive, as $X' + BW$ can equal zero for at most one choice of W.

As an example, we shall stabilize an integrator plant S_1. Its transfer function can be written over $R_H(s)$ as

$$S_1 = \frac{\frac{1}{s+1}}{\frac{s}{s+1}}.$$

Suppose that, using some design procedure, we have found a stabilizing controller for S_1, namely

$$S_2 = -1.$$

This leads to a particular solution X, Y of the equation

$$\frac{s}{s+1}X + \frac{1}{s+1}Y = 1,$$

say $X' = 1$, $Y' = 1$. The solution set in $R_H(s)$ of this equation is

$$X = 1 + \frac{1}{s+1}W, \quad Y = 1 - \frac{s}{s+1}W.$$

Hence all controllers that BIBO-stabilize S_1 are given by those S_2 that are of the form

$$S_2 = -\frac{1 - \frac{s}{s+1}W}{1 + \frac{1}{s+1}W}$$

where W is any function in $R_H(s)$.

1.8 Robust Stabilization

We shall now consider the design of BIBO-stabilizing controllers for imprecisely known plants. Thus a nominal plant description is available, together with a description of the plant uncertainty, and the objective is to design a controller that stabilizes all plants lying within the specified band of uncertainty. Such a controller is said to *robustly stabilize* the family of plants.

The plant uncertainty can be modelled conveniently in terms of its fractional description [48] . Let S_1 be a nominal plant giving rise to the transfer function

$$S_1 = \frac{B}{A}$$

where A and B are coprime elements of $R_H(s)$. We define a *norm* of an element $F(s) \in R_H(s)$ by

$$\|F\| = \sup_{Re\ s \geq 0} |F(s)|.$$

We consider the neighbourhood of S_1 which consists of all plants S_1' having the transfer function

$$S_1' = \frac{B'}{A'}$$

where A' and B' are coprime elements of $R_H(s)$ such that

$$\|A - A'\| < \varepsilon_1, \quad \|B - B'\| < \varepsilon_2$$

for some reals $\varepsilon_1 > 0$ and $\varepsilon_2 > 0$.

Now, let S_2 be a stabilizing controller for S_1 whose transfer function

$$S_2 = -\frac{Y}{X}$$

satisfies

$$AX + BY = 1.$$

Then S_2 stabilizes all plants S_1' in the neighbourhood of S_1 if and only if

$$\varepsilon_1\|X\| + \varepsilon_2\|Y\| \leq 1.$$

See [9], [5], and [48].

When we work in discrete time, we define a norm of an element $F(z) \in R_S(z)$ by

$$\|F\| = \sup_{|z| \geq 1} |F(z)|$$

and proceed along the same lines.

1.9 Parametrization of Closed-Loop Transfer Functions

The utility of the diophantine equation $AX + BY = 1$ for the study of internal BIBO-stability derives not merely from the fact that it provides a parametrization of all controllers that stabilize a given plant in terms of a "free" parameter W, but also from the simple manner in which this parameter enters the resulting (BIBO-stable) closed-loop transfer functions.

In fact,

$$y_1 = B(X' + BW)u_1, \qquad y_1 = -B(Y' - AW)u_2$$

$$y_2 = -B(Y' - AW)u_1, \qquad y_2 = -A(Y' - AW)u_2$$

and we observe that all the transfer functions are *affine* in W.

For example, the integrator stabilized in the last example yields the closed-loop transfer functions

$$y_1 = \frac{1}{s+1}(1 + \frac{1}{s+1}W)u_1, \qquad y_1 = -\frac{1}{s+1}(1 - \frac{s}{s+1}W)u_2$$

$$y_2 = -\frac{1}{s+1}(1 - \frac{s}{s+1}W)u_1, \qquad y_2 = -\frac{s}{s+1}(1 - \frac{s}{s+1}W)u_2$$

where W is arbitrary in $R_H(s)$.

This result serves to parametrize the performance specifications and it is the starting point for the selection of the best controller for the application at hand. The crucial point is that the resulting optimization/selection problem is affine in the parameter W while it is non-linear in the controller S_2.

1.10 Disturbance Rejection

Let us show how further specifications, beyond those of desirability, can be handled. Suppose we are given a plant S_1, with transfer function

$$S_1 = \frac{B}{A}$$

in fractional form over $R_H(s)$, say, whose output is corrupted by a disturbance v of the form

$$v = \frac{Q}{P}$$

where only P is specified. The objective is to design a controller S_2 such that (i) the feedback system shown in the diagram is internally BIBO-stable and (ii) the plant output y is insensitive to the disturbance v.

This means that

$$y = \frac{AX}{AX + BY}\frac{Q}{P}$$

must belong to $R_H(s)$, i.e., both $AX + BY$ and P should cancel. We define coprime elements A_0, P_0 of $R_H(s)$ by

$$\frac{A}{P} = \frac{A_0}{P_0}.$$

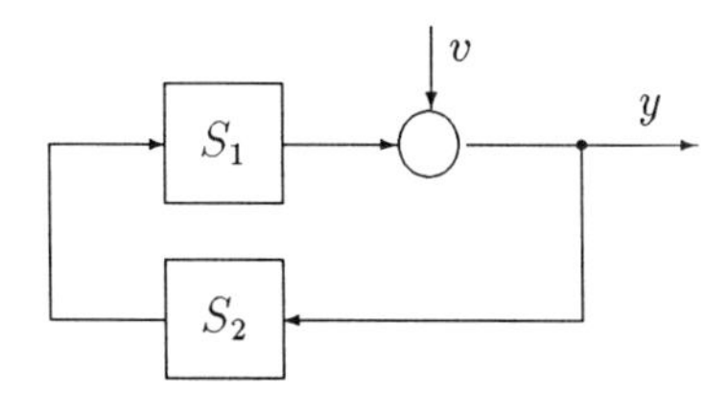

Since $AX + BY$ is a unit for every stabilizing controller, P_0 must divide X in $R_H(s)$. Therefore S_2 exists if and only if P and B are coprime in $R_H(s)$ and has the transfer function

$$S_2 = -\frac{Y}{X_0 P_0},$$

where X_0, Y is any solution pair of the equation

$$AP_0X_0 + BY = 1.$$

The disturbance rejection is said to be *robust* if the conditions (i) and (ii) hold even as the plant S_1 is slightly perturbed, that is to say, if S_1 is replaced by S_1' with transfer function

$$S_1' = \frac{B'}{A'}$$

such that

$$\|A - A'\| < \varepsilon_1, \quad \|B - B'\| < \varepsilon_2$$

for some reals $\varepsilon_1 > 0$ and $\varepsilon_2 > 0$.

Whenever the disturbance can be rejected, it can be rejected in a robust way [11], [48]. All that is needed is to make X divisible by P rather than by P_0. Thus if $\bar{S}_2$ is any controller that stabilizes the plant $\bar{S}_1 = S_1/P$ then $S_2 = \bar{S}_2/P$ achieves the robust disturbance rejection. The presence of the $1/P$ term within the controller is referred to elsewhere [10], [50] as the "internal model principle".

As an example, we shall eliminate the effect of a periodic disturbance

$$v = \frac{as + b}{s^2 + 1}$$

for any real a and b, which is injected at the output of an integrator plant S_1. We take, for instance,

$$A = \frac{s}{s+1}, \quad B = \frac{1}{s+1}, \quad P = \frac{s^2+1}{(s+1)^2}$$

and solve the equation

$$\frac{s}{s+1}\,\frac{s^2+1}{(s+1)^2}X_0 + \frac{1}{s+1}Y = 1$$

to obtain

$$X_0 = 1 + \frac{1}{s+1}W$$

$$Y = \frac{3s^2+2s+1}{(s+1)^2} - \frac{s}{s+1}\,\frac{s^2+1}{(s+1)^2}W$$

for any W from $R_H(s)$.

All stabilizing controllers that reject the disturbance have the transfer function

$$T_2 = -\frac{\frac{3s^2+2s+1}{(s+1)^2} - \frac{s}{s+1}W'}{\frac{s^2+1}{(s+1)^2} + \frac{1}{s+1}W'}$$

where

$$W' = \frac{s^2+1}{(s+1)^2}W.$$

They all achieve robust rejection since A and P are coprime in $R_H(s)$.

1.11 Reference Tracking

By now it is clear how the algebraic design works: the performance specifications call for divisibility conditions in the underlying ring, and these are equivalent to diophantine equations. Another design problem of importance is that of reference tracking.

Suppose we are given a plant S_1, with transfer function

$$S_1 = \frac{B}{A}$$

in fractional form over $R_S(z)$, say, together with a reference r of the form

$$r = \frac{G}{F}$$

where only F is specified. We recall that S_1 contains a one-step delay, hence B is strictly proper. The objective is to design a controller S_2 such that (i) the feedback system

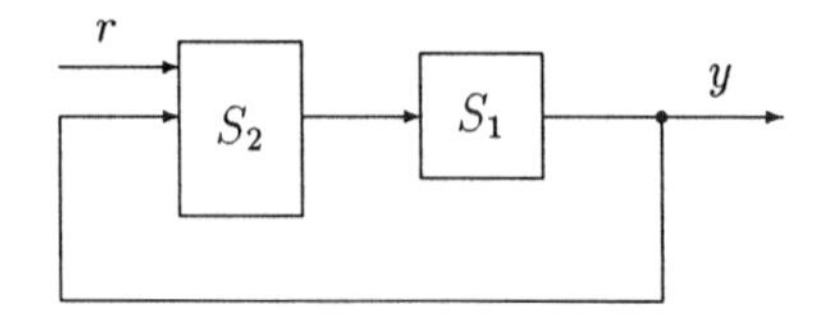

is internally BIBO-stable and (ii) the plant output y tracks the reference signal r. The controller can operate on both r (feedforward) and y (feedback), so it is described by two transfer functions

$$S_{2y} = -\frac{Y}{X}, \quad S_{2r} = \frac{Z}{X}.$$

The requirement of tracking imposes [33] that the tracking error

$$\begin{aligned} e &= r - y \\ &= (1 - \frac{BZ}{AX+BY})\frac{G}{F} \end{aligned}$$

belong to $R_S(z)$. Since $AX + BY$ is a unit for every stabilizing controller, F must divide $1 - BZ$ in $R_S(z)$. Hence there must exist an element V of $R_S(z)$ such that $1 - BZ = FV$. Therefore S_2 exists if and only if F and B are coprime in $R_S(z)$ and its two transfer functions evolve from solving the two diophantine equations

$$AX + BY = 1$$

$$FV + BZ = 1$$

where the element V of $R_S(z)$ serves to express the tracking error as

$$e = VG.$$

The reference tracking is said to be *robust* if both conditions (i) and (ii) hold even as the plant S_1 is slightly perturbed to S_1', with transfer function

$$S_1' = \frac{B'}{A'}$$

factorized over $R_S(z)$ such that B' is strictly proper and

$$\|A - A'\| < \varepsilon_1, \quad \|B - B'\| < \varepsilon_2$$

for some reals $\varepsilon_1 > 0$ and $\varepsilon_2 > 0$.

Since $A'X + B'Y$ is still a unit, say U, of $R_S(z)$, we have

$$e = \left(\frac{A'X}{U} + \frac{B'(Y - Z)}{U}\right)\frac{G}{F}.$$

Hence two conditions are necessary and sufficient for robust reference tracking [11], [48], namely

$$\begin{aligned} &F \text{ divides } X, \\ &F \text{ divides } Y - Z \end{aligned}$$

in $R_S(z)$.

We illustrate on a discrete-time plant S_1 given by

$$S_1 = \frac{1}{1 + 0.5z}$$

whose output is to track every sinusoidal sequence of the form

$$r = \frac{a + bz}{1 + z^2}$$

where a, b are unspecified reals. Taking

$$A = \frac{1 + 0.5z}{z}, \quad B = \frac{1}{z}, \quad F = \frac{1 + z^2}{z^2}$$

and solving the diophantine equations

$$\frac{1+0.5z}{z}X + \frac{1}{z}Y = 1$$

$$\frac{1+z^2}{z^2}V + \frac{1}{z}Z = 1$$

yields the tracking controllers in parametric form

$$S_{2y} = \frac{2 + \frac{1+0.5z}{z}W_1}{2 + \frac{1}{z}W_1}$$

$$S_{2r} = \frac{-\frac{1}{z} - \frac{1+z^2}{z^2}W_2}{2 + \frac{1}{z}W_1}$$

for any elements W_1, W_2 of $R_S(z)$. The resulting error is

$$e = (1 + \frac{1}{z}W_2)(a + bz).$$

Not all of these controllers, however, achieve the robust tracking of the reference. The two divisibility conditions are fulfilled if and only if W_1 is restricted to

$$W_1 = \frac{2}{z} + \frac{1+z^2}{z^2}W$$

where W ranges over $R_S(z)$.

1.12 Model Matching

It is nice to see how the primary requirement of BIBO-stability decouples from the secondary requirements imposed on the control system. To further illustrate this point, we consider the problem of (exact) model matching.

We are given a plant S_1 with transfer function

$$S_1 = \frac{B}{A}$$

in fractional form over $R_H(s)$, say, and a model (proper and stable) rational function M. We suppose that A and B are coprime in $R_H(s)$. The objective is to design a stabilizing controller S_2, having a feedback and a feedforward part, such that the transfer function of the closed-loop system from v to y is M.

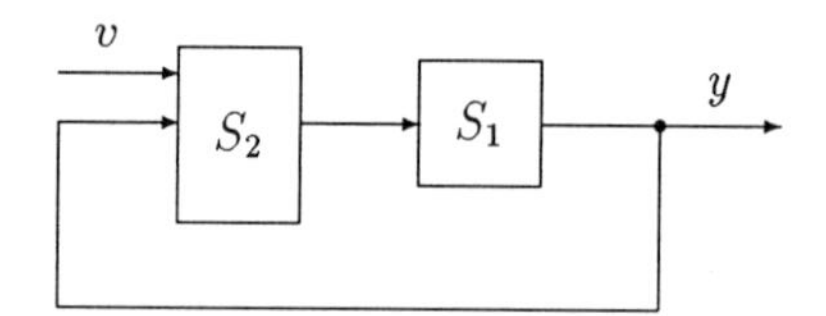

Denoting

$$S_{2y} = -\frac{Y}{X}, \quad S_{2v} = \frac{Z}{X}$$

the transfer functions of the feedback and the feedforward parts of S_2, respectively, the overall transfer function relating v and y reads

$$\frac{BZ}{AX + BY} = M.$$

The requirement of BIBO-stability for the feedback system then decouples the model matching equation into two linear diophantine equations

$$\begin{aligned} AX + BY &= 1 \\ BZ &= M. \end{aligned}$$

It follows that the model matching by a stabilizing controller is possible if and only if B divides M in $R_H(s)$. Any and all controllers are obtained by solving the two equations [33].

As an example, we take again the integrator plant S_1. In order to generate from S_1 a proper and stable system with transfer function

$$M = \frac{s-1}{(s+1)^2}$$

we solve the diophantine equations

$$\begin{aligned} \frac{s}{s+1}X + \frac{1}{s+1}Y &= 1 \\ \frac{1}{s+1}Z &= \frac{s-1}{(s+1)^2} \end{aligned}$$

over $R_H(s)$. All matching controllers are given by

$$S_{2y} = -\frac{1 - \frac{s}{s+1}W}{1 + \frac{1}{s+1}W}, \quad S_{2v} = \frac{\frac{s-1}{s+1}}{1 + \frac{1}{s+1}W}$$

for arbitrary W from $R_H(s)$.

Of course, there are other controllers S_2 that generate M, for example

$$S_{2y} = 0, \quad S_{2v} = sM.$$

They are not stabilizing ones, however.

1.13 $\mathcal{H}_2$ Optimal Control

The performance specifications often involve a norm minimization. Let us consider the ring $R_S(z)$ of rational functions $F(z)$ analytic within the set $|z| \geq 1$ and define a norm by

$$\|F\|_2 = \langle F^* F \rangle^{\frac{1}{2}}$$

where the asterisk denotes conjugation, $F^*(z) = F(z^{-1})$, and $\langle\cdot\rangle$ denotes taking the term independent of z. The corresponding normed space is the rational Hardy space $\mathcal{H}_2$.

We are given a discrete-time plant S_1 having two inputs: the control input u and the exogenous input v. We describe the plant by two transfer functions

$$S_{1u} = \frac{B}{A}, \quad S_{1v} = \frac{C}{A}$$

where A, B, C are elements from $R_S(z)$ and A is a least common denominator. We recall that S_1 contains a one step-delay, hence B is strictly proper.

The objective is to determine a controller S_2 such that (i) the feedback system

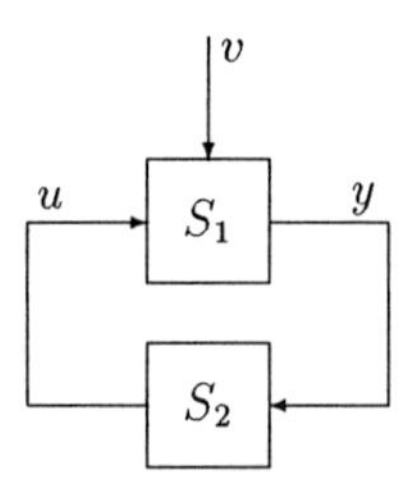

is internally BIBO-stable and (ii) the closed-loop transfer function from v to y, denoted by T, has minimal $\mathcal{H}_2$-norm.

There exists a stabilizing controller for S_1, say

$$S_2 = -\frac{Y'}{X'},$$

if and only if A, B are coprime in $R_S(z)$. Let

$$S_2 = -\frac{Y' - AW}{X' + BW}$$

be a parametrization of all stabilizing controllers in terms of a free parameter W from $R_S(z)$. The closed-loop transfer functions are then parametrized as

$$T = \frac{S_{1v}}{1 + S_{1u}S_2} = C(X' + BW)$$

and the task is to find the parameter W that minimizes $\|T\|_2$.

Suppose there exist two units F_B and F_C of $R_S(z)$, the so-called *spectral factors,* such that

$$B^*B = F_B^*F_B, \quad C^*C = F_C^*F_C.$$

Then

$$\begin{aligned} \|T\|_2^2 &= \langle (X' + BW)^*C^*C(X' + BW)\rangle \\ &= \langle (F_BF_CW + \tfrac{B^*}{F_B^*}X'F_C)^*(F_BF_CW + \tfrac{B^*}{F_B^*}X'F_C)\rangle. \end{aligned}$$

Let P be the proper part of the second term in the parentheses, i.e.

$$\frac{B^*}{F_B^*}X'F_C = P + \frac{Q^*}{F_B^*}$$

for some Q in $R_S(z)$. Then the cross-terms contribute nothing to the norm and we have

$$\|T\|_2^2 = \|\frac{Q}{F_B}\|_2^2 + \|P + F_B F_C W\|_2^2.$$

The first term being independent of W, the minimizing parameter is seen to be

$$W = -\frac{P}{F_B F_C}.$$

By construction, it belongs to $R_S(z)$. It follows that the optimal controller is unique and given by

$$S_2 = -\frac{F_B F_C Y' + AP}{F_B F_C X' - BP}$$

while the minimum norm is

$$\|T\|_{2\min} = \|\frac{Q}{F_B}\|_2.$$

As an example, consider the summator plant

$$\begin{aligned} x_{t+1} &= x_t + u_t - v_t \\ y_t &= x_t + v_t \end{aligned}$$

described by the transfer functions

$$S_{1u} = \frac{1}{z-1}, \quad S_{1v} = \frac{z-2}{z-1}.$$

We take the following elements of $R_S(z)$ to describe the plant

$$A = 1 - z^{-1}, \quad B = z^{-1}, \quad C = 1 - 2z^{-1}$$

and calculate the spectral factors

$$F_B = 1, \quad F_C = 2 - z^{-1}.$$

One particular BIBO-stabilizing controller is given by $S_2 = -1$, hence $X' = 1$ and $Y' = 1$. Since the proper part of

$$\frac{B^*}{F_B^*}X'F_C = 2z - 1$$

is $P = -1$ we obtain the minimizing parameter

$$W = \frac{1}{2 - z^{-1}}.$$

Therefore the $\mathcal{H}_2$ optimal controller is

$$S_2 = -\frac{1}{2}$$

and the corresponding norm $\|T\|_{2\min} = 2$.

The $\mathcal{H}_2$ optimization problem is close to the LQG or minimum variance control problem. In fact, if v is a stationary, zero mean and unit variance white noise sequence, then $\|T\|_2^2$ is the variance of y in the steady state.

1.14 Finite Impulse Response

Transients in discrete-time linear systems can settle in finite time. Systems having the property that a "finite input" produces a "finite output" will be called *finite-input finite-output* (FIFO) *stable.* It is a well-known fact [30] that a system with a proper rational transfer function $H(z)$ is FIFO-stable if and only if $H(z)$ is a polynomial in z^{-1}, i.e. belongs to $R_f(z)$. Equivalently, the unit pulse response of the system is finite.

Let us consider the feedback configuration

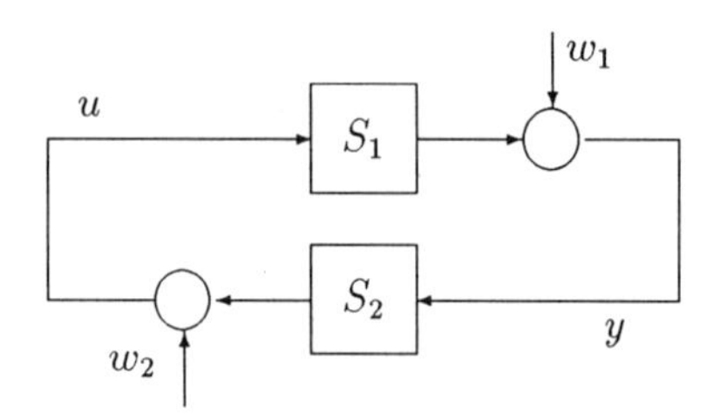

and focus on achieving the desirable property of internal FIFO-stability. To this end, we write the transfer function of the plant as

$$S_1 = \frac{B}{A}$$

where A and B are coprime functions from $R_f(z)$. We recall that S_1 contains a one-step delay, hence B is strictly proper. Repeating the arguments concerning BIBO-stability, we conclude that a FIFO-stabilizing controller exists and that all controllers that internally FIFO-stabilize the given plant have the transfer function

$$S_2 = -\frac{Y}{X}$$

where X, Y is the solution class of the equation

$$AX + BY = 1$$

in $R_f(z)$.

In particular, if X', Y' defines any FIFO-stabilizing controller for S_1, the set of all such controllers can be parametrized as

$$S_2 = -\frac{Y' - AW}{X' + BW}$$

where the parameter W varies over $R_f(z)$. We note that W is completely free because strict properness of B prevents $X' + BW = 0$.

The parameter W can be chosen so as to meet further requirements on the closed-loop system. For reasons of simplicity, economy or aesthetics we may require that S_2 have a smallest McMillan degree. Then W must be taken so that X, Y be the solution having least degrees in z^{-1}. This particular controller achieves also the shortest impulse responses within the closed-loop system.

A well-known example is the *dead-beat* controller. We consider the plant

$$S_1 = \frac{z^{-1}}{1 - z^{-1}}$$

and interpret the exogenous signals w_1 and w_2 as accounting for the effect of the initial conditions of S_1 and S_2. The requirement of FIFO-stability is then equivalent to achieving finite responses u and y for all initial conditions. Since all solutions of the equation

$$(1 - z^{-1})X + z^{-1}Y = 1$$

in $R_f(z)$ read

$$X = 1 + z^{-1}W, \quad Y = 1 - (1 - z^{-1})W$$

we obtain all dead-beat (or FIFO-stabilizing) controllers as

$$S_2 = -\frac{1 - (1 - z^{-1})W}{1 + z^{-1}W}.$$

The dead-beat controller of least McMillan degree is obtained for $W = 0$:

$$S_2 = -1.$$

1.15 Infinite-Dimensional Systems

If the value set of a system is a normed function space, rather than the set of real numbers, we speak of an infinite-dimensional system. The transfer function of such a system is no longer rational.

The diversity of infinite-dimensional systems cannot be captured by a single notion of stability. In the context of discrete-time systems, the BIBO-stability defined earlier is termed the l^∞-stability, because a BIBO-stable system transforms l^∞ input sequences into l^∞ output sequences. The set of transfer functions of causal l^∞-stable systems is the set of the z-transforms of absolutely summable sequences.

In some applications it is more natural to use l^2-stability. A discrete-time causal system is l^2-stable if it transforms l^2 input sequences into l^2 output sequences, where l^2 is the space of infinite sequences $f = (f_0, f_1, ...)$ with a norm defined by

$$\|f\|_2 = \left\{\sum_{i=0}^{\infty} |f_i|^2\right\}^{\frac{1}{2}}.$$

The set of transfer functions of causal l^2-stable systems is the Hardy space $\mathcal{H}_\infty$ of bounded functions on the unit circle with analytic continuation outside the circle.

A good choice for the set of desirable transfer functions is the ring D of functions that are analytic on the open set $|z| > 1$ and continuous on the closed set $|z| \geq 1$. It corresponds to causal l^2-stable systems whose frequency response is continuous. The ring D is not a principal ideal domain, it is not even a Bezout domain [47] . This means that the systems whose transfer functions belong to the field of fractions of D

cannot be assumed a priori to have a coprime fractional representation over D. The transfer functions whose denominators are restricted to having only a finite number of zeros in the closed set $|z| \geq 1$ and none on the unit circle $|z| = 1$, however, do have a coprime fractional representation [48]. The algebraic approach developed for rational systems carries over completely to systems with these transfer functions.

Another desirable property of control systems is related to the ring E of functions that are analytic in the entire complex plane but at $z = 0$; this point can be an essential singularity. A system whose transfer function belongs to E features a unit-pulse response $(h_0, h_1, ...)$ that converges to zero faster than any exponential. This property is a generalization of finite impulse response in finite-dimensional systems. The ring E is a Bezout domain and its field of fractions is the set of meromorphic functions in z^{-1}. Therefore a "rapid descent" control strategy [17] can be formulated and solved for a fairly broad class of infinite-dimensional systems.

1.16 Time-Varying Systems

We shall now outline the extension of the algebraic, transfer function approach to time-varying linear, say discrete-time systems. A class of these systems is defined by unit-pulse responses $(h_0, h_1, ...)$ whose members are infinite sequences in l^∞.

We observe that with pointwise addition and multiplication l^∞ is a ring (commutative, with identity). With the shift operator σ defined for each sequence $x = (x_0, x_1, ...)$ in l^∞ by $(\sigma x)_i = x_{i-1}$ the ring l^∞ is called a difference ring.

When

$$H(z) = \sum_{i=0}^{\infty} z^{-i} h_i$$

exists, it is called the (formal) transfer function of the system. We observe that H resides in $l^\infty(z)$, the ring of rational transfer functions in the indeterminate z over l^∞, with multiplication defined by

$$xz = z(\sigma x), \quad x \text{ in } l^\infty.$$

This non-commutative multiplication captures in a very natural way the time-variance of our systems [22].

We shall study the problem of stabilization: given a plant S_1, design a controller S_2 such that the closed-loop system shown in the diagram is internally uniformly BIBO-stable.

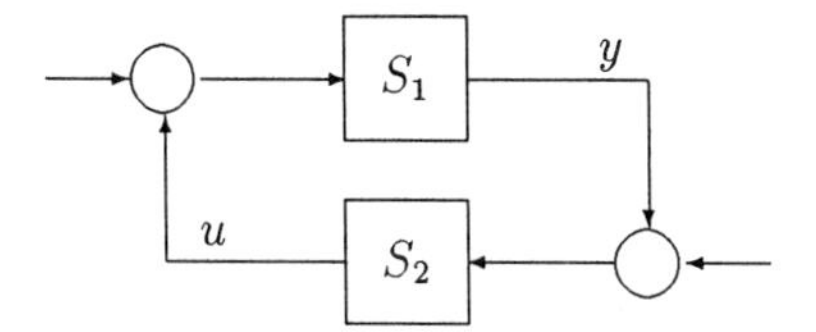

To this effect, we take the set of desirable transfer functions to be $l_S^\infty(z)$, the ring of stable rational functions from $l^\infty(z)$, defined as follows. An element

$$h = \sum_{i=0}^{\infty} z^{-i} h_i$$

of $l^\infty(z)$, with $h_i = (h_{it})_{t=0}^\infty$ is stable if $\|h_i\| \to 0$ as $i \to 0$, where

$$\|h_{it}\| = \sup_t |h_{it}|.$$

Since the ring is non-commutative, we have to distinguish left and right fractions.

Accordingly, we assume that the transfer functions of S_1 and S_2 can be written as

$$S_1 = A^{-1}B = \bar{B}\bar{A}^{-1}, \quad S_2 = -\bar{X}^{-1}\bar{Y} = -YX^{-1}$$

for left coprime elements A, B and $\bar{X}, \bar{Y}$ and for right coprime elements $\bar{A}, \bar{B}$ and X, Y of $l_S^\infty(z)$; otherwise the feedback system could not be stabilized. Then, drawing from the preceding discussion, the feedback system is internally uniformly BIBO-stable if and only if $AX + BY$, or equivalently $\bar{X}\bar{A} + \bar{Y}\bar{B}$, is a unit of $l_S^\infty(z)$.

In order to design a stabilizing controller S_2 for the plant S_1, we solve the equation

$$AX + BY = 1$$

for X, Y in $l_S^\infty(z)$ and put

$$S_2 = -(Y - \bar{A}W)(X + \bar{B}W)^{-1}$$

for an arbitrary parameter W of $l_S^\infty(z)$ such that $X + \bar{B}W \neq 0$. Equivalently, we can proceed by solving the equation

$$\bar{X}\bar{A} + \bar{Y}\bar{B} = 1$$

for $\bar{X}, \bar{Y}$ in $l_S^\infty(z)$ and put

$$S_2 = -(\bar{X} + \bar{W}B)^{-1}(\bar{Y} - \bar{W}A)$$

for an arbitrary parameter $\bar{W}$ of $l_S^\infty(z)$ such that $\bar{X} + \bar{W}B \neq 0$.

To illustrate, we consider the plant [41] given by the recursion

$$x_{t+1} = x_t + a_t u_t, \quad y_t = a_t x_t$$

where $a_t = 1$ if t is even and $a_t = 0$ if t is odd. The transfer function of S_1 is

$$S_1 = z^{-2}a_t + z^{-4}a_t + \ldots$$

and it admits left and right factorizations

$$S_1 = A^{-1}B = \bar{B}\bar{A}^{-1}$$

over $l_S^\infty(z)$, for instance

$$A = 1 - z^{-2}a_t, \quad B = z^{-2}a_t$$
$$\bar{A} = 1 - z^{-2}a_t - z^{-3}a_t, \quad \bar{B} = z^{-2}a_t.$$

Therefore, all stabilizing controllers possess the transfer function

$$S_2 = -[1 - (1 - z^{-2}a_t - z^{-3}a_t)W][1 + z^{-2}a_tW]^{-1}$$

where W ranges over $l_S^\infty(z)$.

1.17 Non-Linear Systems

As a final example we illustrate how the fractional approach can be applied to the stabilization of non-linear systems. Owing to the abundance of unstable non-linear systems, the stabilization problem is probably one of the most commonly encountered problems in engineering science and practice.

We shall work in discrete time, with the signal space l_0, the set of all infinite real sequences $x = (x_0, x_1, ...)$. We can define a norm on l_0 by $\|x\| = \sup_i 2^{-i}|x_i|$. A non-linear system is simply a map $S : l_0 \to l_0$, transforming input sequences into output sequences.

The problem of stabilizing a non-linear plant S_1 consists in finding a (non-linear) controller S_2 such that the feedback system

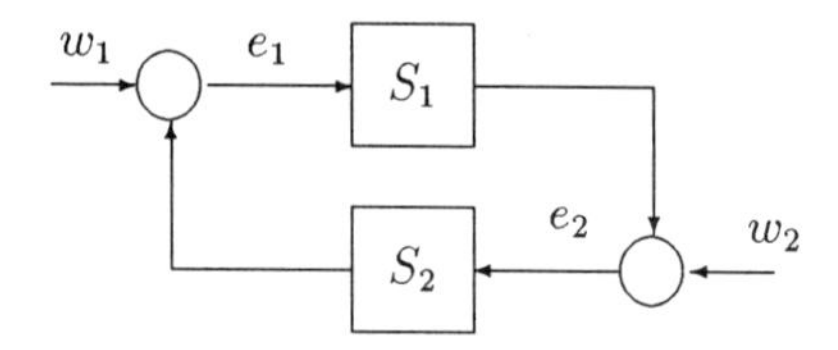

is internally BIBO-stable in that the internal signals e_1 and e_2 of the closed loop are bounded in the presence of suitably bounded, but otherwise arbitrary, signals w_1 and w_2 injected around the loop.

The basic concept that facilitates our development is that of *rationality*. A system S is right-rational if there are BIBO-stable maps A and B with common domain $l \subset l_0$, where A is invertible, such that $S = BA^{-1}$; a system S is left-rational if there are BIBO-stable maps $\bar{A}$ and $\bar{B}$ with common domain $\bar{l} \subset l_0$, where $\bar{A}$ is invertible, such that $S = \bar{A}^{-1}\bar{B}$. We say that A and B are right coprime if the set of all bounded sequences $w \in l_0$ such that Sw is bounded, but $A^{-1}w$ is unbounded, is the empty set. Similarly $\bar{A}$ and $\bar{B}$ are left coprime if the set of all unbounded sequences $w \in l_0$ such that Sw is unbounded and $\bar{B}w$ is bounded is the empty set.

Consider now a strictly causal plant $S_1 : l_0 \rightarrow l_0$ that is right-rational, with a right coprime fraction representation

$$S_1 = BA^{-1}$$

where $A : l \rightarrow l_0$ and $B : l \rightarrow l_0$ are BIBO-stable maps. Then [13] there exist BIBO-stable maps $X : l_0 \rightarrow l$ and $Y : l_0 \rightarrow l$, where X is invertible, such that the following identity holds,

$$XA + YB = I : l \rightarrow l.$$

These maps define the causal controller $S_2 : l_0 \rightarrow l_0$ as

$$S_2 = -X^{-1}Y.$$

Suppose that X and Y satisfy a Lipschitz condition. Then [38] the controller S_2 will internally BIBO-stabilize the plant S_1 in that the signals e_1 and e_2 are bounded for suitably bounded inputs w_1 and w_2, or equivalently all the closed-loop transfer maps are BIBO-stable.

Let the plant S_1 be also left-rational, with a left coprime fraction representation

$$S_1 = \bar{A}^{-1}\bar{B}$$

where $\bar{A} : \bar{l} \rightarrow l_0$ and $\bar{B} : \bar{l} \rightarrow l_0$ are BIBO-stable maps. Then the class of all BIBO-stable maps X_W and Y_W satisfying

$$X_W A + Y_W B = I$$

is characterized in terms of an arbitrary BIBO-stable non-linear map $W : \bar{l} \rightarrow l$ as

$$X_W = X + W\bar{B}, \quad Y_W = Y - W\bar{A}.$$

Therefore, having found one controller S_2 that internally BIBO-stabilizes S_1, the class of all controllers S_{2W} that internally BIBO-stabilize S_1 is given by

$$S_{2W} = -X_W^{-1}Y_W$$

provided that X_W and Y_W satisfy a Lipschitz condition [38].

A broad example of non-linear systems are the *recursive* systems. A system $S : l_0 \rightarrow l_0$ is recursive if there exist non-negative integers α, β and a function f such that, for every input sequence $u \in l_0$, the elements of the output sequence $y = Su$ can be computed recursively from the relation

$$y_{t+\alpha+1} = f(y_t, ..., y_{t+\alpha}, u_t, ..., u_{t+\beta})$$

for all $t = 0, 1, ...$ given the initial conditions $y_0, y_1, ..., y_\alpha$. If S is an injective system with a continuous recursion function f then it is both left and right rational provided it is operated on by bounded input sequences [13]. If the set of bounded input sequences for which S produces bounded output sequences is a rich enough set, in a suitable sense [12], then S can be internally BIBO-stabilized. For example, the system

$$y_{t+1} = e^{y_t} + u_{t+1}$$

is stabilized by

$$u_{t+1} = 10y_t + e^{\phi(y_t)} - \phi(y_{t+1})$$

where

$$\phi(x) = \begin{cases} x, & |x| \leq 1 \\ \text{sign } x, & |x| > 1 \end{cases}$$

provided the input sequences are suitably bounded, whereas no stabilizer exists for the system

$$y_{t+1} = 2y_t + e^{u_t}$$

as no input sequence produces a bounded output sequence.

Bibliography

[1] K. J. Åström , *Introduction to Stochastic Control Theory.* New York: Academic Press, 1970.

[2] H. Blomberg and R. Ylinen, *Algebraic Theory for Multivariable Linear Systems.* London: Academic Press, 1983.

[3] F. M. Callier and C. A. Desoer, "An algebra of transfer functions of distributed linear time-invariant systems," *IEEE Trans. Circuits and Systems*, vol. 25, pp. 651-662, 1978.

[4] F. M. Callier and C. A. Desoer, "Stabilization, tracking and disturbance rejection in multivariable convolution systems," *Annales de la Societé Scientifique de Bruxelles*, vol. 94, pp. 7-51, 1980.

[5] M. J. Chen and C. A. Desoer, "Algebraic theory of robust stability of interconnected systems," *IEEE Trans. Automatic Control*, vol. 29, pp. 511-519, 1984.

[6] C. A. Desoer and W. S. Chan, "The feedback interconnection of linear time-invariant systems," *J. Franklin Inst.*, vol. 300, pp. 335-351, 1975.

[7] C. A. Desoer, R. W. Liu, J. Murray, and R. Saeks, "Feedback system design: the fractional representation approach to analysis and synthesis," *IEEE Trans. Automatic Control*, vol. 25, pp. 399-412, 1980.

[8] C. A. Desoer and M. Vidyasagar, *Feedback Systems: Input-Output Properties.* New York: Academic Press, 1975.

[9] J. C. Doyle and G. Stein, "Multivariable feedback design: concepts for a classical/modern synthesis," *IEEE Trans. Automatic Control*, vol. 26, pp. 4-16, 1981.

[10] B. A. Francis and W. M. Wonham, "The internal model principle of control theory," *Automatica*, vol. 12, pp. 457-465, 1976.

[11] B. A. Francis and M. Vidyasagar, "Algebraic and topological aspects of the regulator problem for lumped linear systems," *Automatica*, vol. 19, pp. 87-90, 1983.

[12] J. Hammer, "Stabilization of nonlinear systems," *Int. J. Control*, vol. 44, pp. 1349-1381, 1986.

[13] J. Hammer, "Fraction representations of nonlinear systems: a simplified approach," *Int. J. Control*, vol. 46, pp. 455-472, 1987.

[14] M. L. J. Hautus, "The formal Laplace transform for smooth linear systems," in *Proc. Internat. Symp. Mathematical Systems Theory*, Udine, Italy, 1975, pp. 29-47.

[15] K. J. Hunt and V. Kučera, "The standard $\mathcal{H}_2$-optimal control problem: a polynomial solution," *Int. J. Control*, to appear.

[16] N. Jacobson, *Lectures in Abstract Algebra, Vol. I.* New York: Van Nostrand, 1953.

[17] J. Ježek, "An algebraic approach to the synthesis of control for linear discrete meromorphic systems," *Kybernetika*, vol. 25, pp. 73-82, 1989.

[18] J. Ježek and J. Nagy, "An algebraic approach to the synthesis of control for linear time-varying systems on a finite time horizon," in *Prep. IFAC Workshop on System Structure and Control*, Prague, Czechoslovakia, 1989, pp. 63-68.

[19] T. Kaczorek, *Two-Dimensional Linear Systems.* Berlin: Springer-Verlag, 1985.

[20] R. E. Kalman, P. L. Falb, and M. A. Arbib, *Topics in Mathematical System Theory.* New York: McGraw-Hill, 1969.

[21] E. W. Kamen, "On an algebraic theory of systems defined by convolution operators," *Math. Syst. Theory*, vol. 9, pp. 57-74, 1975.

[22] E. W. Kamen, P. P. Khargonekar, and K. R. Poola, "A transfer-function approach to linear time-varying discrete-time systems," *SIAM J. Control and Optimization*, vol. 23, pp. 550-565, 1985.

[23] E. W. Kamen, P. P. Khargonekar, and A. Tannenbaum, "Stabilization of time-delay systems using finite-dimensional controllers," *IEEE Trans. Automatic Control*, vol. 30, pp. 75-78, 1985.

[24] V. Kučera, "Algebraic theory of discrete optimal control for single-variable systems, Parts I-III," *Kybernetika*, vol. 9, pp. 94-107, 206-221, 291-312, 1973.

[25] V. Kučera, "Algebraic theory of discrete optimal control for multivariable systems," *Supplement to Kybernetika*, pp. 1-240, 1974.

[26] V. Kučera, "Closed-loop stability of discrete linear single-variable systems," *Kybernetika*, vol. 10, pp. 146-171, 1974.

[27] V. Kučera, "Stability of discrete linear feedback systems," in *Prep. 6th IFAC World Congress*, Boston, 1975, vol. 1, paper 44.1.

[28] V. Kučera, *Algebraic Theory of Discrete Linear Control* (in Czech). Prague: Academia, 1978.

[29] V. Kučera, *Discrete Linear Control: The Polynomial Equation Approach.* Chichester: Wiley, 1979.

[30] V. Kučera, "A dead-beat servo problem," *Int. J. Control*, vol. 32, pp. 107-113, 1980.

[31] V. Kučera, "Linear quadratic control: state space vs. polynomial equations," *Kybernetika*, vol. 19, pp. 185-195, 1983.

[32] V. Kučera, "Stationary LQG control of singular systems," *IEEE Trans. Automatic Control*, vol. 31, pp. 31-39, 1986.

[33] V. Kučera, "Internal properness and stability in linear systems," *Kybernetika*, vol. 22, pp. 1-18, 1986.

[34] V. Kučera, *Analysis and Design of Discrete Linear Control Systems.* London: Prentice-Hall, 1991.

[35] V. Kučera, "Diophantine equations in control," in *Proc. First European Control Conference*, Grenoble, France, 1991, pp. 491-502.

[36] F. L. Lewis, "A survey of linear singular systems," *Circuits Systems Signal Process.*, vol. 5, pp. 3-36, 1986.

[37] A. S. Morse, "Ring models for delay-differential systems," *Automatica*, vol. 12, pp. 529-531, 1976.

[38] A. D. B. Paice and J. B. Moore, "On the Youla-Kucera parametrization for nonlinear systems," *Systems Control Lett.*, vol. 14, pp. 121-129, 1990.

[39] L. Pernebo, "An algebraic theory for the design of controllers for linear multivariable systems, Parts I - II," *IEEE Trans. Automatic Control*, vol. 26, pp. 171-194, 1981.

[40] V. Peterka, "On steady state minimum variance control strategy," *Kybernetika*, vol. 8, pp. 219-232, 1972.

[41] K. R. Poola, "Linear Time-Varying Systems: Representation and Control via Transfer Function Matrices," Ph.D. Thesis, University of Florida, Gainesville, USA, 1984.

[42] H. H. Rosenbrock, *State-space and Multivariable Theory.* New York: Wiley, 1970.

[43] V. Strejc, *Synthese von Regelungssystemen mit Prozessrechner.* Berlin: Akademie-Verlag, 1967.

[44] M. Šebek, "Characteristic polynomial assignment for delay-differential systems via 2-D polynomial equations," *Kybernetika*, vol. 23, pp. 345-359, 1987.

[45] M. Šebek, "n-D polynomial matrix equations," *IEEE Trans. Automatic Control*, vol. 33, pp. 499-502, 1988.

[46] G. C. Verghese, B. C. Lévy, and T. Kailath, "A generalized state-space for singular systems," *IEEE Trans. Automatic Control*, vol. 26, pp. 811-831, 1981.

[47] M. Vidyasagar, H. Schneider, and B. A. Francis, "Algebraic and topological aspects of feedback stabilization," *IEEE Trans. Automatic Control*, vol. 27, pp. 880-894, 1982.

[48] M. Vidyasagar, *Control System Synthesis: A Factorization Approach.* Cambridge, MA: MIT Press, 1985.

[49] L. N. Volgin, *The Fundamentals of the Theory of Controllers* (in Russian). Moscow: Soviet Radio, 1962.

[50] W. M. Wonham, *Linear Multivariable Control: A Geometric Approach, 2nd ed.* New York: Springer-Verlag, 1979.

[51] D. C. Youla, J. J. Bongiorno, and H. A. Jabr, "Modern Wiener-Hopf design of optimal controllers, Part I: The single-input case," *IEEE Trans. Automatic Control*, vol. 21, pp. 3-14, 1976.

[52] D. C. Youla, H. A. Jabr, and J. J. Bongiorno, "Modern Wiener-Hopf design of optimal controlers, Part II: The multivariable case," *IEEE Trans. Automatic Control*, vol. 21, pp. 319-338, 1976.

[53] O. Zariski and P. Samuel, *Commutative Algebra*, Vol. I, New York: Van Nostrand, 1958.

Chapter 2

H_2 Control Problems

K. J. Hunt, M. Šebek and V. Kučera

2.1 Introduction

There are at least three significant approaches to the design of linear $\mathcal{H}_2$ (or LQ-) optimal controllers for multivariable plants. The basic regulator problem has been studied using a time domain approach in the state-space (see, for example, Kwakernaak and Sivan [1]), and a frequency domain approach using transfer function matrices and Wiener-Hopf theory [2]. As an alternative, a *polynomial equation approach* has been developed by Kučera [3] which is based on the algebra of polynomial matrices. As with the other chapters in this volume, the distinguishing feature of our presentation is the employment of the *polynomial equation approach* of Kučera [3]. For LQ type problems controller synthesis reduces to polynomial spectral factorisation and the solution of linear polynomial equations. A deep analysis of the relationship between polynomial (transfer-function) and state-space control synthesis methods may be found in Kučera [4].

Since many different control tasks are encountered in practice, the basic regulator problem solution was later extended to various more complex control structures (such as reference tracking [5, 6] and measurement feedforward [7, 8, 9]) or various types of costing (e.g. including dynamic weights and sensitivity functions [10]). We refer to these problems as *explicit model* problems since the exact form of the model structure must be pre-specified.

During practical design work, it is perhaps necessary to perform the synthesis for every particular structure at hand. However, it is undesirable to have a number of theories each being specific to only one control structure. Instead, one "general" solution is desirable which, of course, can be simplified (adjusted) for every particular practical design. A solution of such generality has been obtained; it is known as the *standard problem* and a polynomial solution can be found in [11, 12].

It is the aim of this chapter to provide an overview of the results mentioned above. In particular, we present the following:

- a brief review of the historical development of *explicit model* problems including pure regulation, reference tracking, and measurable disturbance feedforward,

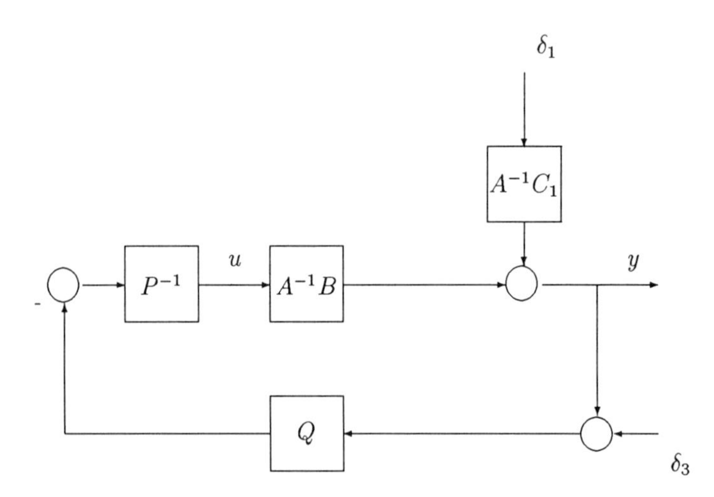

Figure 2.1: The regulator problem

- a detailed presentation of the scalar version of the standard $\mathcal{H}_2$ problem - in the scalar case the solution can be easily followed and deep insights into the problem can be obtained,
- a detailed presentation of the multivariable standard $\mathcal{H}_2$ problem.

Background

The structure of the basic regulator problem is shown in Figure 2.1. The objective here is to regulate the plant output y in the face of an unmeasurable disturbance δ_1 and a measurement noise δ_3. This optimal regulation problem, where we seek the regulator polynomial matrices P and Q, was solved using polynomial techniques by Kučera [3].

The regulator solution was extended by Šebek [6] to the optimal tracking of reference signals. The structure of the tracking problem is shown in Figure 2.2. The reference signal to be tracked is r, and δ_{rn} is a measurement noise associated with r. We require to find one further polynomial matrix R, the tracking compensator. Note that the regulator polynomial matrices P, Q are unchanged by this extension to the problem.

A further significant advance was made with the introduction of feedforward compensation for the optimal rejection of measurable disturbances [8]. The structure of this problem is shown in Figure 2.3. Here, δ_2 is a disturbance signal which we assume can be measured. We seek the feedforward compensator polynomial matrix S. Again, the regulator polynomial matrices P, Q are unchanged with this extension.

The detailed solution of these explicit model problems is given in Section 2.2.

The general structure of Figure 2.4 accounts for all the particular cases mentioned above (regulation, tracking, feedforward) and many others. Such a structure, known as the *standard feedback system* has, in fact, been borrowed from the $\mathcal{H}_\infty$ literature [13], [14]. It is recognized to cover many (if not all) of the linear control problem structures. Thus, the multivariable standard problem solution should be viewed as the most important result presented in this chapter.

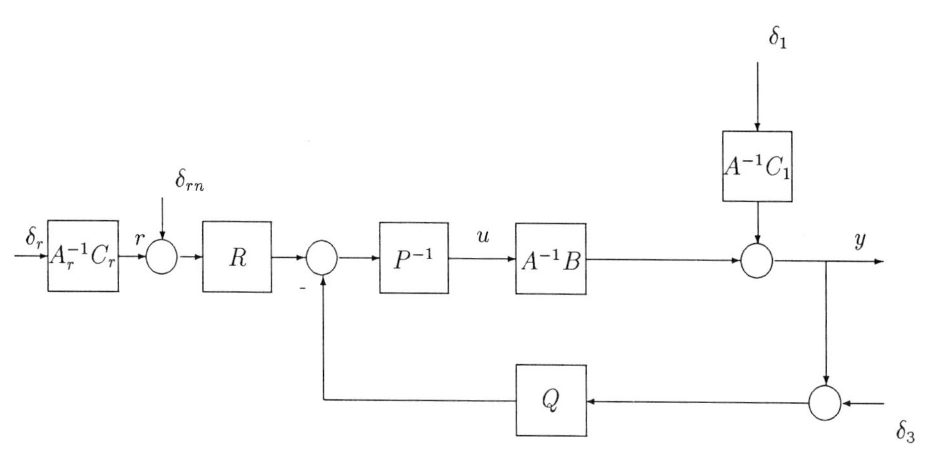

Figure 2.2: Tracking problem

The 'standard' feedback control structure has been widely studied in recent years. It includes a wide range of important control structures as special cases and therefore provides a general framework for control studies. In the standard structure (see Figure 2.4) we denote the plant by G_p. All sensor outputs are incorporated into y, while the regulated variable is z. ζ denotes all exogenous inputs to the system. The regulator K_c processes the measurement y to produce the control signal u. Many control problems can be reformulated within the standard structure (examples include output regulation, reference tracking, feedforward control and model matching).

The focus in this work is on $\mathcal{H}_2$-optimal control problems. In the specific case of the $\mathcal{H}_2$ standard problem our control objective is to minimise the $\mathcal{H}_2$ norm of the closed-loop transfer function between the input ζ and the regulated variable z.

For the solution of the scalar version of this problem the reader is referred to [15, 16, 17]. The multivariable standard problem was recently addressed in [18], but this solution was for a strongly specialised version of the standard structure; the inherent and desired generality was therefore lost. For the solution of the general multivariable $\mathcal{H}_2$ standard problem using polynomial methods the reader is referred to the papers [11, 12]. The results of Hunt *et al* [11] are presented in more detail in this chapter.

Needless to say, various control concepts such as *stability* may have different meanings in different structures. Thus, in solving the problem for one general (standard) system, we must allow some freedom when considering its stability. This becomes clear in the sequel.

Recently, in a parallel development, the other two approaches have also been generalized for more complex structures. The basic Wiener-Hopf theory has been extended to tracking and sensitivity costing [19], measurement feedforward [20, 21] and, finally, to the general structure [22]. The state space approach to $\mathcal{H}_2$-optimal control design has also been generalised to the standard structure in [23].

The standard structure has also been widely studied in the realm of $\mathcal{H}_\infty$ optimisation. Francis [13] and Doyle *et al* [23] tackle the standard $\mathcal{H}_\infty$ problem using a state-space approach, while Kwakernaak [14] employs a polynomial approach.

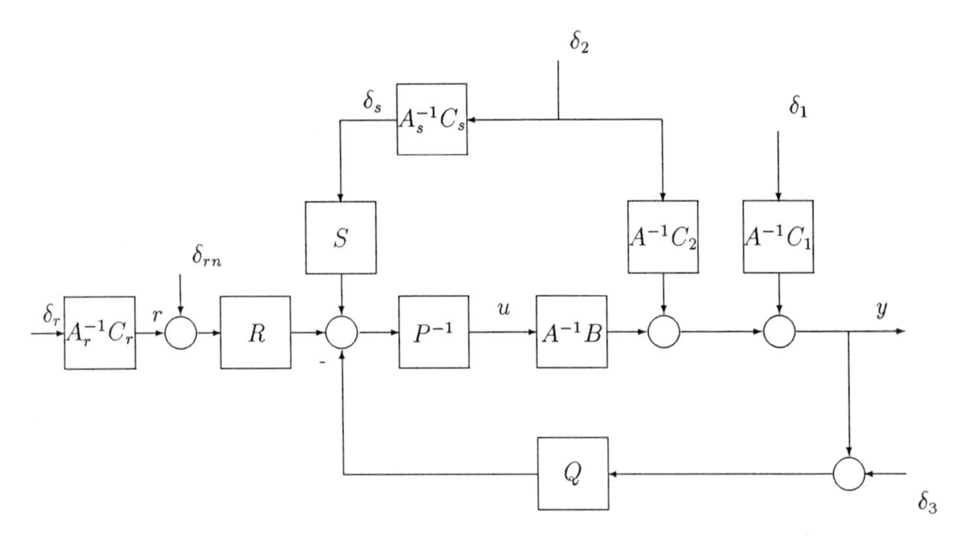

Figure 2.3: Feedforward problem

Notation

All systems considered in this chapter are assumed to be linear, time-invariant and discrete-time. The systems are described by means of real polynomial matrices in an indeterminate d, normally interpreted as the delay operator. The reader is referred to the work by Kučera [3] for details.

For simplicity, the arguments of polynomial matrices are usually omitted; a polynomial matrix $X(d)$ is denoted by X. The adjoint of a polynomial matrix $X(d)$ is written as $X^*(d)$ and defined as $X^*(d) \stackrel{\text{def}}{=} X^T(d^{-1})$. For any polynomial matrix $X(d)$ we define $\langle X \rangle$ as the matrix of terms independent of d. As a shorthand notation we often denote polynomials of the form $(X^{-1})^*$ as X^{-*}. Causal polynomial matrices are those having a non-zero term independent of d. Stable square polynomial matrices are those with all zeros of their determinant having magnitude greater than unity.

2.2 Explicit Model Problems

In this section we review optimal control problems where the plant model under consideration is made explicit (this should be contrasted with the approach of using the "standard" structure where particular problems are special cases of the general problem; see Sections 2.3 and 2.4).

In particular, we consider regulation, tracking and feedforward problems.

2.2.1 Problem formulation

The overall system under consideration here is shown in Figure 2.3; it incorporates regulation (Figure 2.1), tracking (Figure 2.2), and feedforward.

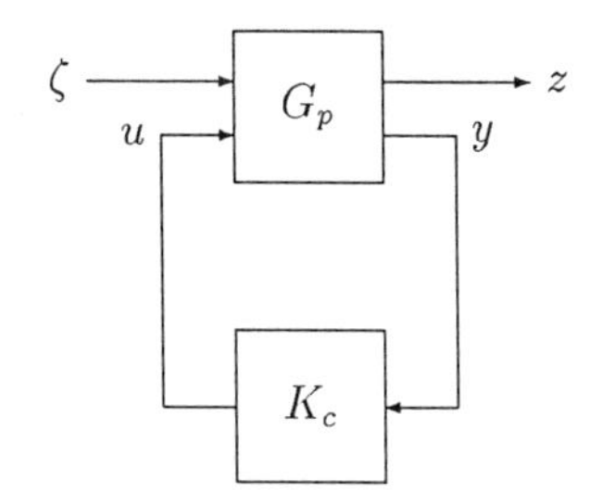

Figure 2.4: Standard feedback system

The discrete multivariable stochastic plant under consideration is governed by

$$y = A^{-1}(Bu + C_1\delta_1 + C_2\delta_2) \tag{2.1}$$

where y is the vector output sequence, u is the vector control sequence, and δ_1 and δ_2 are two vector noise sequences. A, B, C_1 and C_2 are a quadruple of left coprime polynomial matrices, with $\langle A \rangle$ invertible.

The noise component δ_2 passes through a filter to produce a measured disturbance signal δ_s:

$$A_s\delta_s = C_s\delta_2 \tag{2.2}$$

where A_s and C_s are polynomial matrices, with $\langle A_s \rangle$ invertible and C_s square. The filter $A_s^{-1}C_s$ typically represents measurement dynamics.

Further, consider a reference vector sequence r modelled as

$$A_r r = C_r\delta_r \tag{2.3}$$

where A_r and C_r are left coprime polynomial matrices, with $\langle A_r \rangle$ invertible. δ_r is a stochastic generator vector. Under measurement, the reference sequence r is corrupted by an additive observation noise δ_{rn}.

Similarly, the measurement of the plant output y is corrupted by an additive noise sequence δ_3.

The linear controller which operates on the measurement of the plant output, the observation of the reference, and on the measured disturbance, is described by (see Figure 2.3)

$$u = P^{-1}(-Q(y + \delta_3) + R(r + \delta_{rn}) + S\delta_s) \tag{2.4}$$

where P, Q, R and S are the polynomial matrices to be found, with $\langle P \rangle$ invertible. Note that in practice the controller must be realised as a single dynamical system having three vector inputs and one vector output.

All the random sources $\delta_1, \delta_2, \delta_3, \delta_r$ and δ_{rn} are mutually independent stationary white noises with intensities $\phi_1, \phi_2, \phi_3, \phi_r$ and ϕ_{rn}. We further assume, without loss of generality, that $\phi_2 = I$. ϕ_1, ϕ_3, ϕ_r and ϕ_{rn} are real non-negative definite matrices.

The desired optimal controller evolves from minimisation of the quadratic cost function

$$J = tr\langle \Omega\phi_u \rangle + tr\langle \Sigma\phi_{r-y} \rangle \tag{2.5}$$

where ϕ_u and ϕ_{r-y} are the correlation functions of the control sequence u and the tracking error sequence $r-y$ in the steady state, respectively. Ω and Σ are real non-negative definite weighting matrices, not simultaneously zero.

The design problem is to minimise the cost (2.5) subject to the constraint that the closed-loop system defined by Equations (2.1)–(2.4) be asymptotically stable.

Matrix fraction description

The plant input-output transfer matrix is written in terms of left and right coprime matrix fractions:

$$A^{-1}B = A_0^{-1}B_0 = B_1A_1^{-1} \tag{2.6}$$

We then define stable polynomial matrices D_c, D_f and G (the spectral factors) by

$$A_1^*\Omega A_1 + B_1^*\Sigma B_1 = D_c^*D_c \tag{2.7}$$

$$A\phi_3A^* + C_1\phi_1C_1^* = D_fD_f^* \tag{2.8}$$

$$A_r\phi_{rn}A_r^* + C_r\phi_rC_r^* = GG^* \tag{2.9}$$

The following left and right coprime factorisations will also be required:

$$A_0A_r^{-1} = A_4^{-1}A_3 \tag{2.10}$$

$$A^{-1}C_2 = C_aA_c^{-1} \tag{2.11}$$

$$BC_b = C_2B_c \tag{2.12}$$

$$D_f^{-1}A = A_dD_{fa}^{-1} \tag{2.13}$$

$$D_f^{-1}B = B_dD_{fb}^{-1} \tag{2.14}$$

$$(A_3G)^{-1}A_4B_0 = DE^{-1} \tag{2.15}$$

$$G^{-1}A_r = A_5G_2^{-1} \tag{2.16}$$

2.2.2 Regulation

Kučera [3] considered the basic multivariable regulation problem as depicted in Figure 2.1. The feedback part of the controller equation (2.4) may be written in the form

$$P^{-1}Q = p^{-1}q \tag{2.17}$$

where p and q are stable *rational* functions. The optimising solution for the feedback controller is given by

$$p = D_c^{-1}XD_{fb}^{-1} \tag{2.18}$$

$$q = D_c^{-1}YD_{fa}^{-1} \tag{2.19}$$

Here, the polynomial matrices X and Y form the unique solution having the property $\langle V\rangle = 0$ of the polynomial equations

$$D_c^* X + V^* B_d = A_1^* \Omega D_{fb} \tag{2.20}$$

$$D_c^* Y - V^* A_d = B_1^* \Sigma D_{fa} \tag{2.21}$$

2.2.3 Tracking

The first major extension to the regulator solution was given by Šebek [6] who derived a direct polynomial solution for the multivariable tracking problem of Figure 2.2.

The tracking part of the controller equation (2.4) may be written in the form

$$P^{-1}R = p^{-1}n \tag{2.22}$$

where n is a stable *rational* function. The optimising solution for the tracking controller is given by

$$n = D_c^{-1} M G_2^{-1} \tag{2.23}$$

Here, the polynomial matrix M forms the unique solution having the property $\langle N \rangle = 0$ of the polynomial equations

$$D_c^* L + N^* D = A_1^* \Omega E \tag{2.24}$$

$$D_c^* M - N^* A_5 = B_1^* \Sigma G_2 \tag{2.25}$$

We note that the regulator part of the optimal controller is unchanged by the addition of reference tracking; the polynomial matrices P, Q are obtained as above.

2.2.4 Feedforward

The second major extension was the derivation by Hunt and Šebek [8] of a solution to the multivariable feedforward problem; a separate feedforward compensator was derived within the overall optimisation procedure for rejection of measurable disturbances as shown in Figure 2.3.

The feedforward part of the controller equation (2.4) may be written in the form

$$P^{-1}S = p^{-1}s \tag{2.26}$$

where s is a stable *rational* function. The optimising solution for the feedforward controller is given by

$$s = D_c^{-1}(Y D_{fa}^{-1} C_a - Z) A_c^{-1} C_s^{-1} A_s \tag{2.27}$$

Here, the polynomial matrix Z forms the unique solution having the property $\langle W \rangle = 0$ of the polynomial equations

$$D_c^* U + W^* B_c = A_1^* \Omega C_b \tag{2.28}$$

$$D_c^* Z - W^* A_c = B_1^* \Sigma C_a \tag{2.29}$$

Again, the other parts of the controller (regulation and tracking) are unchanged by this extension; the polynomial matrices P, Q and R are obtained as above.

2.2.5 Other developments

Grimble has made several important contributions to the polynomial optimisation theory. These include dynamic cost function weighting elements, explicit sensitivity costing and coloured measurement noise. For an overview of these problems the reader is referred to the book [10].

More recently, Sternad and Ahlén [24] have developed a novel derivation methodology for optimal controller design. This approach is based on a variational argument and stands in contrast to the usual "completing the squares" approach followed by Kučera and the other workers referred to above. The new approach appears to provide a simple route to the solution of a variety of optimal control and filtering problems, including the multivariable measurable disturbance feedforward problem.

2.2.6 Polynomial equations

In a series of papers Roberts gave a deep analysis of the role and necessity of the bilateral polynomial matrix equations (2.20)–(2.21) arising in the solution of the optimal regulator problem. This work began in the reference [25] and has continued up to the recent work [26]. A summary of this work in relation to the regulator problem can be found in [27]. Mosca and co-workers [28] have also contributed to the analysis. Although Kučera's original solution remains unchanged by the analysis, this work has contributed to a deeper understanding of the solution and has quelled the debate which arose on this subject. In a related analysis Hunt *et al* [29] derived the conditions under which a single unilateral equation (related to the original couple) can be used to derive the unique optimal regulator. That equation is obtained by eliminating the coupling term V from (2.20)–(2.21).

Finally, Hunt and Šebek [30] have extended the results of the latter reference and derived for the full control problem, including tracking and feedforward, the conditions under which three single unilateral equations can be used to uniquely determine the optimal controller. These unilateral equations result from the elimination of the coupling terms in each of the pairs of equations (2.20)–(2.21), (2.24)–(2.25) and (2.28)–(2.29).

2.3 Standard Control Structure - Scalar Problem

The *standard control structure* (see Figure 2.4) has become the object of major control studies since 1981. The structure is important because it includes all the various control structures as special cases. The standard structure has particularly been used as the basis of H_∞ optimal control problems [13]. The general standard structure has also been attacked in H_2-optimal control problems; firstly using the Wiener-Hopf approach [22] and then the state-space approach [23].

In Figure 2.4, G_p represents the notional plant while K_c is the controller. All sensor outputs are incorporated into y, while the regulated variable is z. The vector ζ denotes all exogenous inputs to the system including disturbances, references and so on. The vector u represents the control inputs.

For scalar systems the problem of H_2-optimal control for the standard structure was solved using polynomial techniques in [15, 16]. A further solution to this problem, with relaxed stability requirements, was then given in [17].

For full details of these results the reader should consult these references; a summary of the results for the scalar standard problem is presented here.

2.3.1 Problem formulation

Plant

The *causal* plant G_p under consideration (see Figure 2.4) is governed by the polynomial model

$$Ay = Bu + C\zeta \tag{2.30}$$

$$Az = Du + E\zeta \tag{2.31}$$

where y is the measurement, z is the variable to be regulated, u is the control input, and ζ is an external input. A, B, C, D and E is a quintuple of coprime polynomials.

Further, we write A in terms of its stable/unstable factorisation as follows:

$$A = A^+A^- \tag{2.32}$$

where $^+$ denotes a stable polynomial and $^-$ denotes a completely unstable polynomial.

Moreover, it is assumed that

(A.1) The transfer function B/A is strictly causal with $\langle B \rangle = 0$, $\langle A \rangle = 1$ and $\langle A^+ \rangle = 1$.

(A.2) The greatest common divisor of A and B is a stable polynomial.

Regulator

We seek a causal regulator K_c which operates on the plant measurement y described by

$$u = K_c y \tag{2.33}$$

The transfer function K_c is written in rational form as

$$K_c = -\frac{Q}{P} \tag{2.34}$$

where P and Q are polynomials, and $\langle P \rangle = 1$.

Cost

We define the *closed-loop transfer function* as the transfer function between the external input ζ and the regulated variable z. We will denote this transfer function as H i.e.

$$z = H\zeta \tag{2.35}$$

In the $\mathcal{H}_2$ control problem we seek a regulator which minimises the $\mathcal{H}_2$ norm of the closed-loop transfer function H. The desired optimal regulator therefore evolves from minimisation of the cost function

$$J = \| H \|_2 \tag{2.36}$$

When a causal rational transfer function H is expressed as a formal power series in non-negative powers of d via long division:

$$H(d) = \sum_{k=0}^{\infty} h_k d^k \tag{2.37}$$

then its $\mathcal{H}_2$ norm may be expressed as

$$\| H \|_2^2 = \sum_{k=0}^{\infty} h_k^2 \tag{2.38}$$

Thus, the cost function above may be conveniently expressed as

$$J = \langle HH^* \rangle^{\frac{1}{2}} \tag{2.39}$$

Formally, we define the standard $\mathcal{H}_2$-optimal control problem as follows:

Problem Formulation 1 *Determine the regulator K_c such that*

(R.1) *the $\mathcal{H}_2$ norm of the closed-loop transfer function H is minimal.*

Remarks

1. Notice that our problem formulation above laid out no *explicit* requirements on stability of the control system. However, we note that to meet requirement (R.1) it is clear that H must be a causal stable transfer function. It will become apparent in the sequel that this in turn leads to the requirement of stabilising just that part of the system *which can be stabilised*, namely, the 'feedback part' of the system.

   In contrast, the problem of optimal regulation with internal stability has been solved by Hunt *et al* [15]. For this problem to have a solution the absence of unstable hidden modes within the plant and the regulator is required.

2. The $\mathcal{H}_2$ optimisation problem as formulated above is close to the LQ optimisation (or minimum variance) approach (see, for example, Kučera [3]). In fact, if the further assumption that ζ is *a stationary, zero mean, white noise sequence* is added then the $\mathcal{H}_2$ problem becomes identical with the LQ problem. In this case the $\mathcal{H}_2$ norm of the closed-loop transfer function H is related to the variance of the regulated variable in the steady state through

$$\| H \|_2^2 = \langle \phi_z \rangle \tag{2.40}$$

   where ϕ_z denotes the steady state correlation function of z.

3. Assumption (A.1) is not necessary. We may alternatively assume that B/A is merely causal and seek a strictly causal regulator K_c instead. The relevant modifications of the design procedure in this case are described in Kučera [4].

2.3.2 Scalar problem solution

We define the spectral factors F_1 and F_2 by the relations

$$CC^* = F_1 F_1^* \tag{2.41}$$

$$D^* D = F_2^* F_2 \tag{2.42}$$

and note that we are interested in solutions F_1, F_2 which are causal and stable polynomials. It is clear that necessary and sufficient conditions for stable spectral factors to exist are that:

1. C and D have no zeros on the unit circle.

For simplicity, we proceed here under the assumption that the given data make the problem regular (that the above conditions hold) i.e. that

(A.3) There exist stable solutions F_1 and F_2 to Equations (2.41)–(2.42).

We may now summarise the solution to the optimal regulation problem:

Theorem 1 *The $\mathcal{H}_2$-optimal control problem has a solution if and only if*

> **(C.1)** *There exists some polynomial G such that* $CD - BE = GA^-$.

The optimal regulator has the transfer function

$$K_c = -\frac{Y}{X} A^+ \tag{2.43}$$

Here, the polynomials X and Y form the unique solution having the property $\langle Z \rangle = 0$ of the linear polynomial equations

$$F_1^* F_2^* Y + Z^* A = C^* D^* E \tag{2.44}$$

$$F_1^* F_2^* X - Z^* B A^+ = C^* D^* G \tag{2.45}$$

Proof

(a) Stability/Finiteness

Our primary requirement is for minimality of the $\mathcal{H}_2$ norm of the closed-loop transfer function H (requirement R.1). To ensure that the cost is finite it is clear that H must be a causal *stable* rational function.

From the plant and controller equations (2.30)–(2.34) it is straightforward to derive the following expression for H:

$$H = \frac{EAP - (CD - BE)Q}{A(AP + BQ)} \tag{2.46}$$

Now, from Assumption (A.2) we may infer that there exist *stable* rational functions p and q satisfying

$$pA + qB = 1 \tag{2.47}$$

From Equations (2.46)–(2.47) p and q may be defined as

$$p = \frac{P}{AP + BQ} \tag{2.48}$$

$$q = \frac{Q}{AP + BQ} \tag{2.49}$$

Then, stable p and q immediately guarantee stability of the polynomial $AP + BQ$, which appears as a factor in the denominator of (2.46). Stability of H therefore requires the following condition:

1. The unstable part A^- of A must divide the polynomial $CD - BE$.

This necessary and sufficient condition is entirely equivalent to the following:

1. There must exist some polynomial G such that $CD - BE = GA^-$.

This condition is the solvability condition (C.1) of Theorem 1.

Having established the necessary and sufficient conditions for finiteness of the cost (stability of H) it remains to determine which stable rational functions p and q will minimise the cost.

(b) Optimisation

From Equation (2.46) the closed-loop transfer function now reads

$$H = \frac{E - qCD}{A} \tag{2.50}$$

Using this expression for H the cost function (2.39) may now be written as

$$J = \langle A^{-1}(EE^* - EC^*D^*q^* - qCDE^* + qCDC^*D^*q^*)A^{-1^*}\rangle^{\frac{1}{2}} \tag{2.51}$$

Introducing the spectral factorisations (2.41)–(2.42) this equation becomes

$$J = \langle A^{-1}(EE^* - EC^*D^*q^* - qCDE^* + qF_1F_2F_1^*F_2^*q^*)A^{-1^*}\rangle^{\frac{1}{2}} \tag{2.52}$$

After completing the squares, and some algebraic manipulation, the cost becomes

$$J = \langle VV^*\rangle^{\frac{1}{2}} \tag{2.53}$$

where

$$V = \frac{F_1F_2}{A}q - \frac{C^*D^*E}{F_1^*F_2^*A} \tag{2.54}$$

The second term in V contains both causal and anti-causal parts. These parts may be split by introducing the polynomials Y and Z, as follows

$$\frac{C^*D^*E}{F_1^*F_2^*A} = \frac{Y}{A} + \frac{Z^*}{F_1^*F_2^*} \tag{2.55}$$

Note that this equation is just the linear polynomial equation (2.44). V now reads

$$V = \left(\frac{F_1F_2}{A}q - \frac{Y}{A}\right) - \frac{Z^*}{F_1^*F_2^*} \tag{2.56}$$

Under the constraint $\langle Z \rangle = \langle Z^* \rangle = 0$ the contribution of the cross terms in (2.56) to the cost vanishes. To minimise the cost we therefore set the causal part of V to zero:

$$\frac{F_1F_2}{A}q - \frac{Y}{A} = 0 \tag{2.57}$$

or,

$$q = \frac{Y}{F_1F_2} \tag{2.58}$$

From the relation (2.47) the remaining regulator term is

$$p = \frac{1 - qB}{A} \tag{2.59}$$

Substituting firstly for q from (2.58) and then for Y from (2.44) p may be written, after some rearrangement involving (C.1), as

$$p = \frac{C^*D^*G + Z^*BA^+}{F_1^*F_2^*F_1F_2A^+} \tag{2.60}$$

Now, both p and q must be *stable* rational functions. The stability of q is guaranteed by (A.3). For p to be stable the numerator in (2.60) must include the anti-stable polynomials $F_1^*F_2^*$ as factors. In other words, there must exist some polynomial X such that

$$F_1^*F_2^*X = C^*D^*G + Z^*BA^+ \tag{2.61}$$

Note that this is just the linear equation (2.45). Moreover, p now reads

$$p = \frac{X}{F_1F_2A^+} \tag{2.62}$$

and we now see that (A.3) guarantees stability of p also. Finally, from Equations (2.48)–(2.49) the regulator transfer function (2.34) is

$$\frac{Q}{P} = \frac{q}{p} \tag{2.63}$$

Thus, putting together Equations (2.58) and (2.62) the optimal regulator transfer function (2.43) results.

Finally, we note that Assumption (A.2), as well as ensuring stabilisability of the plant, also ensures the existence of a unique solution to the pair of equations (2.44)–(2.45) with $\langle Z \rangle = 0$ (see Kučera [3] for details).

□

Corollary 1 *The polynomials X and Y in Equations (2.44)–(2.45) also satisfy the linear polynomial equation*

$$A^-X + BY = F_1F_2 \tag{2.64}$$

Proof Substitute for p and q from Equations (2.58) and (2.62) into Equation (2.47) (the result can also be proved by eliminating the coupling term Z from the Equations (2.44)–(2.45)).

□

2.3.3 Remarks

It is worth reiterating that our standard $\mathcal{H}_2$ problem formulation included no explicit stability requirements. However, it quickly became clear that minimising the $\mathcal{H}_2$ norm of the closed-loop transfer function H had an implied requirement for stability of H in order to obtain a finite cost. Hence, the first step in the solution derivation consisted in obtaining the necessary and sufficient conditions for stability of H (see (C.1) in Theorem 1).

The second step in the derivation was to find the parameters p and q corresponding to the optimal regulator. The optimisation procedure required the introduction of the linear polynomial equation (2.44) which in fact provided a systematic causal/anti-causal factorisation in the cost.

Finally, we noted that the parameters p and q were required to be *stable* rational functions. While stability of q was immediately evident, one further linear polynomial equation (2.45) was required to show that p is in general stable.

2.4 Standard Control Structure - Multivariable Problem

The standard $\mathcal{H}_2$-optimal control problem applies to a wide class of multivariable control tasks. It involves the minimization of the quadratic norm of the closed-loop transfer matrix of a general linear feedback configuration. The polynomial solution presented here was first presented in abbreviated form in Hunt *et al* [11].

The quadratic norm of a rational transfer matrix H is defined as

$$\| H \|_2 = \sum_{i,j} \| h_{ij} \|_2 \tag{2.65}$$

where the quadratic norm of a causal scalar rational function h is $\| h \|_2{}^2 = \sum_{k=0}^{\infty} h_k^2$ where h is expressed as a formal power series in nonnegative powers of d via long division $h(d) = \sum_{k=0}^{\infty} h_k d^k$.

For a stable rational matrix H, its quadratic norm $\| H \|_2$ reads

$$\| H \|_2 = tr\langle H^* H \rangle^{\frac{1}{2}} \tag{2.66}$$

while for an unstable H the $\| H \|_2$ is infinite.

2.4.1 Problem formulation

The generalised model and control structure is shown in Figure 2.4. Here, the sub-system G_p (the "plant") accounts for all components in the system with the exception of the controller K_c. The input to the controller is the vector y which includes all the measured variables. The vector z, appropriately defined, represents the regulated variables. The vector ζ denotes all exogenous inputs to the system. The vector u represents the control inputs.

The one, two and three degrees of freedom control structures (i.e. regulation, tracking and feedforward) referred to above are all special cases of this standard system; all sensor outputs are incorporated into y, while the regulated variable z may be a weighted dynamical function of u and y, for example, and may also be related to system sensitivity. z is then incorporated into a cost function for optimisation.

The discrete-time plant G_p described above is governed by

$$\begin{bmatrix} z \\ y \end{bmatrix} = \begin{bmatrix} G_{11} & G_{12} \\ G_{21} & G_{22} \end{bmatrix} \begin{bmatrix} \zeta \\ u \end{bmatrix} \tag{2.67}$$

where all the sub-blocks are causal rational matrices. Their dimensions follow from

$$\begin{array}{c} \\ m \\ n \end{array} \begin{array}{c} \begin{array}{cc} k & l \end{array} \\ \begin{bmatrix} G_{11} & G_{12} \\ G_{21} & G_{22} \end{bmatrix} \end{array} . \tag{2.68}$$

The measurement vector y is processed by the controller K_c to produce the control vector u as follows

$$u = K_c y \tag{2.69}$$

where K_c is a causal rational matrix.

Without loss of generality we introduce the following assumptions:

(A.1) The dimension of the external input ζ is at least as great as that of the observed output y, that is, $k \geq n$ and G_{21} has full normal row rank.

(A.2) The dimension of the regulated variable z is at least as great as that of the control input u, that is, $m \geq l$ and G_{12} has full normal column rank.

The assumptions are met for many problems of interest. If they are not satisfied, the problem may be transformed to an equivalent problem for which the assumptions hold (see [13],[14]).

In the configuration of Figure 2.4 the closed-loop transfer function H from the external input ζ to the regulated variable z is easily found to be given by

$$H = G_{11} + G_{12}K_c(I - G_{22}K_c)^{-1}G_{21} \tag{2.70}$$

The problem

The standard $\mathcal{H}_2$-optimal control problem may now be defined as follows:

Problem Formulation 2 *Determine the controller K_c such that*

> **(R.1)** *the quadratic norm $\| H \|_2$ of the closed-loop transfer matrix H is minimal.*

(R.2) *the closed-loop system is stable, in the sense defined below.*

□

To meet the requirement (R.1), clearly H must be a causal stable rational matrix.

The above problem formulation is evidently consistent with the standard $\mathcal{H}_\infty$-optimization (see [14]), the only difference being, of course, the use of $\| \cdot \|_2$ instead of $\| \cdot \|_\infty$.

On the other hand, when adding the assumptions that

(A.3) the vector ζ is a stationary, zero mean, white noise with intensity $\phi > 0$ (without any loss in generality we take $\phi = I$)

(A.4) the plant control - measurement transfer matrix G_{22} is strictly causal

we are consistent with the standard LQ optimisation approach (see [3]) as well since then the quadratic norm of the closed-loop transfer function H equals the trace of the constant term of the regulated variable correlation function Φ_z in the steady state,

$$\| H \|_2{}^2 = tr\langle \Phi_z \rangle \tag{2.71}$$

Stability

It is a generally accepted philosophy to think of the standard structure of Figure 2.4 as an artificial system employed to cover various real control tasks. It is clear, therefore, that the definition of its stability will differ according to the problem at hand.

The requirement of full internal stability of the overall system can be satisfied only under severe restrictions on either the internal structure of G_p or on the stability of its various internal component parts (see Hunt *et al* [15]). This appears to be both unrealistic and unreasonable for most practical cases. For example, consider an unstable reference signal generator in the classical tracking problem or an unstable model system to be matched by a stable one in the model matching problem. Such artificial (in fact, non-existent in reality) parts of G_p are usually *in principle* unstable and remain so under the action of any controller K_c.

Therefore, we relax the requirement of full internal stability of G_p and instead aim to stabilise just *what can be stabilised*, namely, the "feedback part" of G_p, G_{22}. As the realisation of G_{22} is given (often this is the "real plant") we simply look for a controller K_c which stabilises this part of the system while at the same time minimising the cost.

Necessary and sufficient conditions to achieve our stability requirements, which we now introduce as two further assumptions, are

(A.5) The realisation of G_{22} is free of unstable hidden modes.

(A.6) The unstable part of G_{21} is a factor of G_{22} i.e. the polynomial matrices A_y and B_y have no unstable left factor in common.

These conditions are well known from LQ-optimisation theory [3].

2.4.2 Matrix fraction description

The plant

The transfer matrices of the plant are written in the form of left matrix fractions:

$$G_{11} = A_z^{-1} D_z \tag{2.72}$$

$$G_{12} = A_z^{-1} C_z \tag{2.73}$$

$$G_{21} = A_y^{-1} E_y \tag{2.74}$$

$$G_{22} = A_y^{-1} B_y \tag{2.75}$$

where A_y, B_y, E_y and A_z, C_z, D_z are two triples of left coprime polynomial matrices.

For the transfer matrix G_{22} define right coprime polynomial matrices A'_y and B'_y by

$$A_y^{-1} B_y = B'_y A_y'^{-1} \tag{2.76}$$

Further, we write the plant transfer matrices in the form of right matrix fractions:

$$G_{11} = D_\zeta A_\zeta^{-1} \tag{2.77}$$

$$G_{12} = C_u A_u^{-1} \tag{2.78}$$

$$G_{21} = E_\zeta A_\zeta^{-1} \tag{2.79}$$

$$G_{22} = B_u A_u^{-1} \tag{2.80}$$

where A_u, B_u, C_u and $A_\zeta, D_\zeta, E_\zeta$ are two triples of right coprime polynomial matrices.

The plant equation (2.67) may now be written in the compact form

$$\begin{aligned} \begin{bmatrix} z \\ y \end{bmatrix} &= \begin{bmatrix} A_z & \\ & A_y \end{bmatrix}^{-1} \begin{bmatrix} D_z & C_z \\ E_y & B_y \end{bmatrix} \begin{bmatrix} \zeta \\ u \end{bmatrix} \\ &= \begin{bmatrix} D_\zeta & C_u \\ E_\zeta & B_u \end{bmatrix} \begin{bmatrix} A_\zeta & \\ & A_u \end{bmatrix}^{-1} \begin{bmatrix} \zeta \\ u \end{bmatrix} \end{aligned} \tag{2.81}$$

The spectral factors

We define a left spectral factor D_e by the relation

$$E_y E_y^* = D_e D_e^* \tag{2.82}$$

and a right spectral factor D_c by the relation

$$C^* C = D_c^* D_c \tag{2.83}$$

Here, C is obtained from the fraction

$$A_z^{-1} C_z = C A_c^{-1} \tag{2.84}$$

where the polynomial matrices C and A_c are right coprime.

Note that we are interested in solutions D_e, D_c which are *stable* polynomial matrices. The assumptions (A.1)-(A.2) make our problem regular, so that *Schur* spectral factors above exist. To ensure that they are indeed stable we need one more simplifying assumption

(A.7) The polynomial matrices C_z and E_y have no zeros on the unit circle.

The dictionary

In this section we give a concise "dictionary" of polynomial matrix fractions and definitions used in the sequel. The presentation here is brief; the origin of the various fractions will become clear as the proof of the main theorem is followed.

The following left/right coprime polynomial matrix fractions are required:

$$A_z^{-1}D_z = DA_d^{-1} \tag{2.85}$$

$$A_d^{-1}E_y^* = E_2^*A_{d2}^{-1} \tag{2.86}$$

$$D_e^*A_{d2} = A_{d3}D_{e3}^* \tag{2.87}$$

$$LD_{e3}^{-*} = D_{e4}^{-*}L_1 \tag{2.88}$$

In the latter expression the polynomial matrix L is defined by

$$L = C^*DE_2^* \tag{2.89}$$

Next, we factorise

$$D_e^{-1}B_y = B_2D_{e2}^{-1} \tag{2.90}$$

and find right coprime matrices B_3, E_3 such that

$$E_yB_3 = B_yE_3 \tag{2.91}$$

The polynomial matrix A_c is split into its stable and unstable factors as follows:

$$A_c = A_c^+A_c^- = A_c'^-A_c'^+ \tag{2.92}$$

where $^+$ denotes a stable polynomial matrix and $^-$ a completely unstable one.

We then make the factorisation

$$\begin{bmatrix} (A_c^+)^{-1}D_{e2} \\ A_{d3}^{-1}B_2 \end{bmatrix} = \begin{bmatrix} D_{e5} \\ B_5 \end{bmatrix} A_{c5}^{-1} \tag{2.93}$$

which is consistent with

$$D_{e2}^{-1}A_c^+A_c^- = A_{c5}^+D_{e5}^{-1}A_c^- = A_{c5}^+A_{c5}^-D_{e6}^{-1} = A_{c5}D_{e6}^{-1} \tag{2.94}$$

Finally A_y' is split into stable and unstable factors:

$$A_y' = A_y'^-A_y'^+ \tag{2.95}$$

the polynomial matrix K_1 is defined using

$$A_y'^- = A_c'^-K_1 \tag{2.96}$$

and the following coprime fractions are then required:

$$(A_c'^+)^{-1}K_1 = K_2(A_c''^+)^{-1} \tag{2.97}$$

and

$$D_c K_2 = K_0 D_{c2} \tag{2.98}$$

We can well define the following four useful stable rational matrices, p_y, q_y, p'_y, q'_y, along with the two polynomial matrices A'_y and B'_y without unstable right common factors, by the relations

$$K_c = -q_y p_y^{-1} = -p_y'^{-1} q'_y \tag{2.99}$$

and

$$\begin{bmatrix} A_y & B_y \\ q'_y & -p'_y \end{bmatrix} \begin{bmatrix} p_y & B'_y \\ q_y & -A'_y \end{bmatrix} = \begin{bmatrix} p_y & B'_y \\ q_y & -A'_y \end{bmatrix} \begin{bmatrix} A_y & B_y \\ q'_y & -p'_y \end{bmatrix} = I \tag{2.100}$$

2.4.3 Multivariable problem solution

The solution to the general optimisation problem is summarised in the following theorem:

Theorem 2 *Let the assumptions (A.1)-(A.7) be satisfied. Then the $\mathcal{H}_2$-optimal control problem has a solution if and only if*

(C.1) *The rational matrix $A_z^{-1}(D_z B_3 - C_z E_3)$ is stable.*

(C.2) *The rational matrix $A_c^{-1} A'_y$ is stable.*

The optimal controller, if it exists, is a realisation free of unstable hidden modes of the unique transfer matrix

$$K_c = -A_c Q_d F^{-1} A_y \tag{2.101}$$

where

$$F = D_e A_{d3} D_Q - B_y A_c Q_d \tag{2.102}$$

with the right coprime polynomial matrices D_q and Q_d given by

$$Q_d D_q^{-1} = D_c^{-1} Q \tag{2.103}$$

Here, the polynomial matrix Q (along with P' and Z) forms the unique solution having the property $\langle Z \rangle = 0$ of the linear polynomial matrix equations

$$D_{e4}^* D_c^* Q - Z^* A_{d3} = L_1 \tag{2.104}$$

$$D_{e4}^* D_c^* K_0 P' + Z^* B_2 A_{c5}^+ = L_2 \tag{2.105}$$

where the polynomial matrix L_2 is given by

$$L_2 = (D_{e4}^* D_c^* D_c - L_1 A_{d3}^{-1} D_e^{-1} B_y A_c)(A_c^-)^{-1} D_{e5} \tag{2.106}$$

□

Proof

The closed-loop transfer function may be written as

$$H = A_z^{-1}(D_z - C_z q_y E_y) \tag{2.107}$$

Its symmetric product (the corresponding correlation function) is then

$$H^* H = \Phi_z = A_z^{-1}(D_z - C_z q_y E_y)(D_z^* - E_y^* q_y^* C_z^*)A_z^{-*} \tag{2.108}$$

Introducing the spectral factorisations (2.82)–(2.83) and utilising (2.84) we obtain

$$\begin{aligned} H^* H &= D_e^* q_y^* A_c^{-*} D_c^* D_c A_c^{-1} q_y D_e - D_e^* q_y^* A_c^{-*} D_c^* D_c^{-*} C^* A_z^{-1} D_z E_y^* D_e^{-*} \\ &\quad -(D_e^* q_y^* A_c^{-*} D_c^* D_c^{-*} C^* A_z^{-1} D_z E_y^* D_e^{-*})^* + A_z^{-1} D_z D_z^* A_z^{-*} \end{aligned} \tag{2.109}$$

The proof now proceeds in three parts. In part (a) we check the finiteness of $\| H \|_2$, in part (b) we determine the optimal controller and in part (c) we examine the system stability.

(a) Finiteness

To have a finite quadratic norm, H must be a stable rational matrix. When combining (2.107) and (2.91) we get

$$\begin{aligned} H &= A_z^{-1}(D_z - C_z q_y B_y E_3 B_3^{-1}) \\ &= [A_z^{-1}(D_z B_3 - C_z E_3) + C A_c^{-1} A'_y p'_y E_3] B_3^{-1} \end{aligned} \tag{2.110}$$

Now as p'_y can (and will) be found stable by the choice of controller, CA_c^{-1} is coprime and, in fact, B_3 cancels in (2.110) completely, we can conclude that H is stable (and hence $\| H \|_2$ is finite) if and only if (C.1) and (C.2) are satisfied.

(b) Optimisation

When employing (2.109), the cost may be written as

$$\begin{aligned} \| H \|_2{}^2 &= tr\langle D_e^* q_y^* A_c^{-*} D_c^* D_c A_c^{-1} q_y D_e \rangle \\ &\quad - tr\langle D_e^* q_y^* A_c^{-*} D_c^* D_c^{-*} C^* A_z^{-1} D_z E_y^* D_e^{-*} \rangle \\ &\quad - tr\langle (D_e^* q_y^* A_c^{-*} D_c^* D_c^{-*} C^* A_z^{-1} D_z E_y^* D_e^{-*})^* \rangle \\ &\quad + tr\langle A_z^{-1} D_z D_z^* A_z^{-*} \rangle \end{aligned} \tag{2.111}$$

After completing the squares the cost function becomes

$$\| H \|_2{}^2 = tr\langle V_1^* V_1 \rangle + tr\langle V_2 \rangle \tag{2.112}$$

where

$$V_1 = D_c A_c^{-1} q_y D_e - D_c^{-*} C^* A_z^{-1} D_z E_y^* D_e^{-*} \tag{2.113}$$

$$V_2 = -D_c^{-*} C^* A_z^{-1} D_z E_y^* D_e^{-*} D_e^{-1} E_y D_z^* A_z^{-*} C D_c^{-1} + A_z^{-1} D_z D_z^* A_z^{-*} \tag{2.114}$$

We note that only the term V_1 depends upon the controller; the V_2 does not include q_y at all. We therefore proceed by attempting to minimise the term V_1.

The second term in V_1, which we will denote by V_{12}, contains both causal and non-causal parts. To ensure a causal controller results we must isolate the causal and anti-causal parts.

First, we introduce a sequence of left/right coprime fractions to manipulate the second term of V_1 into a form suitable for causal/anti-causal factorisation. Introducing the relation (2.85) obtain

$$V_{12} = D_c^{-*} C^* A_z^{-1} D_z E_y^* D_e^{-*} = D_c^{-*} C^* D A_d^{-1} E_y^* D_e^{-*} \tag{2.115}$$

Then introducing (2.86) obtain

$$V_{12} = D_c^{-*} C^* D E_2^* A_{d2}^{-1} D_e^{-*} \tag{2.116}$$

Utilising (2.87) we get

$$V_{12} = D_c^{-*} C^* D E_2^* D_{e3}^{-*} A_{d3}^{-1} \tag{2.117}$$

Finally, the definition (2.89) together with the fraction (2.88) lead to

$$V_{12} = D_c^{-*} D_{e4}^{-*} L_1 A_{d3}^{-1} \tag{2.118}$$

This term may now be split into its causal and anti-causal parts by introducing the polynomial matrices Q and Z, as follows

$$D_c^{-*} D_{e4}^{-*} L_1 A_{d3}^{-1} = Q A_{d3}^{-1} - D_c^{-*} D_{e4}^{-*} Z^* \tag{2.119}$$

Note that this is just the linear polynomial matrix equation (2.104). V_1 now reads

$$V_1 = D_c A_c^{-1} q_y D_e - Q A_{d3}^{-1} + D_c^{-*} D_{e4}^{-*} Z^* \tag{2.120}$$

Under the constraint $\langle Z \rangle = 0$ the contribution of the final term in (2.120) to the cost is minimised. To minimise the overall cost we therefore set the causal part of V_1 to zero:

$$D_c A_c^{-1} q_y D_e - Q A_{d3}^{-1} = 0 \tag{2.121}$$

or,

$$q_y = A_c D_c^{-1} Q A_{d3}^{-1} D_e^{-1} \tag{2.122}$$

From the relation (2.100) the remaining controller term is

$$p_y = A_y^{-1}(I - B_y q_y) \tag{2.123}$$

Combining (2.122) and (2.123), we result, after some algebraic manipulations, in (2.101) and (2.102).

(c) Stability

Now we require to show that the optimising q_y and p_y defined above are stable.

The relation (2.100) yields the equation

$$A_y' p_y' + q_y B_y = I \tag{2.124}$$

In Appendix 1 it is shown in detail how we proceed from here to the second polynomial matrix equation (2.105). Moreover, it is shown that the p_y' corresponding to the optimal controller is given by

$$p'_y = A'^{+^{-1}}_y A''^{+}_c D^{-1}_{c2} P' D^{-1}_{e5} A^{+^{-1}}_c \tag{2.125}$$

In Appendix 1 we then continue and show that the following relation, *entirely equivalent to (2.124)*, holds:

$$K_0 P' A^-_{c5} + Q B_5 = D_c D_{e5} \tag{2.126}$$

Rearranging,

$$D^{-1}_c K_0 P' A^-_{c5} D^{-1}_{e5} + D^{-1}_c Q B_5 D^{-1}_{e5} = I \tag{2.127}$$

Since both the rational functions on the left side of this equation are stable (due to the definition of the spectral factors and Assumption (A.7)) it is clear by analogy with (2.124) that p'_y and q_y are stable rational functions (indeed, stability of p'_y is already evident from its definition (2.125)). We therefore conclude that all other rational matrices in (2.100) are stable and, in particular, p_y is so.

□

Appendix 1 - Derivation of polynomial equation (2.105)

We proceed by multiplying Equation (2.104) from the right by $A^{-1}_{d3} D^{-1}_e B_y$:

$$D^*_{e4} D^*_c D_c A^{-1}_c A_c D^{-1}_c Q A^{-1}_{d3} D^{-1}_e B_y - Z^* D^{-1}_e B_y = L_1 A^{-1}_{d3} D^{-1}_e B_y \tag{2.128}$$

Substituting from (2.122) this becomes

$$D^*_{e4} D^*_c D_c A^{-1}_c q_y B_y - Z^* D^{-1}_e B_y = L_1 A^{-1}_{d3} D^{-1}_e B_y \tag{2.129}$$

From relation (2.100) we have

$$q_y B_y = I - A'_y p'_y \tag{2.130}$$

Utilising this relation, and introducing the factorisation (2.90) the above expression becomes

$$D^*_{e4} D^*_c D_c A^{-1}_c A'_y p'_y + Z^* B_2 D^{-1}_{e2} = D^*_{e4} D^*_c D_c A^{-1}_c - L_1 A^{-1}_{d3} D^{-1}_e B_y \tag{2.131}$$

Multiplying from the right by $A^+_c D_{e5}$ and introducing the factorisation (2.93) we obtain

$$D^*_{e4} D^*_c D_c A^{-1}_c A'_y p'_y A^+_c D_{e5} + Z^* B_2 A^+_{c5} = (D^*_{e4} D^*_c D_c - L_1 A^{-1}_{d3} D^{-1}_e B_y A_c) A^{-^{-1}}_c D_{e5} \tag{2.132}$$

We denote the right hand side of this equation as the polynomial matrix L_2 i.e.

$$L_2 = (D^*_{e4} D^*_c D_c - L_1 A^{-1}_{d3} D^{-1}_e B_y A_c) A^{-^{-1}}_c D_{e5} \tag{2.133}$$

In Appendix 2 we show that this expression is indeed polynomial, and not a rational function.

Now, from the relations (2.92), (2.95) and (2.96) we have

$$A_c^{-1}A'_y = A_c'^{+^{-1}}A_c'^{-^{-1}}A_y'^{-}A_y'^{+} = A_c'^{+^{-1}}K_1A_y'^{+} \tag{2.134}$$

Introducing the factorisation (2.97) we then obtain

$$A_c^{-1}A'_y = K_2(A_c''^{+})^{-1}A_y'^{+} \tag{2.135}$$

Equation (2.132) may now be written as

$$D_{e4}^{*}D_c^{*}D_cK_2(A_c''^{+})^{-1}A_y'^{+}p'_yA_c^{+}D_{e5} + Z^{*}B_2A_{c5}^{+} = L_2 \tag{2.136}$$

The factorisation (2.98) then gives

$$D_{e4}^{*}D_c^{*}K_0D_{c2}(A_c''^{+})^{-1}A_y'^{+}p'_yA_c^{+}D_{e5} + Z^{*}B_2A_{c5}^{+} = L_2 \tag{2.137}$$

Since the term on the right of this equation and the second term on the left are polynomial, the first term on the left must also be polynomial. Thus, there must exist some polynomial matrix P' such that

$$P' = D_{c2}(A_c''^{+})^{-1}A_y'^{+}p'_yA_c^{+}D_{e5} \tag{2.138}$$

Equation (2.137) can now be written

$$D_{e4}^{*}D_c^{*}K_0P' + Z^{*}B_2A_{c5}^{+} = L_2 \tag{2.139}$$

Note that this is just the polynomial matrix equation (2.105). Moreover, the p'_y corresponding to the optimal controller is seen from (2.138) to be

$$p'_y = (A_y'^{+})^{-1}A_c''^{+}D_{c2}^{-1}P'D_{e5}^{-1}A_c^{+^{-1}} \tag{2.140}$$

We proceed now from the relation

$$A'_yp'_y + q_yB_y = I \tag{2.141}$$

Substituting from the optimal p'_y and q_y we obtain

$$A_y'^{-}A_c''^{+}D_{c2}^{-1}P'D_{e5}^{-1}A_c^{+^{-1}} + A_cD_c^{-1}QA_{d3}^{-1}D_e^{-1}B_y = I \tag{2.142}$$

Utilising the relations (2.96) and (2.90) this becomes

$$A_c'^{-}K_1A_c''^{+}D_{c2}^{-1}P'D_{e5}^{-1}A_c^{+^{-1}} + A_cD_c^{-1}QA_{d3}^{-1}B_2D_{e2}^{-1} = I \tag{2.143}$$

Utilising the relations (2.97) and (2.93) we further obtain

$$A_c'^{-}A_c'^{+}K_2D_{c2}^{-1}P'A_{c5}^{+^{-1}} + A_cD_c^{-1}QA_{d3}^{-1}B_2 = D_{e2} \tag{2.144}$$

From (2.92), (2.98) and (2.93) we get

$$A_cD_c^{-1}K_0P'A_{c5}^{+^{-1}} + A_cD_c^{-1}QB_5A_{c5}^{-1} = D_{e2} \tag{2.145}$$

Rearranging, this can be written

$$K_0 P' A_{c5}^- + Q B_5 = D_c A_c^{-1} D_{e2} A_{c5} \tag{2.146}$$

One final application of (2.93) allows us to write

$$K_0 P' A_{c5}^- + Q B_5 = D_c D_{e5} \tag{2.147}$$

Appendix 2 - Proof that L_2 is polynomial

This section is to show that L_2 is indeed a polynomial matrix. Using its definition (2.106) and various relations from the Dictionary, the following chain of equalities can be derived:

$$\begin{aligned} L_2 &= D_{e4}^* D_c^* D_c (A_c^-)^{-1} D_{e5} - L_1 A_{d3}^{-1} D_e^{-1} B_y A_c^+ D_{e5} && (2.148)\\ &= (D_{e4}^* D_c^* D_c - L_1 A_{d3}^{-1} D_e^{-1} B_y A_c)(A_c^-)^{-1} D_{e5} && (2.149)\\ &= D_{e4}^* (C^* C - C^* D E_2^* A_{d2}^{-1} D_e^{-*} D_e^{-1} B_y A_c)(A_c^-)^{-1} D_{e5} && (2.150)\\ &= D_{e4}^* C^* (C A_c^{-1} - A_z^{-1} D_z B_3 E_3^{-1}) A_c^+ D_{e5} && (2.151)\\ &= D_{e4}^* C^* A_z^{-1} (C_z E_3 - D_z B_3) E_3^{-1} A_c^+ D_{e5} && (2.152) \end{aligned}$$

Comparing now (2.149) and (2.152), the E_3 must cancel and it remains to show that A_z cancels as well. Indeed, its unstable part cancels according to the condition (C.1) and then its stable part cancels against A_c^+.

Bibliography

[1] H. Kwakernaak and R. Sivan, *Linear Optimal Control Systems.* New York: Wiley, 1972.

[2] D. C. Youla, H. A. Jabr, and J. J. Bongiorno, "Modern Wiener-Hopf design of optimal controllers - Part 2: the multivariable case," *IEEE Trans. Automatic Control*, vol. AC-21, pp. 319–338, 1976.

[3] V. Kučera, *Discrete Linear Control: the polynomial equation approach.* Chichester: Wiley, 1979.

[4] V. Kučera, *Analysis and Design of Discrete Linear Control Systems.* Hemel Hempstead: Prentice-Hall, 1992.

[5] M. Šebek, "Polynomial design of stochastic tracking systems," *IEEE Trans. Automatic Control*, vol. 27, pp. 468–470, 1982.

[6] M. Šebek, "Direct polynomial approach to discrete-time stochastic tracking," *Problems of Control and Information Theory*, vol. 12, pp. 293–302, 1983.

[7] M. Šebek, K. J. Hunt, and M. J. Grimble, "LQG regulation with disturbance measurement feedforward," *Int. J. Control*, vol. 47, pp. 1497–1505, 1988.

[8] K. J. Hunt and M. Šebek, "Optimal multivariable regulation with disturbance measurement feedforward," *Int. J. Control*, vol. 49, pp. 373–378, 1989.

[9] M. Sternad and T. Söderström, "LQG-optimal feedforward regulators," *Automatica*, vol. 24, pp. 557–561, 1988.

[10] M. J. Grimble and M. A. Johnson, *Optimal Control and Stochastic Estimation - Vols 1 and 2.* Chichester: Wiley, 1988.

[11] K. J. Hunt, M. Šebek, and V. Kučera, "Polynomial approach to H_2-optimal control: the multivariable standard problem," in *Proc. IEEE Conference on Decision and Control, Brighton, England*, 1991.

[12] A. Casavola and E. Mosca, "Polynomial LQG regulator design for general system configurations," in *Proc. IEEE Conference on Decision and Control, Brighton, England*, 1991.

[13] B. A. Francis, *A Course in H_∞ Control Theory.* Berlin: Springer-Verlag, 1987.

[14] H. Kwakernaak, "The polynomial approach to H_∞-optimal regulation," in *Proc. CIME Course on Recent Developments in H_∞ Control Theory, Como, Italy*, 1990.

[15] K. J. Hunt, V. Kučera, and M. Šebek, "Optimal regulation using measurement feedback: A polynomial approach," *Trans. IEEE on Automatic Control*, vol. 37, pp. 682–685, May 1992.

[16] K. J. Hunt and M. Šebek, "Polynomial LQ optimisation for the standard control structure: scalar solution," *Automatica*, 1992. To appear.

[17] K. J. Hunt and V. Kučera, "The standard H_2-optimal control problem: a polynomial solution," *Int. J. Control*, 1992. To appear.

[18] M. J. Grimble, "Polynomial matrix solution to the standard H_2 optimal control problem," *Int. J. Systems Sci.*, vol. 22, pp. 793–806, 1991.

[19] D. C. Youla and J. J. Bongiorno, "A feedback theory of two degree of freedom Wiener-Hopf design," *Trans. IEEE on Automatic Control*, vol. 30, pp. 652–665, 1985.

[20] K. Park, J. J. Bongiorno, and D. C. Youla, "Wiener-Hopf design of multivariable control systems," *Weber Research Institute Report*, vol. POLY-WRI-1515-1987, 1987.

[21] K. Park and J. J. Bongiorno, "Wiener-Hopf design of servo-regulator-type multivariable control systems including feedforward compensation," *Int. J. Control*, vol. 52, pp. 1189–1216, 1990.

[22] K. Park and J. J. Bongiorno, "A general theory for the Wiener-Hopf design of multivariable control systems," *IEEE Trans. Automatic Control*, vol. 34, pp. 619–626, 1989.

[23] J. C. Doyle, K. Glover, P. P. Khargonekar, and B. A. Francis, "State-space solutions to standard H_2 and H_∞ control problems," *IEEE Trans. Automatic Control*, vol. 34, pp. 831–847, 1989.

[24] M. Sternad and A. Ahlén, "A novel derivation methodology for polynomial-LQ controller design," *submitted for publication*, 1990.

[25] A. Roberts, "Simpler polynomial solutions in stochastic feedback control," *Int. J. Control*, vol. 45, pp. 117–126, 1987.

[26] M. Newmann and A. Roberts, "Polynomial optimisation of stochastic discrete-time control for unstable plants," *Int. J. Control*, vol. 51, pp. 1363–1380, 1990.

[27] K. J. Hunt, "Polynomial optimisation of discrete-time multivariable regulators," *Proc. IEE Pt. D*, vol. 139, pp. 190–196, March 1992.

[28] E. Mosca, L. Giarré, and A. Casavola, "On the polynomial equations for the MIMO LQ stochastic regulator," *IEEE Trans. Automatic Control*, vol. 35, pp. 320–322, 1990.

[29] K. J. Hunt, M. Šebek, and M. J. Grimble, "Optimal multivariable LQG control using a single diophantine equation," *Int. J. Control*, vol. 46, pp. 1445–1453, 1987.

[30] K. J. Hunt and M. Šebek, "Implied polynomial matrix equations in multivariable stochastic optimal control," *Automatica*, vol. 27, pp. 395–398, 1991.

Chapter 3

LQ Controller Design and Self-tuning Control

M. Sternad and A. Ahlén

3.1 Introduction

The present chapter contributes on some aspects of infinite–horizon LQG control for discrete time systems[1]. It takes up four themes, which complement mainly Chapter 2 and Chapter 5 of this volume:

1. A novel *variational technique*, for deriving polynomial equations for LQG controller design, is presented in Section 3.2. It is illustrated on MIMO feedforward design in Section 3.3. (Application of this tool to filtering problems is discussed in Chapter 5.)

2. Rejection of *nonstationary disturbances*, described by marginally stable and non–stabilisable models, is another topic. A control law discussed in Section 3.4 achieves this by utilising the internal model principle (generalised integration) combined with disturbance measurement feedforward.

3. The suitability of polynomial LQG design as a candidate for *self–tuning control* is highlighted, and an algorithm is presented in Sections 3.5.

4. We investigate some of the user choices in LQG control, which influence the *robustness* of both self–tuning and fixed control laws. In particular, the roles of feedforward control and of different choices of observer polynomials are discussed. See Section 3.6.

These contributions are four inter–related pieces of a larger puzzle, which we see as a challenge for research in the coming years. To explain our interest in the discussed issues, a view of the present state of the area is first outlined.

[1]The chapter describes work partially supported by the Swedish National Board for Technical Development (NUTEK), under grants 84-3680 and 87-01573. Parts of Sections 3.4 and 3.5, and most of Section 3.6, have previously appeared in the paper [53] in the International Journal of Control, published by Taylor and Francis Ltd.

3.1.1 The state of the art

Design of regulators for linear time–invariant discrete–time systems is an area where it is natural to use the polynomial approach, This was, historically, the domain where polynomial methods were first developed. Polynomial methods are now included in many standard textbooks on sampled data control. See e.g. [4].

Consider a simple scalar system with input $u(t)$, output $y(t)$ and output disturbance $n(t)$. It is expressed by polynomials in the backward shift operator q^{-1}:

$$y(t) = \frac{B(q^{-1})}{A(q^{-1})}u(t-k) + n(t) \quad ; \quad k \geq 1 \ . \tag{3.1}$$

Control by a feedback regulator

$$R(q^{-1})u(t) = -S(q^{-1})y(t) + r(t) \tag{3.2}$$

results in the closed loop system

$$P(q^{-1})y(t) = B(q^{-1})r(t-k) + A(q^{-1})R(q^{-1})n(t) \tag{3.3}$$

where

$$P(q^{-1}) \triangleq A(q^{-1})R(q^{-1}) + q^{-k}B(q^{-1})S(q^{-1}) \ . \tag{3.4}$$

When a desired closed–loop system denominator $P(q^{-1})$ is given, equation (3.4) constitutes a linear polynomial equation, a *Diophantine equation.* This is one of the three basic types of equations which occur when regulators and filters are designed[2]. It is easily solved in the unknowns $R(q^{-1})$ and $S(q^{-1})$. Two nice features of the polynomial approach are evident in the feedback system above.

1. By manipulating numerator and denominator polynomials separately, the design equation (3.4) becomes linear in the unknowns, in spite of the fact that the regulator enters the closed–loop system in a nonlinear way. (This feature arises whenever a controlled object can be described in fractional form. See [31].)

2. The resulting system (3.3) is in input–output form. This is helpful for gaining immediate insight into some properties of the solution.

When $A(q^{-1})$ and $B(q^{-1})$ have no common factors, the closed–loop denominator $P(q^{-1})$ can, in principle, be chosen in an arbitrary way. For example, the choice $P(q^{-1}) = 1$ results in state deadbeat control, while $P(q^{-1}) = B(q^{-1})$ results in output deatbeat control, if the system is minimum phase. If the disturbance is described by the ARMA model

$$n(t) = \frac{C(q^{-1})}{A(q^{-1})}e(t)$$

---

[2]Polynomial spectral factorisations and the coprime factorisation of polynomial matrices in multivariable problems are the two other types.

where $e(t)$ is white noise, then the choice $P(q^{-1}) = B(q^{-1})C(q^{-1})$ results in minimum variance control for minimum phase systems. See e.g. [4].

With respect to polynomial methods, the level of knowledge around 1970 roughly corresponded to what has been outlined above, including significant multivariable generalisations [45], [57]. Since then, several workers have extended and generalised the polynomial approach to linear controller design, beginning with the minimum variance regulator for non–minimum phase systems of Peterka [42]. Central to this development was, of course, the work by Kučera [27], [28], [29].

During the last decade, considerable interest has been focused on linear quadratic methods, inspired by the solution presented by Kučera in [27] and, more recently, in [32]. In the scalar case above, this LQG solution implies pole placement in

$$P(q^{-1}) = \beta(q^{-1})C(q^{-1}) \tag{3.5}$$

where $\beta(q^{-1})$ is a stable polynomial, obtained from a criterion–related spectral factorisation. For details, see Section 3.4 below.

Compared to the other pole placement rules mentioned above, LQG optimisation is superior as a framework for regulator design. It provides a direct tradeoff between input energy and disturbance rejection. Non–minimum phase dynamics will not cause instability because $\beta(z^{-1})$, as opposed to $B(z^{-1})$, is stable, under mild conditions. Extensions of this solution include feedforward links and the use of dynamic cost weighting. See e.g. [18], [20], [22], [23], [24] [33], [41], [47], [50], [52] and [53]. A recent result is the derivation of polynomial equations for solving the standard $\mathcal{H}_2$ control problem. That framework contains, in principle, all the previously studied LQ problem formulations as special cases. See [11], [25] and Chapter 2 of this volume. Polynomial methods have also found use for $\mathcal{H}_\infty$–optimisation (cf. Chapters 4 and 6), although that field is dominated by the state–space approach of Doyle and co–workers.

3.1.2 So where do we go from here?

Is the theory of polynomial methods for linear controller design complete? No, far from it. It is our strong belief that, in spite of the past progress, some of the most relevant questions for controller design have hardly been posed, and much less answered.

Work on the polynomial approach has so far mainly focused on problems of *systems theory*: deriving design equations, investigating conditions for problem solvability in an idealised sense, and (to some extent) developing numerical algorithms. What is needed is a complementary development, which rather belongs to the field of *control*: we now have some tools, but *how* should we use them? Are they even the right tools for solving the problems we are really interested in?

Take the pole placement equation (3.4) as an example. Algebraically, arbitrary pole placement might be allowed. In reality, a belief that arbitrary pole placement is possible can be outright dangerous. Just consider the extremely high controller gain and sensitivity to model errors which can result from a state deadbeat design. Or think of minimum variance design, which, in addition to the above problems, frequently results in output oscillations between the sampling instants. (The use of LQG is better, but does not guarantee a good control behaviour.) If our goal is to convert the polynomial approach into a sensible design philosophy, much remains to be done. Some important design issues, which should be taken into account, have been discussed in an excellent way by Boyd and Barratt [10] and by Maciejowski [38]. A sample of challenging issues:

• Guidelines are needed on what kind of specifications it is futile to even ask for in a controller design. Today, many limitations on achievable performance are known, but we have no good ways of building them into the polynomial approach. For example, it is well known that non-minimum phase zeros impose upper bounds on the achievable performance of a closed–loop system. These bounds are of engineering nature, rather than algebraic; the roll–off rate could always be chosen so as not to obtain instability, but at the expense of disturbance amplification. The consideration of practical design constraints will often lead to rather different answers than consideration of purely algebraic constraints[3].

• Design involves tradeoffs; we often move in a Pareto–optimal subspace of the design parameter space. Control structures which introduce new degrees of freedom simplify the task for the designer. The study of their properties is of high practical relevance. One such control structure is the use of disturbance measurement feedforward; it is one of the main themes of this chapter.

• An LQG approach, in particular with frequency dependent weightings, might be a fruitful starting point in a quest for a sensible design philosophy. However, the role of design choices, such as dynamic weightings and observer poles, is still insufficiently understood. (A difficulty is that it is rather hard to "see through" Diophantine equations. The task does not exactly become easier when the solution involves a long list of coprime factorisations.) Some guidelines on these issues for SISO systems, based on our experience, are discussed in Section 3.6.

• Furthermore, the models on which control or filtering is based, must be obtained in some way. Here, the connection to system identification [36], [49], recursive identification [37] and adaptive methods is of importance. Such issues are discussed in Section 3.5.

• The question of robustness against modelling errors is perhaps the most difficult challenge. It should be noted that this problem does not disappear just because

[3]An example is the question if one single of two coupled Diophantine equations are required for each degree of freedom in LQG control. For the case of feedback in an adaptive setting, the use of two coupled equations should be preferred. Stable common factors will often appear in identified models, and the use of two equations avoids numerical problems for that reason. See Section 3.4. In feedforward control problems, on the other hand, no *realistic* problem formulation will include strictly unstable models of exogenous signals, which could cause solvability problems. There, the use of one single equation is sufficient. See Section 3.3 and 3.4.

adaptive algorithms are used, cf. Section 3.6. An LQG design can be arbitrarily sensitive, but the conditions when this actually occurs are not well understood[4]. A focus on *performance* robustness, as opposed to just stability robustness, would be desirable. Robust design has largely been ignored in the polynomial literature, with the exception of a recent paper by Grimble [21].

One of our convictions is that, to gain insight and design intuition, one needs to continue to study different special *explicit model* formulations. They often have much simpler and more transparent solutions that the most general problems. (See e.g. the filtering problems discussed in Chapter 5.) For this reason, one of our interests has been in simple derivation methods for obtaining solutions. The result of that work is presented in Section 3.2.

3.1.3 Remarks on the notation

The backward shift operator q^{-1} corresponds to z^{-1} in the frequency domain. Trace and transpose of matrices M will be denoted tr$M$ and M', respectively. $\mathcal{E}$ will denote expectation. For any polynomial matrix $P(z^{-1})$, $P_* = P(z)'$. The arguments will often be omitted, unless there is risk for confusion. A square polynomial matrix, of full normal rank, is called *stable* (or strictly Schur), if its determinant has all zeros in $|z| < 1$. Rational matrices $\mathcal{R}(z^{-1})$ are called stable if all their elements are transfer functions with poles in $|z| < 1$. If $P(q^{-1})$ is a square polynomial matrix, all elements of the rational matrix $P(q^{-1})^{-1}$ are *causal* if and only if the leading coefficient matrix of $P(q^{-1})$, denoted $P(0)$, is nonsingular. The *degree* of $P(q^{-1})$ is the highest degree of any of its polynomial elements.

3.2 Outline of a Variational Procedure for Deriving LQG Control Laws in Polynomial Form

A new way of deriving LQG controller design equations, which requires significantly fewer algebraic steps, compared to the traditional "completing the squares–method" [27], is presented here. Essentially, an old variational argument is utilised in a novel way. Orthogonality between signals and variations is evaluated in the frequency domain, to obtain polynomial equations which define the control law. The resulting equations are, of course, the same as obtained by a "completing the squares"–reasoning. Subsequent discussion of solvability of the equations and stability of the solution remains unaltered.

The method has been presented, and demonstrated on MIMO feedback design, in [54]. Application on control problems with marginally stable (and non–stabilizable) blocks is demonstrated here. That generalisation leaves the reasoning unchanged. It requires mainly a separate verification of finite cost. Examples of such verifications

[4]Much attention has instead been focused on the LQG/LTR modification. The properties, and the drawbacks, of this technique can be analysed by using a polynomial approach. See [9].

are found in Appendix A and B. The method is applied on a multivariable feedforward problem, and a restricted version of the general $\mathcal{H}_2$ problem, in Section 3.3. A scalar feedback and feedforward regulator is derived in Section 3.4. Application to filtering problems can be found in [1] and in Chapter 5 of this volume [2]. Please refer to that chapter for a comparative evaluation of different derivation techniques.

Consider the control of a linear discrete–time system, which is stochastic and time–invariant. Its inputs $u(t) \in R^m$ are to be calculated, based on linear combinations of measurable outputs $z(t) \in R^n$, so that the signals $y(t) \in R^p$ are controlled. Denote the regulator

$$u(t) = -\mathcal{R}(q^{-1})z(t) \tag{3.6}$$

where $\mathcal{R}(z^{-1})$ is a causal and rational $m|n$–matrix. It is to be designed so that the controlled system is stable and the infinite–horizon quadratic criterion

$$J = \mathcal{E}\{\mathrm{tr}Vy(t)(Vy(t))' + \mathrm{tr}Wu(t)(Wu(t))'\} \tag{3.7}$$

is minimised. Above, $V(q^{-1})$ and $W(q^{-1})$ represent polynomial weighting matrices, of dimensions $p|p$ and $m|m$, respectively. (The use of rational weighting matrices is straightforward, but requires additional coprime factorisations in the solution for multivariable problems.) Variational arguments will be used in order to minimize (3.7). For that purpose, introduce the *alternative regulator*

$$u(t) = -\mathcal{R}(q^{-1})z(t) + \nu(t) \tag{3.8}$$

where $\nu(t) \in R^m$ is a linear function of available data up to time t. The use of (3.8) results in the modified signals

$$\begin{aligned} y(t) &= y_o(t) + \delta y(t) \\ u(t) &= u_o(t) + \delta u(t) \end{aligned} \tag{3.9}$$

where $y_o(t)$ and $u_o(t)$ result from control by (3.6), while $\delta y(t)$ and $\delta u(t)$ are caused by the variation $\nu(t)$. The criterion can then be expressed as

$$J = J_o + 2J_1 + J_2 \tag{3.10}$$

where

$$\begin{aligned} J_o &= \mathcal{E}(\mathrm{tr}(Vy_o)(Vy_o)' + \mathrm{tr}(Wu_o)(Wu_o)') \\ J_1 &= \mathcal{E}(\mathrm{tr}(Vy_o)(V\delta y)' + \mathrm{tr}(Wu_o)(W\delta u)') \\ J_2 &= \mathcal{E}(\mathrm{tr}(V\delta y)(V\delta y)' + \mathrm{tr}(W\delta u)(W\delta u)')\ . \end{aligned} \tag{3.11}$$

The goal is now to select $\mathcal{R}$ so that J_1 vanishes. Then, the regulator (3.6) is optimal; no perturbation $\nu(t)$ could improve the performance, since J_o does not depend on $\nu(t)$ and since $J_2 \geq 0$.

So far, this reasoning is well known, see e.g. [56]. We now outline our novel contributions. Assume $Vy_o(t)$, $V\delta y(t)$, $Wu_o(t)$ and $W\delta u(t)$ to be stationary. (This has to be verified, in each particular problem.) Let ℓ be the dimension of the noise vector disturbing the system. By using Parsevals formula, $J_1 = 0$ can be expressed as

$$J_1 = \frac{1}{2\pi i} \oint_{|z|=1} \mathrm{tr}\mathcal{M}(z, z^{-1}) \frac{dz}{z} = 0 \qquad (3.12)$$

where $\mathcal{M}(z, z^{-1})$ is a rational $\ell|\ell$ matrix. The relation (3.12) is fulfilled if *each element of* $\mathcal{M}(z, z^{-1})z^{-1}$ *is made analytic in* $|z| \leq 1$. Then, the scalar integrand $\mathrm{tr}\mathcal{M}(z, z^{-1})z^{-1}$ is also analytic in $|z| \leq 1$, so the integral vanishes. These ℓ^2 element-wise conditions determine $\mathcal{R}$[5]. By using right matrix fraction descriptions (MFD's), they can be satisfied collectively, as will be exemplified in the next section. This results in the polynomial matrix equations to be used in the design.

The choice of variational term $\nu(t)$ is arbitrary, except that the modified control law must remain causal, and the variational term must not destabilise the system. Representations of stationary signals can be written in innovations form[6], where a stationary innovations signal, $\epsilon(t)$, represents the most recent information at time t. Whatever could possibly be achieved by a variation of the regulator (3.6), could just as well be accomplished by adding a *feedforward from the innovations*, $\nu(t) = \mathcal{T}(q^{-1})\epsilon(t)$, to it. Such a control variation preserves stability, since feedback loops are unaffected. With $\mathcal{T}$, of dimension $m|\ell$, being stable and causal and $\epsilon(t)$ being stationary, such a variation will always be admissible.

When there are several separate measurable signal sources, for example reference generators and disturbances, a control structure with several degrees of freedom can be optimised. The trick is to add *several* variational terms in (3.8), one for each signal source. The result is a criterion with several cross terms. These should be set to zero separately. This technique will be exemplified in Section 3.4.

Note that the condition $J_1 = 0$ can be expressed as

$$\mathcal{E}((V\delta y)'(W\delta u)') \begin{pmatrix} Vy_o \\ Wu_o \end{pmatrix} = 0 \ .$$

The vector $((Vy_o)'(Wu_o)')'$ contains signals appearing in the criterion when the regulator (3.6) is used. It is required to be orthogonal to the vector of perturbations $((V\delta y)'(W\delta u)')$, caused by admissible variations of the control law.

The outlined procedure is a *constructive* derivation technique. Its initial steps are related to a proof by contradiction, first presented in [4]. In that approach, the

[5] This is a crucial insight. It would be hard to determine $\mathcal{R}$ from the scalar condition (3.12) directly.

[6] If signals are generated by unstable systems, we call it a generalised innovations form.

optimality of a filter or regulator, obtained by other means, is verified; it is demonstrated that (3.12) is fulfilled, because the integrand is analytic in $|z| \leq 1$.

3.3 Feedforward and Disturbance Decoupling

Assume a stable system to be represented by a model in right MFD form

$$y(t) = BA^{-1}u(t) + DE^{-1}w(t) \tag{3.13}$$

where $w(t) \in R^{\ell}$ is a vector of measurable disturbances. It is described by a model in right MFD form, with poles inside or on the unit circle (describing e.g. a sequence of step disturbances)

$$w(t) = GH^{-1}v(t) \ . \tag{3.14}$$

Here, $v(t) \in R^{\ell}$ is stationary white noise, with zero means and covariance matrix $\psi \geq 0$[7]. The polynomial matrices $B(q^{-1})$, $A(q^{-1})$, $D(q^{-1})$, $E(q^{-1})$, $G(q^{-1})$ and $H(q^{-1})$ have dimensions $p|m$, $m|m$, $p|\ell$, $\ell|\ell$, $\ell|\ell$ and $\ell|\ell$, respectively. Delays are included in the corresponding polynomials of $B(q^{-1})$ and $D(q^{-1})$. The pairs (B, A), (D, E) and (G, H) need not necessarily be right coprime. We assume $w(t)$, but not $y(t)$, to be measurable. Thus, $z(t) = w(t)$. The criterion (3.7) is to be minimised by using feedforward control. Assume the following:

A1. The polynomial matrices A, E and G are all stable and have nonsingular leading coefficient matrices. Thus, they have stable and causal inverses.

A2. There exists a stable $m|m$ right polynomial spectral factor $\beta(q^{-1})$, defined by

$$\beta_* \beta = B_* V_* V B + A_* W_* W A \tag{3.15}$$

with $\beta(0)$ nonsingular[8].

Stability of the system is a natural requirement in an open–loop problem. Stable invertibility of the disturbance description implies that it must have full rank; there should really exist ℓ independent noise sources. The conditions for the existence of a stable spectral factor β are mild, see the footnote below. To assure a finite criterion in the presence of nonstationary disturbances, two additional assumptions are introduced. Let $\{z_j = e^{i\omega_j}\}$ denote all zeros of H on the unit circle. They are

---

[7]The formulation can also be interpreted as a reference feedforward (servo) problem, or a combination of reference and disturbance feedforward. In a reference feedforward problem, $w(t)$ is the command signal and $-DE^{-1}w(t)$ is an ideal response model.

[8]Two conditions are, together, sufficient for A2 to be fulfilled.

1) The matrix $[B_* V_* \ \ A_* W_*]$ has full (normal) row rank m. This is a condition for the existence of a spectral factor, see [27]. It is fulfilled, for example, if all m inputs are penalized.

2) The greatest common left divisor of $B_* V_*$ and $A_* W_*$ has nonzero determinant on $|z| = 1$. This assures $\det \beta(z^{-1}) \neq 0$ on $|z| = 1$. The factor β is unique, up to a left orthogonal factor.

the frequencies where the rank of $H(e^{i\omega_j})$ drops below full rank.

A3. The criterion polynomial W in (3.7) is chosen such that $W(z_j^{-1}) = 0$.

A4. There exist right inverses $B^{\dagger}(z^{-1})$ to $B(z^{-1})$ at all $z = \{z_j\}$.

Note that for $m > 1$, condition A3 is stronger than just requiring W to have a zero at $\{z_j\}$. Assumptions A3 and A4 have interesting physical interpretations, discussed in Appendix A. In order to minimize (3.7), the perturbed feedforward regulator

$$u(t) = -\mathcal{R}w(t) + \nu(t) \tag{3.16}$$

with $\mathcal{R}$ of dimension $m|\ell$, is introduced. Since G is assumed stably invertible, $v(t)$ in (3.14) represents the innovations sequence in this problem. All admissible variations can then be expressed as $\nu(t) = \mathcal{T}v(t)$. The rational matrix $\mathcal{T}(q^{-1})$ must be causal and stable, but is otherwise arbitrary.

When the system is controlled by (3.16), outputs and inputs are given by (3.9), where

$$\begin{aligned} y_o(t) &= (DE^{-1} - BA^{-1}\mathcal{R})w(t) \quad ; \quad \delta y(t) = BA^{-1}\mathcal{T}v(t) \\ u_o(t) &= -\mathcal{R}w(t) \quad ; \quad \delta u(t) = \mathcal{T}v(t) \ . \end{aligned} \tag{3.17}$$

The signals are stationary for any stable $\mathcal{R}$, since E^{-1}, A^{-1} and $\mathcal{T}$, are assumed stable. The use of (3.17) in (3.11) gives the cross–term

$$\begin{aligned} J_1 &= \frac{1}{2\pi i}\oint_{|z|=1} \text{tr}[V(DE^{-1} - BA^{-1}\mathcal{R})GH^{-1}\psi\mathcal{T}_*A_*^{-1}B_*V_*]\frac{dz}{z} \\ &\quad - \frac{1}{2\pi i}\oint_{|z|=1} \text{tr}[W\mathcal{R}GH^{-1}\psi\mathcal{T}_*W_*]\frac{dz}{z} \ . \end{aligned} \tag{3.18}$$

By using $\text{tr}\mathcal{A}\mathcal{B}_* = \text{tr}\mathcal{B}_*\mathcal{A}$, (with $\mathcal{B} = A_*^{-1}B_*V_*$ and W_*, respectively), the matrices in both terms get equal dimension $\ell|\ell$[9]. The use of the spectral factorisation (3.15) to simplify the integrand then gives

$$J_1 = \frac{1}{2\pi i}\oint_{|z|=1} \text{tr}[A_*^{-1}(B_*V_*VDE^{-1}G - \beta_*\beta A^{-1}\mathcal{R}G)H^{-1}\psi\mathcal{T}_*]\frac{dz}{z} \ . \tag{3.19}$$

When (3.19) is set to zero, it corresponds to (3.12). Note that E^{-1}, A^{-1} and H^{-1} have elements with poles only in $|z| \leq 1$ and $1/z$ contributes a pole at $z = 0$. Poles at $z = 0$ can be caused by V, D, G, β, since they contain polynomials in z^{-1}. They must *all* be eliminated. For that purpose, introduce a right coprime factorisation

[9] This trace rotation step is not required in the (otherwise very similar) reasoning for optimising filters in [1] of Chapter 5 [2].

$$G_2 E_2^{-1} = E^{-1} G \tag{3.20}$$

with G_2 and E_2 both of dimension $\ell|\ell$. Since E^{-1} is stable and causal, so is E_2^{-1}. The position of $\mathcal{R}$ in (3.19) enables direct cancellation of βA^{-1}, if $A\beta^{-1}$ is a left factor of $\mathcal{R}$. With $E_2^{-1}G^{-1}$ as a right factor of $\mathcal{R}$, G is also eliminated, while E_2^{-1} has to be factored out to the right, to be cancelled later. The regulator becomes

$$\mathcal{R} = A\beta^{-1} Q E_2^{-1} G^{-1} \tag{3.21}$$

which is stable and causal, since β^{-1}, E_2^{-1} and G^{-1} are stable and causal. The $m|\ell$ polynomial matrix $Q(z^{-1})$ is not yet specified. Thus,

$$J_1 = \frac{1}{2\pi i} \oint_{|z|=1} \mathrm{tr}[A_*^{-1}(B_* V_* V D G_2 - \beta_* Q) E_2^{-1} H^{-1} \psi \mathcal{T}_*] \frac{dz}{z} \quad . \tag{3.22}$$

The elements of $A_*^{-1}(z)$ and $\mathcal{T}_*(z)$ have poles strictly outside $|z| = 1$, since A and $\mathcal{T}$ are stable. Thus, all elements of the integrand become analytic inside $|z| = 1$ if there exists a polynomial matrix $L_*(z)$, of dimension $m|\ell$, such that

$$(B_* V_* V D G_2 - \beta_* Q) E_2^{-1} H^{-1} \frac{1}{z} = L_* \tag{3.23}$$

or

$$B_* V_* V D G_2 = \beta_* Q + L_* z H E_2 \quad . \tag{3.24}$$

This is a bilateral Diophantine equation in $Q(z^{-1})$ and $L_*(z)$. Thus, the regulator can be obtained by solving (3.15) for β, computing G_2 and E_2 from (3.20), solving (3.24) for Q and L_*, and using the control law $u(t) = -A\beta^{-1} Q E_2^{-1} G^{-1} w(t)$.

The reasoning, from (3.17) up to equation (3.24), constitutes a compact derivation of the optimal control law. The finiteness of the minimal cost is verified in Appendix A, under Assumptions A3 and A4.

The single Diophantine equation (3.24) determines the regulator uniquely[10]. Since $\det \beta_*(z)$ has zeros strictly outside $|z| = 1$, while $\det H(z^{-1}) E_2(z^{-1})$ has zeros only in $|z| \leq 1$, the invariant polynomials of β_* are coprime with all those of HE_2. Thus, a solution $(Q^o(z^{-1}), L_*^o(z))$ to (3.24) always exists. See Lemma 1 of [44]. All solutions can be expressed as

$$(Q, L_*) = (Q^o - XzHE_2, L_*^o + \beta_* X)$$

[10]This holds in general for *open-loop* control and estimation problems, if the involved systems are stable or marginally stable. If $\det H(z^{-1})$ had zeros in $|z| > 1$, two coupled Diophantine equations would sometimes be required to determine $\mathcal{R}$, and formally assure a finite criterion value J_0. (See subsection 5.3.9 of Chapter 5 for a discussion.) However, such control laws, designed to cancel exponentially increasing disturbances, are of no practical interest. Actuator limitations and sensitivity problems would defeat any such ambitions.

where the polynomial matrix X is arbitrary [27]. However, causality requires $Q(z^{-1})$ to have only nonpositive powers of z as arguments, while optimality requires $L_*(z)$ to have no negative powers of z as arguments. (If it had, it would contribute zeros at the origin in (3.22).) Thus, $X = 0$ is the only choice, so the solution to (3.24) is unique.

In SISO problems, with $V = 1$, $W = \rho\tilde{\Delta}(q^{-1})$, $G = G_2$ and $A = E = E_2$, (3.21) becomes $\mathcal{R} = Q/\beta G$, while (3.24) reduces to equation (3.12) in [52].

If the weights V and W in the criterion are rational, two additional coprime factorisations are required. It is shown in [7] that this more general problem is *dual to the generalised deconvolution problem* discussed in Section 5.4. It is very simple to demonstrate this duality. Take the block diagram for one of the problems. Reverse all arrows, interchange summation points and node points and transpose all rational matrices. Then, the block diagram for the other problem has been obtained.

The disturbance measurement is often influenced by the control signal. We then have a more general disturbance rejection problem, sometimes called *disturbance decoupling.* If the transfer function between $u(t)$ and the measurement is stable, it can be *subtracted internally*, inside the regulator. Assume the measurement to be given by

$$z(t) = \mathcal{H}_{zu}u(t) + \mathcal{H}_{zw}w(t) \tag{3.25}$$

with $\mathcal{H}_{zu}$ *stable*, rational and causal, $\mathcal{H}_{zw}$ *stably invertible* and $w(t)$ given by (3.14). The use of the control law

$$u(t) = -\mathcal{R}\mathcal{H}_{zw}^{-1}[z(t) - \mathcal{H}_{zu}u(t)] \tag{3.26}$$

reduces this problem to the already solved feedforward control problem, with the optimal $\mathcal{R}$ given by (3.21). This solution is somewhat related to Internal model control (IMC) [40]. A SISO version of it has been discussed in [50] and [51]. The accuracy of the model of $\mathcal{H}_{zu}$ must be good at frequencies where $\mathcal{R}\mathcal{H}_{zw}^{-1}$ has high gain. Otherwise, model imperfections could cause instability, as in all control laws.

This problem can be interpreted as a somewhat specialised standard $\mathcal{H}_2$ problem, represented by

$$\begin{pmatrix} \begin{bmatrix} Vy(t) \\ Wu(t) \end{bmatrix} \\ z(t) \end{pmatrix} = \begin{pmatrix} \begin{bmatrix} VDE^{-1}GH^{-1} \\ 0 \end{bmatrix} & \begin{bmatrix} VBA^{-1} \\ W \end{bmatrix} \\ \mathcal{H}_{zw}GH^{-1} & \mathcal{H}_{zu} \end{pmatrix} \begin{pmatrix} v(t) \\ u(t) \end{pmatrix}$$

with $\| \, (Vy)' \, (Wu)' \, \|_2^2$ being minimised and $z(t)$ measured. We assume $E^{-1}, A^{-1}, \mathcal{H}_{zw}$ and $\mathcal{H}_{zu}$ to be stable, while H^{-1} is stable or marginally stable. While not completely general, the formulation above, with a stable system, includes a large class of practically relevant control problems. The above solution, with *one* spectral factorisation, *one* coprime factorisation and *one* Diophantine equation, is evidently much simpler than the general polynomial $\mathcal{H}_2$ solutions known to date.

3.4 A Scalar Feedback–feedforward LQG Design

The polynomial equations approach to the design of combined feedback–feedforward regulators has received considerable interest. See [43], [47], [52] and [20]. The earliest result, restricted to random–walk disturbance models, was due to Peterka [43].

We will here use the methodology of Section 3.2 to derive a combined feedback–feedforward control law for a scalar plant. This plant may be affected by non–stationary measurable and unmeasurable disturbances, described by systems with poles on the unit circle. The adaptive controller of Section 3.5 will be based on this control law.

3.4.1 The control problem

Let the plant be described by the following linear discrete–time scalar model

$$Ay(t) = Bu(t-k) + Dw(t-d) + Cn(t) \tag{3.27}$$

where the output $y(t)$, input $u(t)$, measurable disturbance $w(t)$ and unmeasurable disturbance $n(t)$ are all scalar signals. All polynomials except $B(q^{-1})$ and $D(q^{-1})$, of degree na, nb etc, are monic. The delays are $k > 0$ and $d \geq 0$.

The disturbances $w(t)$ and $n(t)$ are asumed to be described by

$$w(t) = \frac{G}{H}v(t) = \frac{G}{H_S H_U}v(t) \quad ; \quad n(t) = \frac{1}{F}e(t) \ . \tag{3.28}$$

We assume $v(t)$ and $e(t)$ to be stationary, mutually uncorrelated and zero mean. They are stationary white noises or random spike sequences, with variance λ_v and λ_e, respectively. While $C(z^{-1}), G(z^{-1})$ and $H_S(z^{-1})$ are assumed to be stable, $H_U(z^{-1})$ and $F(z^{-1})$ have all their zeros on the unit circle. The disturbance models thus include

1. *Stationary stochastic disturbances.* (F or $H_U = 1$.)

2. *Drifting stochastic disturbances.* If $w(t)$ e.g. has stationary increments, it is modelled by $H_U = 1 - q^{-1}$ and a white noise $v(t)$.

3. *Shape–deterministic* or piecewise deterministic signals, such as random step sequences, ramp sequences or sinusoids which occasionally change magnitude or phase. A stationary random spike sequence, such as a Bernoulli–Gaussian sequence [11], is then a reasonable model for $v(t)$ or $e(t)$. (Purely deterministic disturbances can also be included in the formulation. See [53]. For an alternative problem formulation, see [35].)

[11] A Bernoulli–Gaussian sequence is given by $v(t) = r(t)s(t)$ where $s(t)$ is a Bernoulli sequence such that $s(t) = 1$ w. p. λ and $s(t) = 0$ w. p. $1 - \lambda$. $r(t)$ is a zero mean Gaussian sequence with variance σ^2 independent of t, cf. [39]. It is then straightforward to show that $v(t)$ is a stationary white sequence with zero mean and variance $\lambda_v = \sigma^2\lambda$.

Assume, for now, all polynomials to be known. The goal is to minimize an infinite horizon criterion

$$J = \mathcal{E}(Vy(t))^2 + \rho\mathcal{E}(WFu(t))^2 \quad . \tag{3.29}$$

The input penalty $\rho \geq 0$ and the polynomials $V(q^{-1})$ and $W(q^{-1})$ are chosen by the designer. Note that the choice of input filters is not completely free: the factor $F(q^{-1})$ must be present whenever $n(t)$ is described by a marginally stable model. If $n(t)$ e. g. is a drifting signal, a drifting input $u(t)$ will be needed. To keep the criterion finite, the input must then be filtered by $F(q^{-1}) = 1 - q^{-1}$ in (3.29).

In this problem, the measurement is $z(t) = (y(t)\ w(t))'$. It will be shown that the optimal linear regulator structure, with feedback and feedforward, is given by

$$RFu(t) = -\frac{Q}{P}w(t) - Sy(t) \quad . \tag{3.30}$$

See Figure 3.1. The polynomial $P(q^{-1})$ is required to be stable. Note that the filter $1/R(q^{-1})F(q^{-1})$ is present in both the feedback and feedforward signal paths. The filtering by $1/F(q^{-1})$ is consistent with the internal model principle [15]. When $F(q^{-1}) = 1 - q^{-1}$, we have an integrating regulator with a feedforward term.

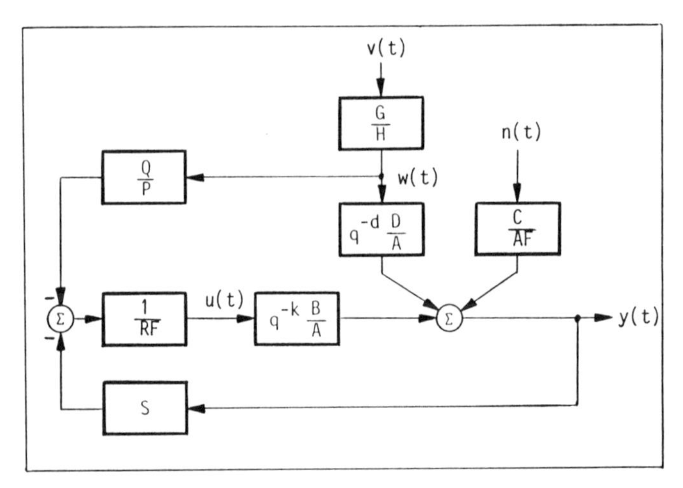

Figure 3.1: The system and regulator structure.

The regulator design will consist of a simple two-step procedure. First, the feedback $\{R, S\}$ is optimised with respect to the unmeasurable disturbance $n(t)$. Then, the feedforward filter $\{P, Q\}$ is calculated so that the measurable disturbance $w(t)$ is rejected in an optimal way. This separability is possible since $w(t)$ and $n(t)$ are assumed uncorrelated. The feedback design equations are well-known [27], but are slightly generalised here to cover $F \neq 1$.

3.4.2 The design equations

Introduce the polynomial spectral factorisation

$$r\beta\beta_* = BVV_*B_* + \rho AFWW_*F_*A_* \tag{3.31}$$

where r is a positive scalar and $\beta(z^{-1})$ is a stable monic polynomial with degree $n\beta$. When $\rho > 0$, stability of β is assured if BV and AFW have no common factors with zeros on $|z| = 1$. If $\rho = 0, BV$ should have no zeros on $|z| = 1$.

The following assumptions are sufficient for the existence of a unique stabilizing solution to the optimisation problem described above.

B1. The polynomials β, C and G are stable.

B2. The polynomials AF and B have no unstable common factors.

B3. H_U is a factor of WFD.

Theorem 3.1

Under the conditions B1–B3 above, the controlled system (3.27),(3.30) attains the global minimum of (3.29), under the constraint of stability, if $\{R, S, P, Q\}$ are calculated as follows.

Let $R(q^{-1}), S(q^{-1})$ and $X_*(q)$ be the unique solution of the coupled linear polynomial equations

$$r\beta_*R - q^{-k+1}BX_* = \rho WW_*F_*A_*C \tag{3.32}$$

$$r\beta_*S + qAFX_* = q^kCVV_*B_* \ . \tag{3.33}$$

Let

$$P = G$$

and let $Q(q^{-1})$, together with $L_*(q)$, be the unique solution of

$$q^{-d+1}DFGX_* = r\beta_*Q + qCHL_* \ . \tag{3.34}$$

□

3.4.3 Proof of Theorem 3.1

We add a variational term $\nu(t)$ to a linear regulator with general structure

$$u(t) = -\frac{S_1}{R_1}y(t) - \frac{Q_1}{P_1}w(t) + \nu(t) \tag{3.35}$$

As mentioned in Section 3.2, the variational term is set to a sum of potential feed-forward contributions from the two innovation signals in the problem structure

$$\nu(t) = \mathcal{T}_1 v(t) + \mathcal{T}_2 e(t) \ . \tag{3.36}$$

The use of (3.35) and (3.36) on the system (3.27), (3.28) results in the following expressions for the nominal closed–loop signals and their perturbations (3.9)

$$\begin{aligned} y_o(t) &= \frac{NG}{P_1\alpha H}v(t) + \frac{R_1C}{\alpha F}e(t) \ ; \ \delta y(t) = q^{-k}\frac{BR_1}{\alpha}(\mathcal{T}_1 v(t) + \mathcal{T}_2 e(t)) \\ u_o(t) &= -\frac{UG}{P_1\alpha H}v(t) - \frac{S_1C}{\alpha F}e(t) \ ; \ \delta u(t) = \frac{AR_1}{\alpha}(\mathcal{T}_1 v(t) + \mathcal{T}_2 e(t)) \end{aligned} \tag{3.37}$$

where

$$N \triangleq (q^{-d}DP_1 - q^{-k}BQ_1)R_1 \ ; \ \alpha \triangleq AR_1 + q^{-k}BS_1 \ ; \ U \triangleq AQ_1R_1 + q^{-d}DS_1P_1 \ .$$

The polynomial α must be stable, and will be required to be monic. Using (3.37) in (3.11), the condition for optimality can be expressed as

$$\begin{aligned} J_1 &= \mathcal{E}(Vy_o)(V\delta y) + \rho\mathcal{E}(WFu_o)(WF\delta u) = \\ &= \mathcal{E}\left\{V\left(\frac{NG}{P_1\alpha H}v(t) + \frac{R_1C}{\alpha F}e(t)\right)Vq^{-k}\frac{BR_1}{\alpha}(\mathcal{T}_1 v(t) + \mathcal{T}_2 e(t))\right\} \\ &+ \rho\mathcal{E}\left\{WF\left(-\frac{UG}{P_1\alpha H} - \frac{S_1C}{\alpha F}e(t)\right)WF\frac{AR_1}{\alpha}(\mathcal{T}_1 v(t) + \mathcal{T}_2 e(t))\right\} = 0 \end{aligned}$$

In order to obtain a two degree of freedom controller, we set the contributions from the two variational terms $\mathcal{T}_2 v(t)$ and $\mathcal{T}_1 e(t)$ to zero *separately*, by tuning feedback and feedforward filters, respectively [12]. By using the assumption that $v(t)$ and $e(t)$ are uncorrelated, and by using the spectral factorisation (3.31), we obtain

Term 1:

$$\mathcal{E}\left\{\left(\frac{VNG}{P_1\alpha H}v(t)\right)Vq^{-k}\frac{BR_1}{\alpha}\mathcal{T}_1 v(t)\right\} - \rho\mathcal{E}\left\{\left(\frac{WFUG}{P_1\alpha H}v(t)\right)W\frac{FAR_1}{\alpha}\mathcal{T}_1 v(t)\right\} =$$

$$\frac{\lambda_v}{2\pi i}\oint \frac{G[V(z^{-d}DP_1 - z^{-k}BQ_1)R_1V_*z^kB_* - \rho WF(AQ_1R_1 + z^{-d}DS_1P_1)W_*F_*A_*]R_{1*}}{P_1\alpha H\alpha_*} . \mathcal{T}_{1*}\frac{dz}{z} =$$

$$\frac{\lambda_v}{2\pi i}\oint \frac{G[z^{-d+k}VV_*DR_1P_1B_* - r\beta\beta_*Q_1R_1 - \rho WW_*FF_*z^{-d}DS_1P_1A_*)]R_{1*}}{P_1\alpha H\alpha_*}\mathcal{T}_{1*}\frac{dz}{z} = 0 \tag{3.38}$$

[12]Note that the part affected by $e(t)$ only (term 2 below) can vanish only by choosing the feedback properly.

Term 2:

$$\mathcal{E}\left\{\left(\frac{VR_1C}{\alpha F}e(t)\right)Vq^{-k}\frac{BR_1}{\alpha}\mathcal{T}_2e(t)\right\}-\rho\mathcal{E}\left\{\left(\frac{WFS_1C}{\alpha F}e(t)\right)WF\frac{AR_1}{\alpha}\mathcal{T}_2e(t)\right\}=$$

$$\frac{\lambda_e}{2\pi i}\oint\frac{(R_1CVV_*z^kB_*-\rho WFS_1CW_*F_*A_*)R_{1*}}{\alpha\alpha_*F}\mathcal{T}_{2*}\frac{dz}{z}=0\ . \qquad (3.39)$$

In order to find the *optimal feedback*, we shall set term 2, which depends on $e(t)$, to zero. All poles of the integrand of (3.39) within $|z|\leq 1$ are cancelled by zeros if there exists a polynomial $X_*(z)$, such that

$$R_1CVV_*z^kB_*-\rho WW_*F_*A_*CFS_1=X_*z\alpha F=X_*z(AR_1+z^{-k}BS_1)F \qquad (3.40)$$

or

$$R_1(z^kCVV_*B_*-zX_*AF)=(\rho WW_*F_*A_*C+z^{-k+1}BX_*)FS_1\ .$$

If R_1 and S_1 are assumed coprime, there must exist a polynomial $K_*(z,z^{-1})$, such that

$$F(\rho WW_*F_*A_*C+z^{-k+1}BX_*)\ =\ K_*R_1 \qquad (3.41)$$

$$z^kCVV_*B_*-zAFX_*\ =\ K_*S_1\ . \qquad (3.42)$$

Multiply (3.41) by A and (3.42) by $z^{-k}B$ and add them. This gives

$$Cr\beta\beta_*=K_*\alpha\ .$$

Thus, we obtain $\alpha=\beta C$ (stable) and $K_*=r\beta_*$, since α is required to be stable and monic. Note that F in (3.41) must be a factor of $K_*R_1=r\beta_*R_1$. Since β_* has no zeros on the unit circle, F must be a factor of R_1. Thus, set

$$R_1=RF\ \ ;\ \ S=S_1\ . \qquad (3.43)$$

Use (3.43) and $K_*=r\beta_*$ in (3.41), (3.42) and substitute q for z. After F has been cancelled in (3.41), they are seen to be identical to the feedback design equations (3.32), (3.33). This completes the feedback derivation.

In order to find the *optimal feedforward* polynomials, we set the first term (3.38), which depends on $v(t)$, to zero. The integrand is analytic in $|z|\leq 1$, if there exists a polynomial $L_*(z)$, such that

$$z^{-d+k}GVV_*DR_1P_1B_*-Gr\beta\beta_*Q_1R_1-\rho GWW_*FF_*z^{-d}DS_1P_1A_*=zL_*P_1\alpha H\ .$$

A rearrangement of the terms gives, with $R_1=RF$ and $S_1=S$,

$$P_1(z^{-d+k}GVV_*DRFB_*-\rho GWW_*FF_*z^{-d}DA_*S-zL_*\alpha H)=Gr\beta\beta_*Q_1RF \quad (3.44)$$

The feedforward denominator P_1 must be a factor of the right–hand side. To preserve stability, factors of P_1 which are not included in the feedback loop i.e. as factors of

$R_1 = RF$, must be stable. With Q_1 being a filter polynomial and β_* unstable, we may set

$$P_1 = RFG\beta \qquad (3.45)$$

By cancelling P_1, (3.44) reduces to a Diophantine equation

$$(VV_*RB_* - \rho WW_*F_*A_*Sz^{-k})z^{-d+k}DGF = r\beta_*Q_1 + zL_*\alpha H \quad . \qquad (3.46)$$

This is the feedforward desing equation, to be solved for $Q_1(z^{-1})$ and $L_*(z)$. The equation depends on the feedback polynomials R and S. It can be used for optimising a feedforward filter which works in conjunction with an *arbitrary* feedback regulator, e.g. PID–type. (For $V = 1$ and $F = 1$, it corresponds to (3.7) in [52].)

When used in conjunction with an LQG feedback, which results in pole placement in $\alpha = \beta C$, the equation can be further simplified. Note that by using $R_1 = RF, S_1 = S$ and $\alpha = \beta C$ in (3.40), that equation can be written

$$(VV_*RB_* - \rho WW_*F_*A_*Sz^{-k})z^kCF = X_*z\beta CF \quad .$$

By cancelling CF above, we obtain a new expression for the left–hand side of (3.46). That equation now becomes

$$z^{-d+1}DGFX_*\beta = r\beta_*Q_1 + zL_*\beta CH \quad . \qquad (3.47)$$

Since β is a factor of two terms, it must be a factor of Q_1 as well, i.e. $Q_1 = Q\beta$. Thus,

$$\frac{Q_1}{P_1} = \frac{Q\beta}{RFG\beta} = \frac{Q}{RFG} \quad .$$

Therefore, $P = G$, (3.47) equals (3.34) and the controller structure (3.30) is verified. By using (3.43), (3.45), $Q_1 = \beta Q$ and $\alpha = \beta C$ in (3.37), the closed loop is found to be stable, except for the poles belonging to H_u and F. They belong to the external, non–stabilisable, disturbance–generating systems. In Appendix B, it is verified that the minimal cost is finite, under condition B3. □

3.4.4 Remarks and interpretations

Degrees of the feedback equations.

The variables in (3.32) and (3.33) have degrees

$$\begin{aligned} nx &= n\beta + k - 1 \\ ns &= \max\{nf + na - 1, nc - k\} \\ nr &= \begin{cases} \max\{nb + k - 1, nc + nw\} & \text{if } \rho \neq 0 \\ nb + k - 1 & \text{if } \rho = 0 \end{cases} \end{aligned} \qquad (3.48)$$

For a discussion of the unique solvability of (3.32) and (3.33), see [27].

Interpretation of the conditions B1–B3.

In the proof above, optimal feedback was found to imply pole placement in βC:

$$AFR + q^{-k}BS = \beta C \quad . \tag{3.49}$$

In addition, the feedforward filter introduces poles in the zero locations of $P = G$. This explains condition B1. If AF and B have no common factors, an optimal feedback can be calculated from the implied pole placement equation (3.49). With common factors, this will not be possible [30]. (We cannot cancel common factors and utilise e.g. the minimum degree solution. That solution would not satisfy the original equations.) The equations (3.32) and (3.33) do, however, give the correct solution, as long as the common factors of AF and B are stable.

Condition B2, the absence of unstable common factors of AF and B, is the condition for solvability of (3.32) and (3.33). It also corresponds to stabilizability and detectability of an equivalent system, with $Fu(t)$ as input, obtained by multiplication of (3.27) by F:

$$(AF)y(t) = q^{-k}B(Fu(t)) + q^{-d}(DF)w(t) + Ce(t) \quad .$$

The condition B3, that H_U divides WFD, guarantees that the minimal cost is finite. See Appendix B. It has a natural explanation. To be more specific, let $w(t)$, with $H_U(q^{-1}) = 1 - q^{-1} \triangleq \Delta(q^{-1})$, be a drifting stochastic disturbance, i.e. $w(t) = w(t-1) + (G/H_S)v(t)$. Let $V(q^{-1}) = 1$. The controller must then eliminate drifts from the signals $y(t)$ and $WFu(t)$, which appear in the criterion. Otherwise, the criterion would be infinite.

- If $H_U = \Delta$ is a factor of D, the non-stationary mode of $w(t)$ is blocked before it reaches the output. Consequently, it does not result in a drifting $y(t)$. Since Δ is a factor of D and H in (3.34), it will also be a factor of Q. Thus, the feedforward control signal $(Q/P)w(t)$ remains stationary.

- If $H_U = \Delta$ is a factor of F, we have an integrating feedback regulator. It eliminates drifts in $y(t)$ caused by drifts in $w(t)$ (or in $n(t)$), regardless of the presence of any feedforward filter. The signal $Fu(t)$ appearing in the criterion (but not $u(t)$ itself), remains stationary.

- If $H_U = \Delta$ is not a factor of DF, the responsibility for eliminating drifts in $y(t)$ is placed on the feedforward filter. To accomplish this task, the filter generates a drifting control signal $(Q/P)w(t)$. Consequently, Δ must be a factor of W, so that $Wu(t)$ gives a finite contribution to the criterion.

When the disturbance $w(t)$ is described by a model with poles on the unit circle, such effects on $y(t)$ (e.g. static errors, drifts or undamped sinusoids) can be controlled either by the feedback or by the feedforward action. Use of the feedback, when H_U

divides F, results in a "robust" disturbance rejection. It is robust in the sense that the criterion is finite if modelling errors in A, B, C or D are present, as long as the closed–loop system is stable. However, F itself must be known exactly. When the feedforward action is used, the magnitude of static errors, drifts or sinusoids can be reduced, but can not be eliminated completely in practice. Modelling errors will cause imperfect cancellation.

The feedforward control.

Note that the solution of only one additional Diophantine equation, namely (3.34), is needed for optimising a feedforward filter. Since β (stable) and $C_* H_*$ (unstable) cannot have common factors, (3.34) is always solvable. The degrees of $Q(z^{-1})$ and $L_*(z)$ are defined uniquely by the requirement that they should cover the maximal occuring powers of z^{-1} and z, respectively, in (3.34):

$$\begin{aligned} nQ &= \max\{nd + nf + ng + d, nc + nh\} - 1 \\ nL &= \max\{0, k - d\} + n\beta - 1 \ . \end{aligned} \qquad (3.50)$$

(See also Theorem 5.3.1 in Chapter 5 of this volume, or Lemma 1 in [1].)

The addition of a feedforward link simplifies the control task remaining for the feedback. (See also Section 3.6.) The delay d affects the achievable control quality significantly. It can be shown that application of feedforward can always improve the control performance when $d > 0$, compared to feedback from $y(t)$ only. The improvement is a nondecreasing function of d. It is advantageous to place the $w(t)$–sensor so that the disturbance is captured as early as possible, i. e. d is large.

Complete elimination of the measurable disturbance can be achieved if and only if $d \geq k$, and all unstable factors of $B(q^{-1})$ are factors also of $D(q^{-1})$. Any part of the system with non–minimum phase dynamics is then located beyond the point where the disturbance meets the control action. See [51], Section 3.3.

It is straightforward to generalise the solution to multiple measurable disturbances. One additional scalar Diophantine equation (3.34) is then obtained for each disturbance. The general multivariable result corresponding to Theorem 3.1 is presented in the report version of [54]. An alternative solution is derived in [24].

Numerical aspects.

The regulator should be realised minimally, as a single dynamical system having two inputs and one output. A common special case is when the measurable disturbance is drifting or of random step type, and an integrating regulator is used. Then, $H_U = F = \Delta$. Since Δ becomes a factor of both the left-hand side and rightmost term in (3.34), it must also be a factor of Q. With $Q = Q_2\Delta$, equation (3.34) is reduced to

$$z^{-d+1} D G X_* = r\beta_* Q_2 + z C H_S L_* \qquad (3.51)$$

In this case, the controller (3.30) must be modified slightly. It can be implemented in differential form, using an explicit differentiation of the measurable disturbance:

$$
\begin{aligned}
R(\Delta u(t)) &= -\frac{Q_2}{G}(\Delta w(t)) - Sy(t) \\
u(t) &= u(t-1) + \Delta u(t) \ . \qquad (3.52)
\end{aligned}
$$

Alternatively, one can use a structure with the feedforward filter separated from the integration:

$$
Ru(t) = -\frac{Q_2}{G}w(t) - \frac{S}{\Delta}y(t) \ . \qquad (3.53)
$$

If (3.34) were used, small numerical errors and finite word–length effects would cause $Q \neq Q_2\Delta$. This could lead to large errors in the low-frequency gain of the feedforward filter $-Q/R\Delta P$ in (3.30). Design from (3.51), with $nQ_2 = nQ - 1$, and realisation according to (3.52) or (3.53), avoids such problems. Equation (3.34) must, however, be used in the general case, when $H_U \neq F$.

Reliable algorithms for polynomial spectral factorisation can be found in [27] and [26]. They are iterative, requiring typically 3-10 iterations, when starting from $\beta = 1$. In adaptive control, β from the previous controller calculation can be used as initial value. Then, normally only 1–2 iterations per updating are required.

The coupled equations (3.32),(3.33) represent an over–determined set of simultaneous equations in the coefficients of R, S and X. The system will, however, have a unique solution [27]. (Some equations are linear combinations of the others.) This (exact) solution can be found by computing the least–squares solution to the overdetermined system. Equation (3.34), with polynomial degrees (3.50), corresponds to a square system of linear equations, with full rank.

3.4.5 A numerical example

Example 3.1.

Consider the system

$$
(1 - 0.9q^{-1})y(t) = (0.1 + 0.08q^{-1})u(t-2) + (0.2 + 0.4q^{-1})w(t-2) + e(t)
$$

with $w(t) = w(t-1) + v(t)$ i.e. $H = 1 - q^{-1}$. Unit step disturbances $w(t)$ cause output deviations with amplitude ≈ 6 in this system. We design a regulator (3.30), such that the criterion (3.29), with $V = 1$ and with differential input penalty $\rho = 0.1, W = 1 - q^{-1}$, is minimized.

The spectral factorisation (3.31) becomes

$$r(1+\beta_1 q^{-1}+\beta_2 q^{-2})(1+\beta_1 q+\beta_2 q^2) = (0.1+0.08q^{-1})(0.1+0.08q) + 0.1(1-0.9q^{-1})(1-q^{-1})(1-q)(1-0.9q)$$

with solution $r=0.2670$, $\beta_1=-0.9887$ and $\beta_2=0.3370$.

The feedback part of the regulator is calculated from (3.32) and (3.33)

$$r(1+\beta_1 q+\beta_2 q^2)R(q^{-1})-q^{-1}(0.1+0.08q^{-1})X_*(q) = 0.1(1-q^{-1})(1-q)(1-0.9q)$$

$$r(1+\beta_1 q+\beta_2 q^2)S(q^{-1})+q(1-0.9q^{-1})X_*(q) = q^2(0.1+0.08q) \ .$$

The variables have degree $nx=3, ns=0$ and $nr=2$, given by (3.48) Multiply the first equation by $q^{-n\beta}=q^{-2}$, the second by $q^{-nx-1}=q^{-n\beta-k}=q^{-4}$, and let $\bar{X}(q^{-1})=q^{-3}X_*(q)$. We then obtain equations in powers of q^{-1} only.

$$r(\beta_2+\beta_1 q^{-1}+q^{-2})R(q^{-1})-(0.1+0.08q^{-1})\bar{X}(q^{-1}) = 0.1(1-q^{-1})(q^{-1}-1).\,.(q^{-1}-0.9)$$

$$r(\beta_2+\beta_1 q^{-1}+q^{-2})q^{-2}S(q^{-1})+(1-0.9q^{-1})\bar{X}(q^{-1}) = q^{-1}(0.1q^{-1}+0.08) \ .$$

By considering terms with equal power of q^{-1}, a system of simultaneous equations, with block–Toepliz structure and with 10 equations and 8 unknowns, is obtained.

$$\left(\begin{array}{ccc|c|cccc} r\beta_2 & 0 & 0 & & -0.1 & 0 & 0 & 0 \\ r\beta_1 & r\beta_2 & 0 & 0 & -0.08 & -0.1 & 0 & 0 \\ r & r\beta_1 & r\beta_2 & & 0 & -0.08 & -0.1 & 0 \\ 0 & r & r\beta_1 & & 0 & 0 & -0.08 & -0.1 \\ 0 & 0 & r & & 0 & 0 & 0 & -0.08 \\ \hline & & & 0 & 1 & 0 & 0 & 0 \\ & & & 0 & -0.9 & 1 & 0 & 0 \\ & 0 & & r\beta_2 & 0 & -0.9 & 1 & 0 \\ & & & r\beta_1 & 0 & 0 & -0.9 & 1 \\ & & & r & 0 & 0 & 0 & -0.9 \end{array}\right) \begin{pmatrix} r_o \\ r_1 \\ r_2 \\ s_o \\ x_3 \\ x_2 \\ x_1 \\ x_0 \end{pmatrix} = \left(\begin{array}{c} 0.09 \\ -0.28 \\ 0.29 \\ -0.10 \\ 0 \\ \hline 0 \\ 0.08 \\ 0.1 \\ 0 \\ 0 \end{array}\right)$$

The (exact) solution, computed by least squares, is

$$\begin{aligned} R(q^{-1}) &= 1-0.08871q^{-1}+0.1210q^{-2} \\ S(q^{-1}) &= s_0 = 1.3617 \\ X_*(q) &= 0.4040+0.04945q+0.08q^2+0q^3 \ . \end{aligned}$$

In specific examples, the polynomial degrees may be lower than the values indicated by (3.48) and (3.50). This is evident here, where X_* actually only has degree 2. Note that, in general, the obtained $R(q^{-1})$ will be nonmonic, i.e. $r_0 \neq 1$.

The feedforward polynomial $Q(q^{-1})$ is obtained from the equation (3.34)

$$q^{-1}(0.2 + 0.4q^{-1})X_*(q) = r(1 + \beta_1 q + \beta_2 q^2)Q(q^{-1}) + q(1 - q^{-1})L_*(q)$$

with $X_*(q)$ from above, and with degrees $nQ = 2$, $nL = 1$ from (3.50). The solution is

$$Q(q^{-1}) = 1.8609 + 0.9750q^{-1} + 0.6052q^{-2} \quad ; \quad L_*(q) = 0.2521 - 0.1675q \ .$$

Thus, the optimal regulator (3.30) is

$$\begin{aligned} u(t) \quad &= \quad 0.08871u(t-1) - 0.1210u(t-2) - 1.3617y(t) \\ &\quad -1.8609w(t) - 0.9750w(t-1) - 0.6052w(t-2) \end{aligned}$$

This regulator eliminates a unit step disturbances $w(t)$, after a small initial transient with peak value 0.16, without excessive input variations □

3.5 An LQG Self–tuner

The polynomial approach to adaptive LQG feedback control was first investigated a decade ago by Åström and Zhao–Ying and by Grimble [3] [17], for feedback only. LQG self–tuners based on state–space methods, using an iterative solution of the Riccati equation, have been proposed by different investigators. See e.g. [5], [8], [14] and [46]. There now exist commercially available adaptive LQG controllers, from the company First Control in Västerås, Sweden [6].

Compared to a pole placement algorithm, the "tuning knobs" of an LQG scheme are fewer and more closely related to design objectives. A frequency-weighted input penalty is more easy work with than a number of pole locations. Furthermore, stable common factors may occur in identified models. They cause no problems if the coupled equations (3.32) and (3.33) are used.

3.5.1 The adaptive algorithm

We will now present an adaptive LQG feedback–feedforward regulator in polynomial form [13]. It is based on results in the previous section. An explicit algorithm, based on the certainty equivalence principle, is considered. Models of $y(t)$ and $w(t)$ are updated recursively, using the Recursive Prediction Error Method (RPEM) [37]. The

[13] It has been developed and studied in [51]. A similar algorithm has been suggested in [23].

controller is redesigned periodically, assuming these models to be correct. Regulator polynomials are computed from Theorem 3.1. The regulator may be re–designed at each sample, or with larger intervals.

The regulator is designed to optimise the performance for zero set point, according the criterion (3.29). In the model structure (3.27)–(3.28), upper bounds on all polynomial degrees are assumed known, together with the unstable disturbance model factor $F(q^{-1})$. The input penalty ρ and the polynomials $V(q^{-1})$ and $W(q^{-1})$ in the criterion are user choices, given by the desiger.

The regulator, complemented with a servo filter, is summarised below.

1. Read new samples of $y(t), w(t)$ and a set-point $r(t)$.
2. Update models of $y(t)$ and $w(t)$ with the structure

   $$\hat{A}y(t) = \hat{B}u(t) + \hat{D}w(t) + \hat{C}\varepsilon_y(t) \quad ; \quad \hat{H}w(t) = \hat{G}\varepsilon_w(t)$$

   using two RPEM routines for single output systems.
3. Compute r and $\beta(q^{-1})$ from the spectral factorisation (3.31).
4. Solve for $R(q^{-1}), S(q^{-1})$ and $X_*(q)$ in (3.32),(3.33).
5. Calculate $Q(q^{-1})$ (and $L_*(q)$) from (3.34).
6. If needed, design a servo filter $T(q^{-1})/E(q^{-1})$.
7. Compute the control action:
   $RFu(t) = -(Q/\hat{G})w(t) - Sy(t) + (T/E)r(t)$
8. Shift all data vectors, and go to step 1.

3.5.2 Remarks on the algorithm

Step 2: Identification. The main model and the disturbance model are estimated by two separate recursive prediction error algorithms for single output systems. They constitute two Gauss–Newton algorithms, with the following structure

$$K(t) = P(t)\psi(t) = \frac{P(t-1)\psi(t)}{\lambda + \psi'(t)P(t-1)\psi(t)}$$

$$[\hat{\theta}(t) = \hat{\theta}(t-1) + K(t)[\varepsilon(t)]_{\text{sat}}]_{\text{proj}} \tag{3.54}$$

$$P(t) = \frac{1}{\lambda}(P(t-1) - K(t)\psi'(t)P(t-1)) \ .$$

Here, $\varepsilon(t)$ is a scalar prediction error, $\psi(t)$ is the sensitivity vector $-\partial\varepsilon(t)/\partial\hat{\theta}$, and the matrix $P(t)$ is an approximation of the inverse Hessian. The updating of $P(t)$ should be implemented in factorised form, for numerical reasons. (The code in [37], which is based on UD-factorisation, has been used in the simulation study.) Above, λ is a forgetting factor $0 < \lambda \leq 1$. Time–varying systems and disturbance dynamics are tracked using $\lambda < 1$. The parameter vectors in the two algorithms are

$$\hat{\theta}_y = (\hat{a}_1 \dots \hat{a}_{na}\ \hat{b}_1 \dots \hat{b}_{nb}\ \hat{d}_0 \dots \hat{d}_{nd}\ \hat{c}_1 \dots \hat{c}_{nc})' \quad ; \quad \hat{\theta}_w = (\hat{h}_1 \dots \hat{h}_{nh}\ \hat{g}_1 \dots \hat{g}_{ng})'$$

respectively. The model polynomials $\hat{A}, \hat{H}, \hat{C}$ and $\hat{G}$ have fixed leading coefficients 1. Unknown and time–varying delays k and d are handeled by selecting degrees of $\hat{B}$ and $\hat{D}$ which cover the maximal expected delay. The terms $\hat{b}_1 \dots \hat{b}_{k-1}$ and $\hat{d}_0 \dots \hat{d}_{d-1}$ will then converge to zero[14].

As one–step prediction errors

$$\varepsilon_y(t) = y(t) - \varphi_{fy}(t)'\hat{\theta}_y(t-1) \quad ; \quad \varepsilon_w(t) = w(t) - \varphi_{fw}(t)'\hat{\theta}_w(t-1) \tag{3.55}$$

are used, based on filtered regressor vectors

$$\varphi_{fy}(t) = \frac{F(q^{-1})}{N_1(q^{-1})}\varphi_y(t) \quad ; \quad \varphi_{fw}(t) = \frac{1}{N_2(q^{-1})}\varphi_w(t) \tag{3.56}$$

where

$$\begin{aligned} \varphi_y(t) &= (-y(t-1) \dots - y(t-na)\ u(t-1) \dots u(t-nb) \\ &\quad\ w(t) \dots w(t-nd)\ \bar{\varepsilon}_y(t-1) \dots \bar{\varepsilon}_y(t-nc))' \end{aligned} \tag{3.57}$$

$$\varphi_w(t) = (-w(t-1) \dots - w(t-nh)\ \bar{\varepsilon}_w(t-1) \dots \bar{\varepsilon}_w(t-ng))'$$

$$\bar{\varepsilon}_y(t) = y(t) - \varphi_{fy}(t)'\hat{\theta}_y(t) \quad ; \quad \bar{\varepsilon}_w(t) = w(t) - \varphi_{fw}(t)'\hat{\theta}_w(t) \ .$$

The use of prediction errors (3.55) instead of residuals $\bar{\varepsilon}_y$,$\bar{\varepsilon}_w$ as regressors would slow the convergence. Regressor filtering by (3.56), where $N_1(q^{-1})$ and $N_2(q^{-1})$ are stable polynomials, is utilised. The filtering of $\varphi_y(t)$ by $F(q^{-1})$ avoids biased estimates in the case of non–stationary or non–zero mean disturbances. With $N_1(q^{-1})$, the filter can be modified to improve the estimation accuracy in important freqency regions. In the simulations below, however, we use $N_1(q^{-1}) = N_2(q^{-1}) = 1$.

The sensitivity functions are obtained by filtering the regressor vectors (3.56):

$$-\partial\varepsilon_y(t)/\partial\hat{\theta}_y = \psi_y(t) = \frac{1}{\hat{C}(q^{-1})}\varphi_{fy}(t) \quad ; \quad -\partial\varepsilon_w(t)/\partial\hat{\theta}_w = \psi_w(t) = \frac{1}{\hat{G}(q^{-1})}\varphi_{fw}(t) \ . \tag{3.58}$$

In (3.54), $[\ \cdot\ ]_{\text{proj}}$ represents stability monitoring and projection into stable regions of the polynomials $\hat{G}$ and $\hat{C}$. This is vital, not only because of the role of these polynomials in the control law, but also because they are used for generating the

[14] The algorithm would, however, have to be modified to cope with time delays above 10–15 sampling periods.

sensitivity vectors ψ_y, ψ_w above.

Furthermore, the function $[\varepsilon(t)]_{\text{sat}}$ in (3.54) represents a linear function, except for a dead–zone for small prediction errors, and a saturation for large $|\varepsilon(t)|$. While the former guards against estimator wind–up, the latter modification makes the adaptation more robust against rare but very large noise samples ("outliers"). Monitoring of the regressor energy has also been implemented as a precaution against identification based on insufficient data.

Step 6: Servo design. Since it is problematical to obtain good stochastic models of e.g. manually generated setpoints, we have *not* included servo desing in the optimization. Instead, we design a servo filter T/E by cancelling poles and stable zeros, so that the controlled system approximates a pre–specified response model $y_m(t) = (q^{-k}B_m/A_m)r(t)$. If the response model is not too extreme, this works well, but results in a rather high–order filter. Other approaches, such as including the servo design in the optimisation, are discussed in [51].

Step 7: Control computation. When appropriate, the regulator (3.52) or (3.53), based on equation (3.51), should be used.

Convergence. Global convergence of explicit LQG self–tuners can be demonstrated, under idealised conditions. See, for example, [12], [16] or [19]. In general, a linear model structure cannot be expected to describe the true system exactly. Mismodelling is inevitable, to some extent. A good estimate of $q^{-k}B/A$, in the frequency ranges where the input has significant energy, is needed to assure stability. See Example 3.4. Errors in the estimates of the transfer functions $q^{-d}D/A, C/A$ or G/H will affect the control performance, but they cannot cause instability. (Since the stability of $\hat{C}$ is monitored, pole placement in $\beta\hat{C}$ results in a stable system, if A and B are estimated correctly.)

Complexity. The computational burden of this algorithm is significantly higher than for simple direct self–tuners, such as generalised minimum variance control [13]. See Table 1. With the increasing speed of computers, this should be no significant restriction in most control applications. There is no need to recalculate the regulator at each sample. Steps 3-6 can be placed in a background process, which provides a new regulator every m'th sample. For $\lambda > 0.95$, the use of $m = 5 - 10$ results in only a small degradation of the adaptation transient when the system dynamics changes. (It has been shown by Shimkin and Feuer [48] that it may be important to update the regulator infrequently, to assure convergence.)

2:	Identification		$37n^2$	$+\,36n$
3:	Spectral factorisation (per iteration)		$3n^2$	$+\,3n$
4:	Feedback optimisation	$36n^3$	$+\,87n^2$	$+\,135n$
5:	Feedforward optimisation	$9n^3$	$+\,14n^2$	
7:	Control			8n

Table 3.1. The approximate number of mult–add operations required per sample, assuming all model polynomials to have equal degree n. A least squares solution is computed in Step 4.

3.5.3 Simulation examples

Example 3.2

Let $[1/(1-q^{-1})]v(t)$ be a square wave disturbance, with unit amplitude and period 60. It disturbs the system

$$(1-0.5q^{-1})y(t) = (b_2 + b_3q^{-1})u(t-2) + (1+2q^{-1})w(t-1)$$

$$w(t) = \frac{1-0.3q^{-1}}{1-0.9q^{-1}}\left(\frac{1}{1-q^{-1}}v(t)\right) .$$

The polynomial $b_2 + b_3q^{-1}$ changes from $1+0.1q^{-1}$ to $0.5+0.05q^{-1}$ at time 300. The LQG self–tuner, with correctly parametrized models, is applied. An input penalty $\rho = 0.5$ and $V = 1, W = 1 - q^{-1}, F = 1$ is used. Thus, we use no integration. The forgetting factor λ is 0.98 in both RPEM algorithms. After an initial open loop identification period of 20 samples, the regulator quickly converges □

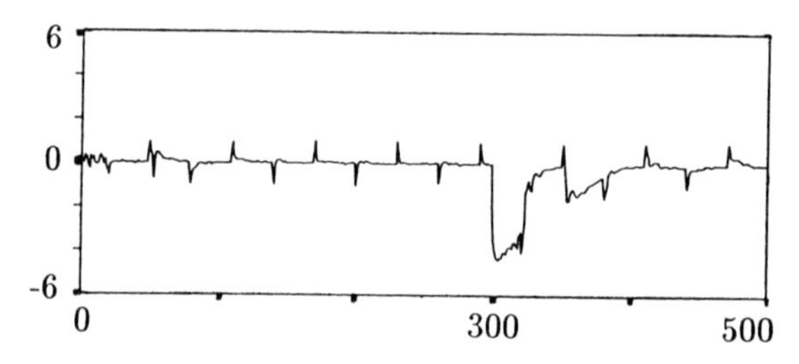

Figure 3.2: The controlled output $y(t)$ in Example 3.2. The disturbance $w(t)$ is cancelled almost completely, although the delay difference $k-d=1$ prevents perfect cancellation. At $t = 300$, the system gain is halved. At $t = 400$, the control performance has recovered to the off–line optimal one.

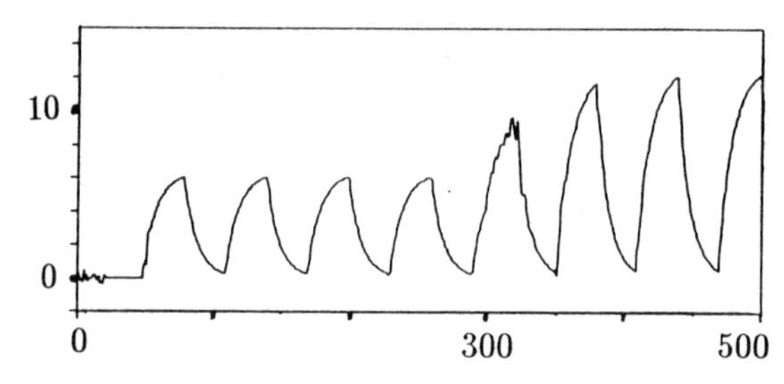

Figure 3.3: The input $u(t)$ in Example 3.2. When the system gain is halved at time 300, the regulator modifies itself, so that its gain is doubled.

Example 3.3

Consider the unstable and non–minimum phase system

$$(1-2q^{-1}+1.5q^{-2})y(t) = (1+2q^{-1}+2q^{-2})u(t-1) + (1+0.5q^{-1})w(t-2)$$

where $w(t)$ is white noise with standard deviation 0.1. As reference for the controlled output,

$$y_m(t) = \frac{0.7}{1 - 0.3q^{-1}} r(t)$$

was used, with $r(t)$ being a square wave. Adaptation, with a correctly parametrized model, and with $V = 1, W = 1, \rho = 0$, started at $t = 1$. The regulator had essentially converged to off–line optimal control after 30 samples. [15]. See Figure 3.4. Because the system is non–minimum phase, and the unstable part of B is not a factor of D, complete cancellation of the disturbance cannot be achieved □

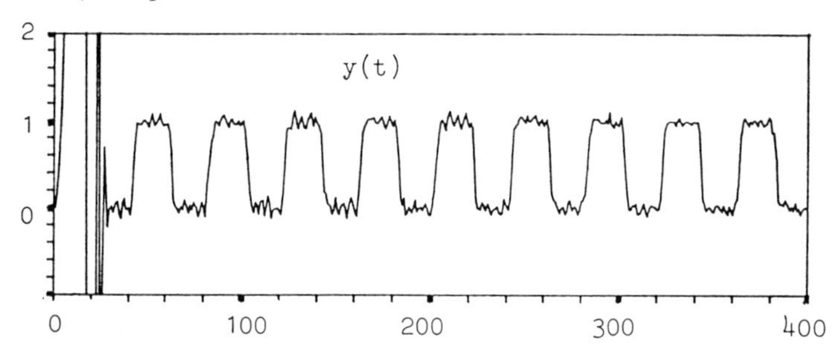

Figure 3.4: Control of the unstable non–minimum phase system in Example 3.3.

In further simulation studies, the adaptive algorithm has been found to behave very well in general. It has been compared to the Explicit criterion minimisation approach, cf. [55], which minimises the same criterion. Compared to that algorithm, the convergence rate of LQG self–tuners is much faster. See [51].

One (seldomly occuring) remaining problem is that over–parametrised models may contain *unstable* common factors of $\hat{A}$ and $\hat{B}$. The test of schemes, such as that of [34], to avoid this, is a problem for further research. A simple alternative can be used for the large class of systems which are open–loop stable or marginally stable. By using stability monitoring and projection on $\hat{A}$, in a similar way as is already done for $\hat{C}$ and $\hat{G}$, the risk of unstable common factors is eliminated.

3.6 User Choices Affecting the Robustness

The robustness against unmodelled dynamics of an LQG self–tuner is affected by properties of both the estimator and the control law. Simple considerations regarding the LQG control strategy, which in general improve the robustness of both off–line and self–tuning designs, are illustrated by the following example.

---

[15] Of course, it would not be advisable to use an adaptive regulator, without a prior model, during the start–up of such a system in practice; there is no guarantee that the signals behave acceptably in the transient phase. The instability at the beginning might cause too large input amplitudes, input saturation and thus divergence.

Example 3.4.

The system

$$(1-1.2q^{-1}+0.52q^{-2})y(t) = q^{-2}(1+0.8q^{-1})u(t) + q^{-2}(1-0.2q^{-1})w(t) + (1-0.2q^{-1})n(t)$$

is affected by measurable and unmeasurable drifting stochastic disturbances

$$w(t) = w(t-1) + v(t) \quad ; \quad n(t) = n(t-1) + e(t) \quad .$$

The white noises $v(t)$ and $e(t)$ have standard deviations 0.3 and 0.1, respectively. Thus, the largest disturbance is measurable, and $H_U = F = \Delta$.

The control error standard deviation was measured (after convergence) in simulation runs with four different self–tuners. Integrating regulators with the structure (3.52), with $r(t) = 0$ and $V = W = 1$, were used. The results are shown in Figure 3.5, as functions of the input penalty ρ. Curve (1) represents the performance of LQG feedback and feedforward. When $\rho \to 0$, the disturbance $w(t)$ is cancelled completely by the feedforward control action. When only feedback is used, curve (2) is obtained. The disturbances $w(t)$ and $n(t)$ are then treated as one unmeasurable noise. The performance is obviously degraded without disturbance measurement. Correctly parametrised models were used in these experiments. The performance in each case was indistinguishable from the off–line optimum.

Curve (3) and (4) result when an *underparametrised* $\hat{B}$ is used. (Degree 2 instead of 3, including the delay.) For input penalties $\rho \leq 1$, the closed loop is unstable.

The reason for this behaviour is explained by Figures 3.6 and 3.7. Figure 3.6 shows Bode magnitude plots of some under–parametrised models, obtained at the end of the simulation runs. Compare them with the true system. The high–frequency properties of the system are badly estimated. For low ρ, the regulators have large feedback gains at high frequencies, cf. Figure 3.7. (This is often the case for minimum variance regulators.) The combination of large feedback gain and an incorrect model at high frequencies leads to instability.

One way of reducing the high–frequency feedback gain is to modify the polynomial $C(q^{-1})$, used in (3.32)–(3.33). Instead of the estimate $\hat{C}$, a fixed polynomial $C_O = (1-0.5q^{-1})^2$ was used. This decreased the feedback high–frequency gain (cf. (4) in Figure 3.7) □

Based on the experience from this and other experiments, let us summarize some robustness–enhancing user–choices:

• By *increasing the input penalty* ρ *from zero*, the control signal variations, and the high–frequency gains of both feedback and feedforward filters, are reduced. Large

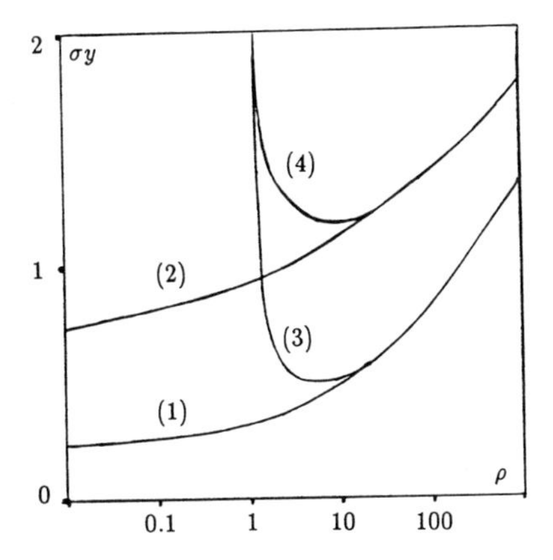

Figure 3.5: The output standard deviation σy in Example 3.4, as a function of the input penalty ρ.
(1): Feedback and feedforward. $\hat{B}$ of correct order 3.
(2): Feedback only. $\hat{B}$ of correct order 3.
(3): Feedback and feedforward. $\hat{B}$ of order 2.
(4): Feedback only. $\hat{B}$ of order 2.

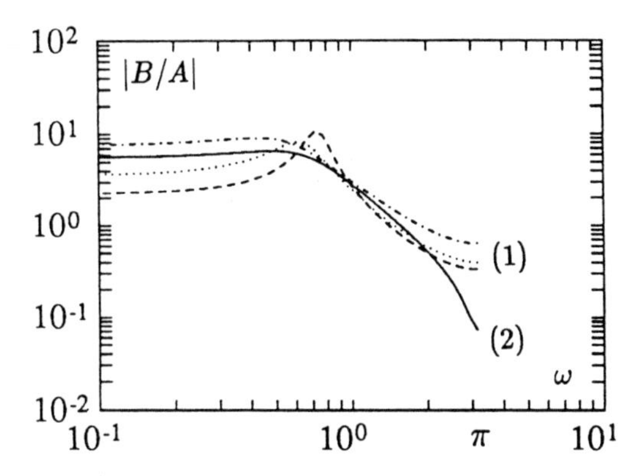

Figure 3.6: Bode magnitude plots of system and adapted models in Example 3.4.
(1): Transfer function magnitudes for some under-parametrized models.
(2): The true system.

reductions can often be achieved, with only a minor deterioration of the disturbance rejection. This increases the robustness against unmodelled high–frequency dynamics for open–loop stable systems. Problems with hidden inter–sample output oscillations are also avoided[16].

[16] "Hidden oscillations" are caused by pole placement on the negative real axis. The corresponding oscillative modes are unobservable in discrete time, at the sampling instants, but are evident in continuous time. See e.g. [4], Section 5.4. This often occurs in minimum variance control, with pole placement in BC, since B–polynomials of sampled data systems often have zeros on the negative real axis. See e.g. [4], Section 3.6.

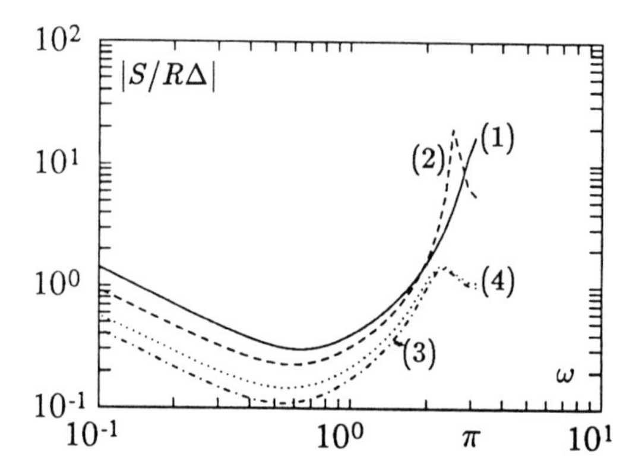

Figure 3.7: Transfer function magnitudes of feedback filters in Example 3.4.
(1): $\rho = 0$. (2): $\rho = 0.5$. (3): $\rho = 10$.
(4): $\rho = 0.5$, with pole placement in C_O.

• *Use of feedforward* can increase the stability robustness. This is possible when high disturbance rejection is required, and the main system disturbance is measurable. For example, consider Figure 3.5: if $\sigma y < 1$ is required, this could be attained, in the ideal case (2), by a high gain (low ρ) feedback. Instability would, however, result in the underparametrised case (4). With both feedback and feedforward, a low gain regulator ($\rho = 3 - 200$) can be used. It easily attains the required performance, also in the underparametrised case. Thus, when most of a disturbance can be eliminated by feedforward, the high-frequency gain of the feedback can be reduced. The feedback can more easily be designed to achieve robust stability. Also, more of the feedback control action can be allocated to robust performance issues, rather than to high (nominal) disturbance rejection.

• With LQG control, poles are placed at the zero locations of βC, cf. (3.49). The polynomial C corresponds to the observer dynamics in a state space formulation. Use of *a fixed prespecified observer polynomial* C_O, with $1/C_O$ being low-pass, has several advantages. First, uncertain estimates of C are avoided in the pole placement. The C–polynomials are *often hard to estimate* accurately, in particular if they have zeros close to the unit circle. Improved control performance may actually be achieved by not using them in the control law. Secondly, the estimated $\hat{C}$ may contain a factor close to $1 - q^{-1}$. (That happens when regressors are differentiated in (3.56), but the disturbance $n(t)$ is not generated by a system with $F = 1 - q^{-1}$.) It is obviously not desirable to force such a slow mode into the closed loop pole placement. Thirdly, unlike spectral factors β, we have no control over the zeros of C. A fixed C_0-polynomial may be used by the designer to obtain a safe and robustifying pole placement.

Appendix A. Conditions for, and verification of, finite cost in Section 3.3.

The conditions A3 and A4 are best understood if we consider drifting stochastic disturbances, generated by systems with poles at $z_j = 1$. (In other words, $\omega_j = 0$). To cancel these disturbances, we must have at least as many inputs as outputs. (If two nonstationary disturbances $w(t) = (w_1(t)\ w_2(t))'$ cause two outputs $y(t) = (y_1(t)\ y_2(t))'$ to drift in different directions, we could never hope to compensate for both of these drifts with just a scalar input $u(t)$.) Furthermore, the control action at the critical frequencies $\{\omega_j\}$ must not be blocked by zeros of the system BA^{-1}. These requirements are summarized by the condition A4.

To cancel a drifting disturbance, at least some components of the input vector $u(t)$ must also be drifting. The requirement A3 prevents such control modes, with infinite power, from appearing in the criterion. (Conditions analogous to A3 and A4 appear also in the scalar example of Section 3.4.)

Let us now verify that the optimal feedforward control law results in a finite criterion value. Using (3.17) and (3.21), the optimal criterion value is

$$\mathcal{E}(\mathrm{tr}(y_f y_f') + \mathrm{tr}(u_f u_f'))$$

where

$$y_f(t) = V(DE^{-1} - BA^{-1}\mathcal{R})GH^{-1}v(t) = V(DE^{-1}G - B\beta^{-1}QE_2^{-1})H^{-1}v(t)$$

$$u_f(t) = W\mathcal{R}GH^{-1}v(t) \ .$$

The transfer functions above have finite magnitude, except possibly at the frequencies $\{z_j\}$ which correspond to poles on the unit circle of GH^{-1}. They must be shown to be finite there also. For the input term $u_f(t)$, this is immediate from condition A3. That condition furthermore implies that

$$\beta_*\beta|_{z=z_j} = B_*V_*VB|_{z=z_j} \ .$$

The use of (3.23) gives

$$\beta^{-1}QE_2^{-1} = \beta^{-1}\beta_*^{-1}(B_*V_*VDG_2E_2^{-1} - L_*zH) \ .$$

Let us evaluate the transfer function from $v(t)$ to $y_f(t)$ at $z = \{z_j\}$. Using the two above relations and (3.20), it can be expressed as

$$V(DE^{-1}G - B\beta^{-1}\beta_*^{-1}(B_*V_*VDG_2E_2^{-1} - L_*zH))H^{-1}\Big|_{z=z_j} =$$

$$V(DE^{-1}G - B(B_*V_*VB)^{-1}B_*V_*VBB^{\dagger}DE^{-1}G)H^{-1} + VB\beta^{-1}\beta_*^{-1}L_*z\Bigg|_{z=z_j} =$$

$$V(DE^{-1}G - BB^{\dagger}DE^{-1}G)H^{-1} + VB\beta^{-1}\beta_*^{-1}L_*z \Big|_{z=z_j}$$

Since $BB^{\dagger} = I_p$, the first parenthesis is zero at $z = z_j$, so the first term cancels out. The second term is finite, since neither β^{-1} nor β_*^{-1} has poles on the unit circle. Thus, they can have no poles at z_j. Consequently, the non-stationary modes cancel at the output. This is, of course, a non–robust property, in that it requires perfect model knowledge at the critical frequencies $\{\omega_j\}$.

Appendix B. Verification of finite cost in Section 3.4

Let J_0 denote the cost (3.29) when the optimal regulator is applied. Using $R_1 = RF, S_1 = S, Q_1 = \beta Q$ and $P_1 = G\beta RF$ in (3.37), it can be expressed as

$$J_0 = \mathcal{E}(y_v(t) + y_e(t))^2 + \rho\mathcal{E}(z_v(t) + z_e(t))^2$$

where $WFu(t) \triangleq z_v(t) + z_e(t)$, $Vy(t) \triangleq y_v(t) + y_e(t)$ and

$$y_v(t) = \frac{VM}{\alpha G}\frac{G}{H_S H_U}v(t) \quad ; \quad y_e(t) = \frac{VFRC}{\alpha}\frac{1}{F}e(t)$$

$$z_v(t) = -\frac{WF(q^{-d}DSG + AQ)}{\alpha G}\frac{G}{H_S H_U}v(t) \quad ; \quad z_e(t) = -\frac{WFSC}{\alpha}\frac{1}{F}e(t)$$

Above, $M \triangleq q^{-d}DRFG - q^{-k}BQ$. Both $y_e(t)$ and $z_e(t)$ are stationary and have finite variance, since the cancellation of the unstable factor F is assumed to be exact.

The signal $y_v(t)$ has finite variance if (and only if) the unstable denominator factor H_U divides M. Consider the polynomial $r\beta_* M$. By using first (3.34) and then (3.32), it can be expressed as

$$\begin{aligned} r\beta_* M &= q^{-d}DFGr\beta_* R - q^{-k}Br\beta_* Q \\ &= q^{-d}DFGr\beta_* R - q^{-d}DFGq^{-k+1}BX_* + q^{-k+1}BCHL_* \\ &= q^{-d}DFG\rho WW_* F_* A_* C + q^{-k+1}BCH_S H_U L_* \end{aligned}$$

Since H_U is assumed to be a factor of DFW, it will be a factor of $\beta_* M$. But β_* has no zeros on the unit circle. So H_U, which has all its zeros on the unit circle, must be a factor of M. Consequently, $y_v(t)$ has finite variance.

Factors of H_U which are factors of WF are cancelled in the expression for $z_v(t)$ above. Common factors of H_U and D must, according to (3.34), also be factors of Q. (They cannot be factors of β_*, since β_* is assumed to have no zeros on the unit circle.) Thus, such factors are factors of $q^{-d}DSG + AQ$. Consequently, $z_v(t)$ is stationary, with finite variance, if H_U is a factor of WFD.

Bibliography

[1] A. Ahlén and M. Sternad, "Wiener filter design using polynomial equations", *IEEE Transactions on Signal Processing,* vol. 39, pp 2387–2399, 1991.

[2] A. Ahlén and M. Sternad, "Optimal filtering problems", in K. Hunt. ed. *Polynomial methods in Optimal Contol and Filtering.* Control Engineering Series, Peter Peregrinus, London, 1992.

[3] K. J. Åström and Z. Zhao–Ying, "A linear quadratic gaussian self–tuner," *Richerche di Automatica*, vol 13, pp. 106–122, 1982.

[4] K. J. Åström and B. Wittenmark, *Computer–Controlled Systems.* Prentice–Hall, Englewood Cliffs, NJ, second ed. 1990.

[5] G. Bartolini et. al.,"The ICOF approach to infinite horizon LQG adaptive control," *Richerche di Automatica*, vol 13, pp. 123, 1982.

[6] G. Bengtsson, "New regulators for industrial use," *Swedish Control Conference "Reglermöte"* , Linköping, Sweden, Oct. 1990.

[7] B. Bernhardsson and M. Sternad, "Feedforward control is dual to deconvolution." To appear in the *International Journal of Control*, 1992.

[8] R. R. Bitmead, M. Gevers and V. Wertz, *Adaptive Optimal Control; the Thinking Man's GPC.* Prentice Hall International, London, 1990.

[9] H. Bourles and E. Irving, "LQG/LTR: a polynomial approach," *European Control Conference ECC*, Grenoble, France, 1991.

[10] S. P. Boyd and C. H. Barratt, *Linear Controller Design: Limits of Performance.* Prentice–Hall, Englewood Cliffs, NJ, 1991.

[11] A. Casavola and E. Mosca, "Polynomial LQG regulator design for general system configurations," Proc. *30th CDC*, Brighton, England, pp 2307–2312, 1991.

[12] H. F. Chen and L. Guo, "Optimal adaptive control with consistent parameter estimates for ARMAX models with quadratic cost," *SIAM Journal on Control and optimisation*, vol 25, pp. 845–867, 1987.

[13] D. W. Clarke and P. J. Gawthrop, "Self–tuning control," *IEE Proceedings, Pt D*, vol 126, pp. 633-640, 1979.

[14] D. W. Clarke, P. P. Kanjilal and C. Mohtadi, "A generalized LQG approach to self–tuning control," *International Journal of Control*, vol 41, pp. 1509–1543, 1985.

[15] B. A. Francis and W. M. Wonham, "The internal model principle of control theory," *Automatica*, vol 12, pp. 457-465, 1976.

[16] L. Guo, "Robust stochastic adaptive control for non–minimum phase systems," *IFAC workshop on Robust Adaptive Control*, Newcastle, Australia. Preprints, pp. 185–189, 1988.

[17] M. J. Grimble, "Implicit and explicit LQG self–tuning controllers," *Automatica*, vol 20, pp. 661–669, 1984.

[18] M. J. Grimble, "Multivariable controllers for LQG self–tuning application with coloured measurement noise and dynamic cost weighting," *International Journal of System Science*, vol. 17, no. 4, pp. 543–557, 1986.

[19] M. J. Grimble, "Convergence of explicit LQG self–tuning controllers," *IEE Proceedings, Pt D*, vol 135, no 4, 1988.

[20] M. J. Grimble, "Two-degrees of freedom feedback and feedforward optimal control of multivariable stochastic systems," *Automatica*, vol 24, pp. 809-817, 1988.

[21] M. J. Grimble, "LQG optimal controller design for uncertain systems," *Proc. IEE, part D*, vol 139, pp. 21–30, 1992.

[22] M. J. Grimble and M. A. Johnsson, *Optimal Control and Stochastic Estimation.* Wiley, Chichester, 1988.

[23] K. J. Hunt, *Stochastic Optimal Control Theory with Application to Self–tuning Control.* Springer–Verlag, Berlin, 1989.

[24] K. J. Hunt and M. Šebek, "Optimal multivariable regulation with disturbance measurement feedforward," *International Journal of Control*, vol. 49, pp. 373–378, 1989.

[25] K. J. Hunt, M. Sebek and V. Kučera, "Polynomial approach to $\mathcal{H}_2$–optimal control: the multivariable standard problem," Proc. 30th CDC, Brighton, England, pp 1261–1266, 1991.

[26] J. Ježek and V. Kučera, "Efficient algorithm for matrix spectral factorization," *Automatica* vol. 21, pp. 663–669, 1985.

[27] V. Kučera, *Discrete Linear Control: The Polynomial Equations Approach.* Wiley, Chichester, 1979.

[28] V. Kučera, "Stochastic multivariable control: a polynomial equations approach", *IEEE Transactions on Automatic Control,* vol. 25, pp. 913–919, 1980.

[29] V. Kučera, "New results in state estimation and regulation," *Automatica*, vol. 17, pp. 745–748, 1981.

[30] V. Kučera, "The LQG control problem: A study of common factors," *Problems in Control and Information Theory*, vol 13, pp. 239-251, 1984.

[31] V. Kučera, "Diophantine equations in control," *European Control Conference ECC*, Grenoble, France, 1991.

[32] V. Kučera, *Analysis and Design of Linear Control Systems.* Academia, Prague and Prentice Hall International, London, 1991.

[33] V. Kučera and M. Šebek, "A note on stationary LQG control," *IEEE Transactions on Automatic Control*, vol 30, pp. 1242–1245, 1985.

[34] P. De Laminat, "On the stabilizability condition in indirect adaptive control," *Automatica*, vol 20, pp. 793-795, 1984.

[35] B. K. Lee, B. S. Chen and Y. P. Lin, "Extensions of linear quadratic optimal control theory for mixed backgrounds," *International Journal of Control*, vol 54, pp. 943–972, 1991.

[36] L. Ljung, *System Identification – Theory for the User.* Prentice–Hall, Englewood Cliffs, NJ, 1987.

[37] L. Ljung and T. Söderström, *Theory and Practice of Recursive Identification.* MIT Press, Cambridge, MA, 1983.

[38] J. M. Maciejowski, *Multivariable Feedback Design.* Addison Wesley, Reading, Mass, 1989.

[39] J. M. Mendel, *Optimal Seismic Deconvolution.* Academic Press, New York, 1983.

[40] M. Morari and E. Zafiriou, *Robust Process Control.* Prentice–Hall, Englewood Cliffs, NJ, 1989.

[41] E. Mosca and G. Zappa, "Matrix fraction solution to the discrete–time LQ stochastic tracking and servo problems," *IEEE Transactions on Automatic Control*, vol. 34, pp. 240–242, 1989.

[42] V. Peterka, "On steady–state minimum variance control strategy," *Kybernetika*, vol 8, pp. 219–232, 1972.

[43] V. Peterka, "Predictor–based self–tuning control," *Automatica*, vol 20, pp. 39-50, 1984.

[44] A. P. Roberts and M. N. Newmann, "Polynomial approach to Wiener filtering," *International Journal of Control*, vol. 47, pp. 681–696, 1988.

[45] H. H. Rosenbrock, *State–Space and Multivariable Theory.* Thomas Nelson and Sons, London, 1970

[46] C. Samson, "An adaptive LQ controller for non–minimum phase systems," *International Journal of Control*, vol 35, pp 1–28, 1982.

[47] M. Šebek, K. J. Hunt and M. J. Grimble, "LQG regulation with disturbance measurement feedforward," *International Journal of Control*, vol 47, pp. 1497-1505, 1988.

[48] N. Shimkin and A. Feuer, "On the necessity of 'block–invariance' for the convergence of adaptive pole–placement algorithm with persistently exciting input," *IEEE Trans. on Automatic Control*, vol 33, pp. 775-780, 1988.

[49] T. Söderström and P. Stoica, *System Identification.* Prentice Hall International, London, 1989.

[50] M. Sternad, "Disturbance decoupling adaptive control," *IFAC ACASP-86, Lund, Sweden.* Preprints, pp. 399-404, 1986.

[51] M. Sternad, *Optimal and Adaptive Feedforward Regulators.* PhD Thesis, Department of Technology, Uppsala University, Sweden, 1987.

[52] M. Sternad and T. Söderström, "LQG–optimal feedforward regulators," *Automatica* vol. 24, pp. 557–561, 1988.

[53] M. Sternad, "The use of disturbance measurement feedforward in LQG self–tuners," *International Journal of Control*, vol 52, pp. 579–596, 1991.

[54] M. Sternad and A. Ahlén, "A novel derivation methodology for polynomial–LQ controller design," *IEEE Transactions on Automatic Control*, vol 37, October 1992. Report UPTEC 90058R, Dept of Technology, Uppsala University.

[55] E. Trulsson and L. Ljung, "Adaptive control based on explicit criterion minimization," *Automatica*, vol 21, pp. 385-399, 1985.

[56] J. E. Weston and J. A. Bongiorno, "Extension of analytical desing techniques to multivariable feedback control systems," *IEEE Transactions on Automatic Control*, vol 17, pp. 613–620, 1972.

[57] V. A. Wolovich, *Linear Multivariable Systems.* Springer–Verlag, New York, 1974.

Chapter 4

Mixed H_2/H_∞ Stochastic Tracking and Servo Problems

A. Casavola and E. Mosca

4.1 Introduction

Multiple objective control problems, involving different types of norms on closed-loop transfer functions, are motivated by tradeoffs among competing objectives, e.g. stability robustness versus performance. Well established results exist for performance optimization for given plant and disturbance models in an H_2-norm setting. On the other hand, H_∞ control is finalized to robustify closed-loop stability in the face of plant unstructured uncertainties. When both performance and robust stability are considered, mixed H_2/H_∞ optimization may turn out to be more physically justifiable.

Though a general solution to mixed H_2/H_∞ control seems hard to be found, in special cases the solution can be obtained by solving independently H_2 and H_∞ problems. In this chapter we address one of this special cases, viz. the minimax LQ stochastic servo and tracking problem.

The treatment is for multivariable linear discrete-time plants and stochastic exogenous signals (reference and disturbance) of finite variance, with possible uncertainty on the stochastic properties of the disturbances.

In our formulation the reference to be tracked is assumed to be known up to the generic time $t+\tau$, τ being an arbitrary positive or negative integer and t the current time. Then, the servo problem ($\tau > 0$) and the tracking one ($\tau \leq 0$) are unified.

One finds that a two-degrees-of-freedom (2DOF) controller is the optimal structure for this problem and that a suitable separation property exists between the feedforward and the feedback actions, in that both of them can be designed independently one from the other. In particular, the feedforward loop is optimized with respect to an H_2 criterion costing the output tracking error, while the feedback part is determined by solving an H_∞ mixed-sensitivity pure regulation (1DOF) problem.

The above approach can be used to add a model following capability to the above optimal control law. In fact, since the Model Matching (MM) problem presumes the existence of a reference signal to be tracked, the problem can be approached as an optimal servo or tracking problem.

4.2 Signals and Operators

All the stochastic signals considered are assumed to be zero-mean, weakly-stationary (finite covariance), $\Re^m$-valued sequences with rational spectral density matrices. This space, denoted hereafter by RL_2^m, is a Hilbert space when the following inner product is introduced

$$\prec x(t), y(t) \succ = \varepsilon\{x'(t)y(t)\} \tag{4.1}$$

where $\varepsilon\{\cdot\}$ stands for the expectation operator. Two random vectors are orthogonal if their inner product is zero, i.e. if they are uncorrelated.

The norm of a causal operator $\mathcal{T}(d)$, has the following definition

$$\|\mathcal{T}\|_{RL_2^m} := sup_{\varepsilon\{\|x(t)\|_2\}=1}\varepsilon\{\|\mathcal{T}(d)x(t)\|_2\} \tag{4.2}$$

with $\{x(t)\} \in RL_2^m$ and $\mathcal{T}(d) \in \Re_{pm}(d)$ a stable transfer matrix, being $\Re_{pm}(d)$ the set of the $p \times m$ causal rational transfer matrices in the unit backward shift operator d. When $\{x(t)\}$ in (4.2) reduces to a white sequence of random vector with unitary variance, the following well known result holds.

Lemma 1 - Provided that $\{x(t)\}$ is white

$$\varepsilon\{\|\mathcal{T}x(t)\|_2^2\} = \|\mathcal{T}\|_2^2 := tr\left\{\frac{1}{2\pi}\int_{-\pi}^{\pi}\mathcal{T}^*(e^{j\omega})\mathcal{T}(e^{j\omega})d\omega\right\}$$

Proof - Since $x(t) \in \Re^m$ is white, $\varepsilon\{x'(t+i)x(t+j)\} = \delta_{i,j}$, being $\delta_{i,j}$ the kronocker operator (1 for $i = j$, 0 otherwise). Then $\varepsilon\{\|x'(t)x(t)\|_2^2\} = 1$, and by direct substitution, one finds

$$\begin{aligned}\|\mathcal{T}\|_{RL_2^m}^2 &= \varepsilon\{\|\sum_{i=0}^{\infty}T_ix(t-i)\|_2^2\} = \varepsilon\{\sum_{i,j}^{\infty}x'(t-i)T_i'T_jx(t-j)\} = tr\sum_{i=j}^{\infty}T_i'T_j \\ &= \|\mathcal{T}\|_2^2 := tr\left\{\frac{1}{2\pi}\int_{-\pi}^{\pi}\mathcal{T}^*(e^{j\omega})\mathcal{T}(e^{j\omega})d\omega\right\}\end{aligned}$$

where $\mathcal{T}(d) = \sum_{i=0}^{\infty}T_id^i$ and $\mathcal{T}^*(d) := \mathcal{T}'(d^{-1})$. □

This result enables one to consider H_2-norm optimization problems when the statistical properties of the exogenous signals are fully known.

A key result of the theory of stochastic realization is that any stochastic signal $\{x(t)\} \in RL_2^m$ with power constraint

$$\varepsilon\{\|x(t)\|_2^2\} = 1 \tag{4.3}$$

can be thought, as the output of some causal, stable linear system with transfer matrix $\mathcal{Q}(d) \in \Re_{mm}(d)$ excited by a zero-mean, white random sequence $\{e(t)\}$ with unitary variance. In this way the class of admissible exogenous signal is defined by the set of stable transfer matrix whose spectrum, $\mathcal{Q}^*(d)\mathcal{Q}(d)$, has component variances whose sum is equal to one.

As a consequence, the induced operator norm (4.2), can be rewritten as

$$\|\mathcal{T}\|_{RL_2^m} = \sup_{\|\mathcal{Q}\|_2=1}\varepsilon\{\|\mathcal{T}(d)\mathcal{Q}(d)e(t)\|_2\} = \sup_{\|\mathcal{Q}\|_2=1}\|\mathcal{T}\mathcal{Q}\|_2 \tag{4.4}$$

where $\mathcal{Q}(d)$ is in RH_∞. Finally, the following lemma provides some justification for minimizing the H_∞-norm when the knowledge of statistic properties of the exogenous signals is limited.

Lemma 2 [1] - Let $\{x(t)\}$ be a stochastic sequence in RL_2^m such that $\varepsilon\{\|x(t)\|_2^2\} = 1$. Then,

$$\|\mathcal{T}(d)\|_{RL_2^m} = \|\mathcal{T}(d)\|_\infty^2 := \sup_{\omega\in[0,2\pi]} \sigma_{max}(\mathcal{T}(e^{-j\omega})). \tag{4.5}$$

with $\sigma_{max}(\cdot)$ the maximum singular value of $\mathcal{T}(e^{-j\omega})$.

Proof - For all $\omega \in [-\pi,\pi]$, $\mathcal{T}(e^{j\omega})^*\mathcal{T}(e^{j\omega})$ is a Hermitian, positive definite matrix with rank m. Thus, there exist m real eigenvalues $\lambda_i(e^{j\omega}) > 0$, $i \leq m$, and m distinct eigenvectors $\mu_i(e^{j\omega})$ which form an ortonormal basis in $\Re^m$. Further, $\mathcal{T}(e^{j\omega})^*\mathcal{T}(e^{j\omega})$ is diagonalizable by means of a unitary transformation $\mathcal{U}^*(e^{j\omega})\mathcal{T}(e^{j\omega})^*\mathcal{T}(e^{j\omega})\mathcal{U}(e^{j\omega}) = \Lambda(e^{j\omega})$, with $\Lambda = \mathcal{D}iag\{\lambda_i\}$ Hermitian and diagonal and $\mathcal{U} = [\mu_1 \mid \mu_2 .... \mid \mu_m]$ such that $\mathcal{U}^*\mathcal{U} = I$. It follows that

$$\begin{aligned}
\|\mathcal{T}\mathcal{Q}\|_2^2 &= tr\left\{\frac{1}{2\pi}\int_{-\pi}^{\pi} \mathcal{Q}^*(e^{j\omega})\mathcal{T}^*(e^{j\omega})\mathcal{T}(e^{j\omega})\mathcal{Q}(e^{j\omega})d\omega\right\} \\
&= tr\left\{\frac{1}{2\pi}\int_{-\pi}^{\pi} \mathcal{Q}(e^{j\omega})\mathcal{Q}^*(e^{j\omega})\mathcal{U}(e^{j\omega})\Lambda(e^{j\omega})\mathcal{U}^*(e^{j\omega})d\omega\right\} \\
&\leq tr\left\{\frac{1}{2\pi}\int_{-\pi}^{\pi} \mathcal{Q}(e^{j\omega})\mathcal{Q}^*(e^{j\omega})\mathcal{U}(e^{j\omega})\mathcal{D}iag\{\sup_{\omega\in[-\pi,\pi]} \max_{i\leq m} \lambda_i(e^{j\omega})\}\mathcal{U}^*(e^{j\omega})d\omega\right\} \qquad (4.6) \\
&= \sup_{\omega\in[-\pi,\pi]} \max_{i\leq m} \lambda_i(e^{j\omega})\} tr\left\{\frac{1}{2\pi}\int_{-\pi}^{\pi} \mathcal{Q}^*(e^{j\omega})\mathcal{Q}(e^{j\omega})d\omega\right\} \\
&= \|\mathcal{T}\|_\infty^2
\end{aligned}$$

It can be proved that the bound in (4.6) is sharp, in that is there exist $\{x(t)\} \in RL_2^m$ with $\frac{\varepsilon\{\|\mathcal{T}x\|_2^2\}}{\varepsilon\{\|x\|_2^2\}}$ as close as one wants to $\|\mathcal{T}\|_\infty^2$ (but not equal, unless $\mathcal{T}$ has constant modulus on the unit circle) [2]. □

The conclusion is that in the absence of information about the statistical properties of the exogenous signals, one can consider the H_∞-norm of the operator. For this reason such a strategy is sometimes referred to as the "worst case analysis", meaning that is fully equivalent to the H_2-norm approach for the "worst" exogenous signal in RL_2^m.

4.3 Minimax LQ Stochastic Tracking and Servo Problem

This section is devoted to the minimax LQ stochastic tracking and servo problem, where disturbance regulation and robust stability are optimized in H_∞ sense while tracking performance in an H_2 sense. Consequently a mixed H_2/H_∞ optimization problem results.

Unlike previous relevant contributions [3], here the reference to be tracked by the output at time t is assumed to be known up to time $t+\tau$, τ positive or negative.

In particular, if τ is positive and large enough, the optimal feedforward command can be tightly approximated without requiring any stochastic dynamic model of the reference. However, if this is available, a closed-form solution is given in terms of a bilateral matrix polynomial Diophantine equation.

4.3.1 Problem formulation and results

Consider a *set* of discrete-time multivariable stochastic linear *plants*

$$y(t) = \mathcal{P}(d)u(t) + \mathcal{Q}(d)e(t) \tag{4.7}$$

where: $y(t) \in \Re^p$ is the output; $u(t) \in \Re^m$ is the input; $\{e(t)\}$ is a stationary white sequence of random vectors in $\Re^p$ with zero-mean and unit covariance; $\mathcal{P}(d) \in \Re_{pm}(d)$ and $\mathcal{Q}(d) \in \Re_{pp}(d)$ are causal *rational* transfer matrices in the unit backward shift operator d. $\Re_{pm}(d)$ denotes the set of $p \times m$ matrices with elements in $\Re(d)$, the set of polynomial fractions in the indeterminate d. Moreover:

(A.3.1)
- The plant with transfer matrix $\mathcal{P}(d)$ is strictly causal and its actual realization is free of unstable hidden modes.
- Every unstable root of the characteristic polynomial (c.p.) of the overall system (4.7) is also a root with the same multiplicity of the c.p. of the subsystem with transfer matrix $\mathcal{P}(d)$.

The second condition in (**A.3.1**) states that all the unstable modes of (4.7) are due solely to $\mathcal{P}(d)$. This does not imply that $\mathcal{Q}(d)$ in (4.7) needs to be stable, since the disturbances can act on the plant in a whatever point between its inputs and outputs. The situation is depicted in Figure 4.1, where $\delta(t) := \bar{\mathcal{Q}}(d)e(t)$, $\bar{\mathcal{Q}} \in \Re_{pp}(d)$, is any disturbance having bounded arbitrary rational power spectrum. Moreover, the following normalization condition over the class of admissible disturbances is assumed

(A.3.2)
- $\bar{\mathcal{Q}}(d)$ is any stable transfer matrix such that $\|\bar{\mathcal{Q}}\|_2^2 := tr\left\{\frac{1}{2\pi}\int_{-\pi}^{\pi} \bar{\mathcal{Q}}^*(e^{j\omega})\bar{\mathcal{Q}}(e^{j\omega})d\omega\right\} = 1,$

Here, if $\mathcal{F}(d)$ represents any rational matrix, $\mathcal{F}^*(d) := \mathcal{F}'(d^{-1})$, with the *prime* denoting transpose and $\{\delta(t)\}$ represents a stochastic sequence with arbitrary rational spectrum $\bar{\mathcal{Q}}^*\bar{\mathcal{Q}}$ and component variances whose sum is equal to one.

Let $w(t) \in \Re^p$ be the output *reference* variable. It is assumed that:

(A.3.3)
- $\{w(t)\}$ is a second-order stationary process uncorrelated with $\{e(t)\}$.

Let $I^t := \{y^t, u^{t-1}, w^{t+\tau}\}$, $y^t := \{y(t), y(t-1), ..., \}$ be the available information up to time t, and $RL_2^m(z)$ denote the subspace of $\Re^m$-valued random vectors with finite covariance and rational power density function spanned by $\{z\}$. The problem that is addressed hereafter is to find among the *linear non anticipative control laws*

$$u(t) = \mathcal{L}(I^t) \in RL_2^m(I^t) \tag{4.8}$$

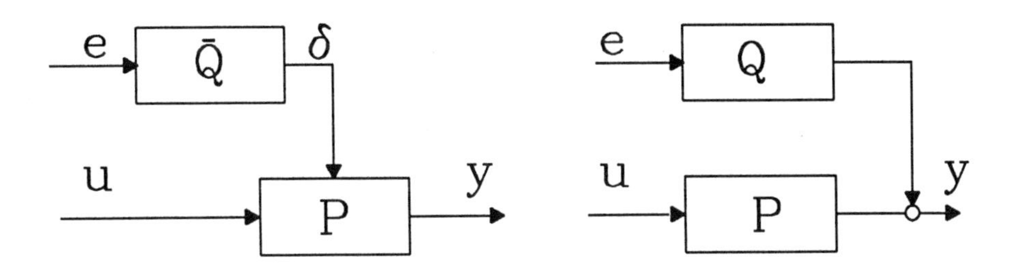

Figure 4.1: Equivalent plant and disturbance descriptions

the ones that, in stochastic steady-state, stabilize the plant (4.7) and solve the following *minimax Linear-Quadratic (LQ) stochastic tracking or servo problem*

$$\inf_{\mathcal{L}} \sup_{\bar{Q}} \varepsilon\{\|\mathcal{W}_y(d)y(t) - w(t)\|^2_{\psi_y} + \|\mathcal{W}_u(d)u(t)\|^2_{\psi_u}\} \tag{4.9}$$

In (4.9), $\psi_y = \psi_y^T \geq 0, \psi_u = \psi_u^T \geq 0$, $\|v\|^2_\psi := v^T\psi v$ and $\mathcal{W}_y(d) \in \Re_{pp}(d)$ and $\mathcal{W}_u(d) \in \Re_{mm}(d)$ are suitable dynamic weights that are assumed to be of full rank on the unit circle. In particular, it is assumed that they have the following polynomial representation

$$\mathcal{W}_y(d) = B_y(d)A_y(d)^{-1}, \qquad \mathcal{W}_u(d) = B_u(d)A_u(d)^{-1} \tag{4.10}$$

with $B_y(d)$ and $A_y(d)$ polynomial matrices in $\Re_{pp}[d]$, $\Re_{pp}[d]$ denoting the set of $p \times p$ matrices with elements in $\Re[d]$, the set of polynomial in d, and $B_u(d)$ and $A_u(d)$ polynomial matrices in $\Re_{mm}[d]$. It is further assumed that $A_y(d)$ and $A_u(d)$ are strictly Hurwitz. The whole feedback system structure is shown in Figure 4.2.

Remark 3.1 - If $w(t) \equiv 0$, problem (4.7)-(4.10) is the same as finding a stabilizing controller $u(t) = -\mathcal{K}(d)y(t)$, $\mathcal{K}(d) \in \Re_{mp}(d)$, which minimizes

$$\sup_{0 \leq \omega < 2\pi} \lambda_{max}\{[\mathcal{S}^*(e^{j\omega})\mathcal{W}_y^*(e^{j\omega})\psi_y\mathcal{W}_y(e^{j\omega})\mathcal{S}(e^{j\omega}) + \mathcal{M}^*(e^{j\omega})\mathcal{W}_u^*(e^{j\omega})\psi_u\mathcal{W}_u(e^{j\omega})\mathcal{M}(e^{j\omega})]\} \tag{4.11}$$

where: $\mathcal{S} := (I_p + \mathcal{P}\mathcal{K})^{-1} = I_p - \mathcal{P}(I_p + \mathcal{K}\mathcal{P})^{-1}\mathcal{K}$ and $\mathcal{M} := \mathcal{K}\mathcal{S}$ are, respectively, the *sensitivity matrix* and the *power transfer matrix* of the feedback system. The H_∞ *mixed sensitivity problem* (4.11) has been considered in a number of papers: [4, 5, 6, 7, 8, 9]. In particular, [6, 9, 8] showed how solutions can be obtained by polynomial equations.□

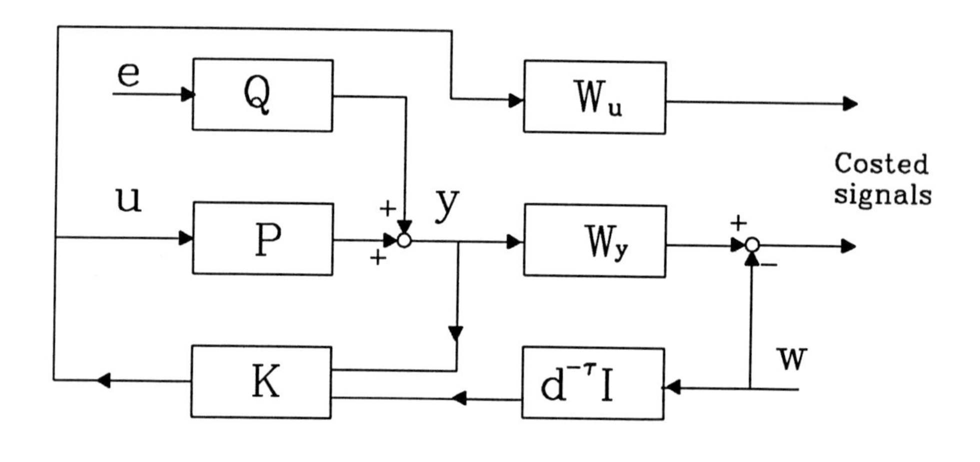

Figure 4.2: Feedback system structure

Write $\mathcal{P}$ and $\mathcal{Q}$ in the form of a left matrix fraction descriptions (m.f.d.)

$$\mathcal{P}(d) = A^{-1}(d)B(d), \qquad \mathcal{Q}(d) = A^{-1}(d)D(d) \tag{4.12}$$

where $A(d) \in \Re_{pp}[d]$, $B(d) \in \Re_{pm}[d]$ and $D(d) \in \Re_{pp}[d]$ and $A(d)$ is a least common divisor of $\mathcal{P}(d)$ and $\mathcal{Q}(d)$. It is further assumed:

(A.3.4) $\{$ • D is strictly Hurwitz, i.e. E^{-1} is analytic on the closed unit disk.

Further, define the polynomial matrices B_2, $B_3 \in \Re_{pm}[d]$ and A_2, A_{y3}, A_{2o}, A_{uo} all $\in \Re_{mm}[d]$, by the following coprime polynomial matrix fraction descriptions

$$A^{-1}B = B_2A_2^{-1}, \quad A_y^{-1}B_2 = B_3A_{y3}^{-1}, \quad A_u^{-1}A_2A_{3y} = A_{2o}A_{uo}^{-1} \tag{4.13}$$

It is easily to assess that both A_{y3} and A_{uo} are strictly Hurwitz since A_u and A_y are so. Further, define a right spectral factor $E(d) \in \Re_{mm}[d]$, solution of the following spectral factorization problem

$$A_{2o}^*B_u^*\psi_uB_uA_{2o} + A_{uo}^*B_3^*B_y^*\psi_yB_yB_3A_{uo} = E^*E \tag{4.14}$$

It is further assumed that :

(A.3.5) $\{$ • E is strictly Hurwitz, i.e. E^{-1} is analytic on the closed unit disk.

E.g., **(A.3.5)** is fulfilled if $\psi_y > 0$ and $\psi_u > 0$ [10].

Let a solution of the mixed sensitivity problem, viz.(4.7)-(4.10) with $w(t) \equiv 0$, be

$$u(t) = -\mathcal{M}_1^{-1}\mathcal{N}_1 y(t) \tag{4.15}$$

where $\mathcal{M}_1 \in \Re_{mm}(d)$ and $\mathcal{N}_1 \in \Re_{pp}(d)$ are suitable stable causal transfer matrices such that [12]

$$\mathcal{M}_1 A_2 + \mathcal{N}_1 B_2 = I_m \tag{4.16}$$

Notice that the solution exists and it is stabilizing if the above assumptions **(A.3.1)-(A.3.5)** are fulfilled [6, 8]. Consider now the following equation

$$\mathcal{M}_1 u(t) = -\mathcal{N}_1 y(t) + v(t). \tag{4.17}$$

Since $\mathcal{M}_1$ and $\mathcal{N}_1$ are specific stable and causal transfer matrices, for each $v(t) \in RL_2^m(I^t)$, (4.17) yields an admissible control law. Conversely, any admissible control law (4.8) can be obtained as in (4.17) for a suitable $v(t) \in RL_2^m(I^t)$. Thus, the stated minimax LQ stochastic tracking or servo problem amounts to finding a process $\{v(t)\}$, $v(t) \in RL_2^m(I^t)$, such that (4.17) stabilizes the plant and solves (4.11).

Since, according to **(A.3.3)** and (4.8), $RL_2^m(I^t) = RL_2^m(e^t) \bigoplus RL_2^m(w^{t+\tau})$, a convenient orthogonal decomposition for $v(t)$ is

$$v(t) = v_e(t) + v_w(t) \tag{4.18}$$

with $v_e(t) \in RL_2^m(e^t)$ and $v_w(t) \in RL_2^m(w^{t+\tau})$. Further, under the control law (4.17), the regulated $y(t)$ and $u(t)$ can be decomposed as follows:

$$\begin{aligned} u(t) &= (I_p + \mathcal{KP})^{-1} v(t) - (I_p + \mathcal{KP})^{-1}\mathcal{KQ}e(t) \\ &= A_2 v(t) - A_2\mathcal{N}_1\mathcal{Q}e(t) := u_v(t) + u_e(t). \qquad (4.19) \\ y(t) &= \mathcal{P}(I_p + \mathcal{KP})^{-1} v(t) + (I_p - \mathcal{P}(I_p + \mathcal{KP})^{-1}\mathcal{K})\mathcal{Q}e(t) \\ &= B_2 v(t) - B_2\mathcal{N}_1\mathcal{Q}e(t) := y_v(t) + y_e(t). \qquad (4.20) \end{aligned}$$

With reference to the decomposition (4.19)-(4.20) , the quadratic term

$$J := \varepsilon\{\|\mathcal{W}_y(d)y(t) - w(t)\|^2_{\psi_y} + \|\mathcal{W}_u(d)u(t)\|^2_{\psi_u}\} \tag{4.21}$$

can be split as follows

$$J = J_{ee} + J_{ev} + J_{ew} + J_{ww} + J_{vw} + J_{vv} \tag{4.22}$$

where:

$$\begin{aligned} J_{ee} &:= \varepsilon\{\|\mathcal{W}_y y_e(t)\|^2_{\psi_y} + \|\mathcal{W}_u u_e(t)\|^2_{\psi_u}\}; \\ J_{ev} &:= 2\varepsilon\{y_e'(t)\mathcal{W}_y^*\psi_y\mathcal{W}_y y_v(t) + u_e'(t)\mathcal{W}_u^*\psi_u\mathcal{W}_u u_v(t)\}; \\ J_{ww} &:= \varepsilon\{\|w(t)\|^2_{\psi_y} = 0 \ \ [\mathbf{A.3.3}]; \\ J_{ew} &:= -2\varepsilon\{w'(t)\psi_y\mathcal{W}_y y_e(t)\}; \\ J_{vw} &:= -2\varepsilon\{v'(t)B_2^* A_y^{-*} B_y^* \psi_y w(t)\}; \\ J_{vv} &:= \varepsilon\{\|\mathcal{W}_y y_v(t)\|^2_{\psi_y} + \|\mathcal{W}_u u_v(t)\|^2_{\psi_u}\} = \varepsilon\{\|EA_{uo}^{-1}A_{y3}^{-1}v(t)\|^2\} \\ &= \varepsilon\{\|EA_{uo}^{-1}A_{y3}^{-1}v_e(t)\|^2 + \|EA_{uo}^{-1}A_{y3}^{-1}v_w(t)\|^2\} := J_{v_e v_e} + J_{v_w v_w} \end{aligned}$$

where $E^{-*} := (E^*)^{-1}$. The main result of the section can now be stated :

Theorem 3.1 [11] - Let **(A.3.1)-(A.3.5)** be fulfilled. Then, the *minimax LQ stochastic tracking or servo control law* is given by

$$\mathcal{M}_1(d)u(t) = -\mathcal{N}_1(d)y(t) + A_{y3}(d)A_{uo}(d)E^{-1}(d)u_c(t) \tag{4.23}$$

where : $\mathcal{M}_1$ and $\mathcal{N}_1$ are the transfer matrices in (4.15) and $u^c(t)$ is the *feedforward command* defined by

$$u_c(t) = \varepsilon\{E^{-*}A^*_{uo}B^*_3B^*_y\psi_y w(t) \mid w^{t+\tau}\}. \tag{4.24}$$

Proof - By **(A.3.3)**, $J_{ev} = J_{ev_e}$, $J_{vw} = J_{v_w w}$ and $J_{ew} = 0$. Further, J_{ww} is not affected both by $\mathcal{Q}$ and v. Consequently,

$$\begin{aligned}\inf_{v_e,v_w}\sup_{\bar{\mathcal{Q}}} J &= \inf_{v_e,v_w}\sup_{\bar{\mathcal{Q}}}(J_{ee} + J_{ev_e} + J_{v_ev_e} + J_{v_ww} + J_{v_wv_w}) \\ &= \inf_{v_e}\sup_{\bar{\mathcal{Q}}}(J_{ee} + J_{ev_e} + J_{v_ev_e}) + \inf_{v_w}(J_{v_ww} + J_{v_wv_w})\end{aligned}$$

Last equality follows since both J_{v_ww} and $J_{v_wv_w}$ do not depend on v_e and $\mathcal{Q}$, whereas J_{ee}, J_{ev_e} and $J_{v_ev_e}$ are not affected by v_w . Now

$$\inf_{v_e}\sup_{\bar{\mathcal{Q}}}(J_{ee} + J_{ev_e} + J_{v_ev_e})$$

is the same as the underlying minimax LQ stochastic regulation problem (the H_∞ mixed sensitivity problem of Remark 3.1), viz.(4.7)-(4.10) with $w(t) \equiv 0$, of which the pair $(\mathcal{M}_1, \mathcal{N}_1)$ in (4.17) is a solution. Thus $v_e = 0$. Next, the minimum of

$$\begin{aligned}&J_{v_ww} + J_{v_wv_w} = \\ &\varepsilon\{\|EA^{-1}_{uo}A^{-1}_{y3}v_w(t) - E^{-*}A^*_{uo}B^*_3B^*_y\psi_y w(t)\|^2\} - \varepsilon\{\|E^{-*}A^*_{uo}B^*_3B^*_y\psi_y w(t)\|^2\}\end{aligned}$$

is attained, among all $v_w(t) \in RL^m_2(w^{t+\tau})$, at $v_w(t) = A_{y3}A_{uo}E^{-1}u^c(t)$, with $u_c(t)$ as in (4.24). □

It is straightforward to verify that $u_c(t)$ is a mean-square (ms) bounded process, being a conditional expectation of a strictly anticausal ms bounded process. Hence, it consists in general of a linear combination of conditional expectation of future samples of $w(t)$ w.r.t. to $w^{t+\tau}$, i.e.

$$u_c(t) = \sum_{i=1}^{\infty}\alpha_i\varepsilon\{w(t+i) \mid w^{t+\tau}\} \tag{4.25}$$

where $\{\alpha_i\}$ is a convergent sequence of constant matrices such that $E^{-*}A^*_{uo}B^*_3B^*_y\psi_y = \sum_{i=1}^{\infty}\alpha_i d^{-i}$. Further, if $\tau > 0$, (4.25) can be rewritten as

$$u_c(t) = u^\tau_c(t) + \tilde{u}_c(t) = \sum_{i=1}^{\tau}\alpha_i w(t+i) + \sum_{i=\tau+1}^{\infty}\alpha_i\varepsilon\{w(t+i) \mid w^{t+\tau}\} \tag{4.26}$$

where only the second term in the right-side of (4.26) depends on the statistic properties of $w(t)$.

One of the features of expression (4.26) is that it gives, if τ is positive and large enough, a tight approximation of the optimal command input when a stochastic model for the reference is not available. In fact, this can be obtained simply by replacing $u_c(t)$ with $u_c^\tau(t)$ and it is generally found that the increment of the cost due to this approximation decay exponentially as τ increases. Consider the following example:

Example 3.1 - Consider a SISO plant with $B = B_2 = d(1 - \alpha d)$ and a minimum sensitivity H_∞ control problem ($\psi_u = 0$ and $\psi_y = 1$ in (4.9)) with constant dynamic weights $\mathcal{W}_y = \mathcal{W}_u = 1$. Then, for $|\alpha| > 1$ (non minimum-phase plant), we get from (4.14)

$$E(d) = k(1 - \alpha^{-1}d), \qquad k := \left(\frac{1+\alpha^2}{1+\alpha^{-2}}\right)^{\frac{1}{2}} \tag{4.27}$$

and the optimal feedforward input becomes

$$u_c(t) = \frac{1}{k}\varepsilon\{\frac{d^{-1}(1-\alpha d^{-1})}{1-\alpha^{-1}d^{-1}}w(t) \mid w^{t+\tau}\} \tag{4.28}$$

As in (4.26), if $\tau > 0$, it can be decomposed as $u_c^\tau(t) + \tilde{u}_c(t)$:

$$u_c^\tau(t) = \frac{1}{k}\left(w(t+1) + (1-\alpha^2)\sum_{i=1}^{\tau-1}\alpha^{-i}w(t+i+1)\right) \tag{4.29}$$

$$\tilde{u}_c(t) = \frac{1}{k}(1-\alpha^2)\alpha^{-\tau+1}\sum_{i=1}^{\infty}\alpha^{-i}\varepsilon\{w(t+\tau+i) \mid w^{t+\tau}\} \tag{4.30}$$

A tight approximation of the optimal feedforward input can be obtained simply by replacing u_c with u_c^τ. In order to valuate the approximation error, let $w(t)$ be a white sequence with covariance σ_r^2. The increment in the cost functional due to this approximation amounts to

$$\begin{aligned} J_{u_c^\tau u_c^\tau} - J_{u_c u_c} &= \varepsilon\{\tilde{u}_c(t)'\tilde{u}_c(t)\} \\ &= k^{-2}(1-\alpha^2)^2\alpha^{-2\tau+2}\gamma_r, \end{aligned} \tag{4.31}$$

$$\gamma_r := <(\sum_{i=1}^{\infty}\alpha^{-i}d^{-i})(\sum_{i=1}^{\infty}\alpha^{-i}d^{i}) > \sigma_r^2 \tag{4.32}$$

where $<\sum_{\infty}^{\infty}\beta_i d^i> := \beta_o$ extracts the zero-power coefficient of a two-sided sequence. From (4.31) follows that the approximation error decreases exponentially with τ. □

Remark 3.2 - It is to be pointed out that the optimal feedforward action (4.24) is independent of the output disturbance transfer matrix $\mathcal{Q}$. This is an important feature of the solution and gives a clear justification of the fact that u_c as determined in (4.24) is also the optimal solution for the LQG stochastic tracking and servo problem discussed in [13], that is the problem (4.7)-(4.10) for a fixed $\mathcal{Q}$. Consider the following example for a better understanding.

Example 3.2 - Here the aim is to compare the solution of the LQG stochastic tracking or servo problem as reported in [13] with the one given in (4.23). A

SISO plant is considered with $\mathcal{P} = \frac{d}{(1-3d)}$ and a quadratic cost as in (4.21) with $\psi_y = \psi_u = \mathcal{W}_y = \mathcal{W}_u = 1$. Consequently, from (4.14), we get $E = \alpha(3.370 - d)$, where $\alpha = 0.943$.

In the LQG stochastic servo problem, we assume for simplicity that the output disturbance is white, viz. $\mathcal{Q} = 1$. Thus, the stabilizing control law minimizing (4.21) can be obtained as shown in [12] and it is given by

$$(9.538 - 0.314d)u(t) = -22.920y(t) + (3-d)u^c(t) \tag{4.33}$$

with

$$\begin{aligned} u_c(t) &= \varepsilon\left\{\frac{d^{-1}}{E(d^{-1})}w(t) \mid w^{t+\tau}\right\} \\ &= (3.18)^{-1}\varepsilon\{\sum_{j=1}^{\infty}(3.370)^{-j+1}w(t+j) \mid w^{t+\tau}\} \end{aligned} \tag{4.34}$$

In the minimax LQ context, the underlying H_∞ mixed sensitivity problem (4.11)) can be readily solved by the polynomial approach of [8] or [6] and one gets:

$$\mathcal{M}_1 = \frac{9(1-0.29\overline{6}d)}{\alpha(3-d)(3.370-d)} \quad ; \quad \mathcal{N}_1 = \frac{24(1-0.29\overline{6}d)}{\alpha(3-d)(3.370-d)} \tag{4.35}$$

The above $\mathcal{M}_1$ and $\mathcal{N}_1$ are also the solution to the stochastic LQG pure regulation problem for $\psi_y = \psi_u = \mathcal{W}_1 = \mathcal{W}_u = 1$, the given $\mathcal{P}$, and the "worst" noise colouring filter $\mathcal{Q} = \frac{25.78}{(27-8d)}$. Consequently, the control law solving the minimax stochastic tracking or servo problem is given by

$$9(1-0.29\overline{6}d)u(t) = -24(1-0.29\overline{6}d)y(t) + (3-d)u^c(t) \tag{4.36}$$

where $u^c(t)$ is again given by (4.34).

4.3.2 ARMA and recurrent reference models

Expression (4.24) can be further elaborated when a stochastic model for the reference realization $w(t)$ is available. In this connection, assume that

$$w(t) = G_2(d)F_2^{-1}(d)\nu(t), \tag{4.37}$$

where $\nu(t) \in \Re^p$, is a white sequence of random vectors with zero-mean and unitary covariance, $G_2F_2^{-1} \in \Re_{pp}(d)$ is a causal rational transfer matrix and G_2 and F_2 polynomial matrices in $\Re_{pp}[d]$. Moreover, the following is assumed

(A.3.6) $\{$ • F_2 and G_2 are both strictly Hurwitz.

Let

$$p := \max\{\partial E, \partial B_y B_3 A_{uo} - \min\{\tau, 0\}\} \tag{4.38}$$

where ∂E denotes the degree of E, viz. the greatest degree of the polynomial entries of E. Further, let

$$\bar{E}(d) = d^p E^*(d), \quad \bar{B}_4 = d^{p+\min\{\tau,0\}} A_{uo}^* B_3^* B_y^* \tag{4.39}$$

Corollary 3.1 [13] - Let **(A.3.1)-(A.3.6)** be fulfilled and the reference be modelled as in (4.37). Then, the optimal *feedforward command* is given by

$$u_c(t) = \Delta(d)G_2^{-1}(d)w(t+\tau) \tag{4.40}$$

where $(\Delta(d), \Gamma(d)) \in \Re_{mp}[d]$ is the unique solution of minimum degree w.r.t. $\Gamma(d)$, i.e. $\partial\Gamma < \partial\bar{E}$, of the following bilateral Diophantine equation

$$\bar{E}\Delta + \Gamma F_2 = d^{\max\{\tau,0\}}\bar{B}_4\psi_y G_2 \tag{4.41}$$

Proof - First, since $\bar{E}$ is nonsingular and anti-Hurwitz while F_2 is strictly Hurwitz, there exist unique polynomial matrices (Δ, Γ), $\partial\Gamma < \partial\bar{E}$, satisfying (4.41) [12]. Next, taking into account (4.37), (4.24) becomes

$$u_c(t) = \varepsilon\{\bar{E}^{-1}\bar{B}_4\psi_y G_2 F_2^{-1} d^\tau \nu(t+\tau) \mid \nu^{t+\tau}\} \tag{4.42}$$

We consider separately the case $\tau \geq 0$ and $\tau < 0$. For $\tau \geq 0$ one finds

$$\begin{aligned} u_c(t) &= \varepsilon\{\bar{E}^{-1}\bar{B}_4\psi_y G_2 d^\tau F_2^{-1}\nu(t+\tau) \mid \nu^{t+\tau}\} \\ &= \varepsilon\{\bar{E}^{-1}(\bar{E}\Delta + \Gamma F_2)F_2^{-1}\nu(t+\tau) \mid \nu^{t+\tau}\} \\ &= \varepsilon\{(\Delta F_2^{-1} + \bar{E}^{-1}\Gamma)\nu(t+\tau) \mid \nu^{t+\tau}\} \\ &= \Delta F_2^{-1}\nu(t+\tau). \end{aligned} \tag{4.43}$$

The last equality follows from the degree constraint on Γ. In fact, $\bar{E}^{-1}\Gamma = E^* d^{-p}\Gamma$ is a *strictly anticausal* transfer matrix. Finally, for $\tau < 0$

$$\begin{aligned} u_c(t) &= \varepsilon\{\bar{E}^{-1}\bar{B}_4\psi_y G_2 F_2^{-1}\nu(t+\tau) \mid \nu^{t+\tau}\} \\ &= \varepsilon\{(\Delta F_2^{-1} + \bar{E}^{-1}\Gamma)\nu(t+\tau) \mid \nu^{t+\tau}\} \\ &= \Delta F_2^{-1}\nu(t+\tau). \quad \square \end{aligned} \tag{4.44}$$

The last case considered concerns a *recurrent* or *predictable* reference. This is the case when the reference realization is governed by

$$W(d)w(t) = 0_p \tag{4.45}$$

where $W(d) \in \Re_{pp}[d]$ is proper, viz. the matrix coefficient of the highest power of d is nonsingular, and marginally stable with all roots of $\det W(d)$ of unit modulus.

Corollary 3.2 [13] - Let (**A.3.1**)-(**A.3.5**) be fulfilled and the reference predictable as in (4.45). Then, the optimal *feedforward command* is given by

$$u_c(t) = \Delta(d)w(t) \tag{4.46}$$

where $(\Delta(d), \Gamma(d)) \in \Re_{mp}[d]$ is the unique solution of minimum degree w.r.t. $\Delta(d)$, i.e. $\partial\Delta < \partial W$, of the following bilateral Diophantine equation

$$\bar{E}\Delta + \Gamma W = \bar{B}_4\psi_y \tag{4.47}$$

Proof - W.l.o.g., one can set $\tau = 0$. Thus

$$\begin{aligned} u_c(t) &= \varepsilon\{\bar{E}^{-1}\bar{B}_4\psi_y w(t) \mid w^t\} \\ &= \Delta w(t) + \bar{E}^{-1}\Gamma W w(t) \mid w^t\} \\ &= \Delta w(t). \end{aligned} \tag{4.48}$$

That (4.47) can be uniquely solved follows from the properness of W and the fact that $\det E$ and $\det W$ do not share common roots. □

Example 3.3 - Considering the following non minimum-phase system defined by

$$B_2 = \begin{bmatrix} d(3d+1) & 0 \\ d & d \end{bmatrix}, \qquad A_2 = I_2 \tag{4.49}$$

Here $A_2 = I_2$ only for simplicity. The theory applies to the general case. By choosing $\psi_u = 0_2$ and $\psi_y = \mathcal{W}_y = \mathcal{W}_u = I_2$ in (4.9), E in (4.14) equals $E = \begin{bmatrix} 3+d & 0 \\ 1 & 1 \end{bmatrix}$. Further, one has $B_y = A_{uo} = I_2$, $B_3 = B_2$, $B_2^* = \begin{bmatrix} d^{-1}(3d^{-1}+1) & d^{-1} \\ 0 & d^{-1} \end{bmatrix}$ and $E^{-*} = \frac{1}{3+d^{-1}} \begin{bmatrix} 1 & -1 \\ 0 & 3+d^{-1} \end{bmatrix}$ Thus, the following feedforward input $u^c(t) = \begin{bmatrix} u_c^{(1)}(t) & u_c^{(2)}(t) \end{bmatrix}'$ defined in (4.24) results after straightforward steps

$$\begin{aligned} u_c^{(1)}(t) &= \varepsilon\{\frac{d^{-1}(3d^{-1}+1)}{(3+d^{-1})} w^{(1)}(t) \mid w^{t+\tau}\} \\ &= \frac{1}{3}\varepsilon\{w^{(1)}(t+1) \mid w^{t+\tau}\} - \frac{8}{3}\sum_{j=1}^{\infty}(-3)^{-j}\varepsilon\{w^{(1)}(t+j+1) \mid w^{t+\tau}\} \qquad (4.50) \\ u_c^{(2)}(t) &= \varepsilon\{w^{(2)}(t+1) \mid w^{t+\tau}\}. \qquad (4.51) \end{aligned}$$

with $w(t) = \begin{bmatrix} w^{(1)}(t) & w^{(2)}(t) \end{bmatrix}'$.

Finally, the closed-loop output $y_v(t)$ in (4.20) is given, after simplification steps, by

$$y_v^{(1)}(t) = \frac{1}{(3+d)} u_c^{(1)}(t) \tag{4.52}$$

$$y_v^{(2)}(t) = -\frac{1}{(3+d)} u_c^{(1)}(t) + u_c^{(2)}(t) \tag{4.53}$$

If a stochastic model for the reference is given the results of the Corollary 3.1 may than be applied. To this end, consider the following model

$$G_2 = I_2, \quad F_2 = \begin{bmatrix} 1 & 0 \\ 0 & 3+d \end{bmatrix} \tag{4.54}$$

and select $\tau = 2$. Then $p = 2$, $\bar{E} = \begin{bmatrix} d(3d+1) & d^2 \\ 0 & d^2 \end{bmatrix}$ and $\bar{B}_4 = \begin{bmatrix} d(3+d) & d \\ 0 & d \end{bmatrix}$ follows from (4.38) and (4.39) and the required solution for the Diophantine equation (4.41) is given by

$$\Delta = \begin{bmatrix} d + \frac{1}{3}d^2 & 0 \\ 0 & d \end{bmatrix}, \qquad \Gamma = 0_2 \tag{4.55}$$

and the feedforward command, carried out by (4.40), results

$$u_c^{(1)}(t) = w^{(1)}(t) \tag{4.56}$$

$$u_c^{(2)}(t) = w^{(2)}(t+1). \tag{4.57}$$

The closed-loop expressions (4.52) and (4.53) apply with $u^c(t)$ given by (4.56) and (4.57).

If a predictable or recurrent reference sequence is provided the result of Corollary 3.2 may then be applied. To this end, consider the following recurrent model

$$W(d)w(t) = \begin{bmatrix} 1-d+d^2 & 0 \\ 0 & 1+d+d^2 \end{bmatrix} w(t) = 0_2. \tag{4.58}$$

The required solution of (4.47) is given by

$$\Delta = \begin{bmatrix} 3-8d & 0 \\ 0 & -(1+d) \end{bmatrix}, \quad \Gamma = \begin{bmatrix} 0 & d \\ 0 & d \end{bmatrix}, \tag{4.59}$$

and the feedforward command, carried out by (4.46), results

$$u_c^{(1)}(t) = w^{(1)}(t) \tag{4.60}$$
$$u_c^{(2)}(t) = -(1+d)w^{(2)}(t). \tag{4.61}$$

As above, the closed-loop expressions (4.52) and (4.53) apply with $u^c(t)$ given by (4.60) and (4.61).

4.4 Minimax LQ Exact and Inexact Model Matching Problems

The problem of model matching control has encountered considerable attention in the past decades [14, 15]. It consists of designing a stabilizing controller in such a way that the compensated system performs closely to an assigned model. Conditions under which this can be exactly accomplished are well known. In particular, it is shown in [16] that the unstable part of the plant numerator is critical in designing *zero-cancelling* strategies as model matching control is. In fact, exact model matching is achievable only when the numerator of the given model contains all the unstable roots of the plant numerator. If this is not the case, an inexact model matching problem results, and, for internal stability, at most a suitable approximation of the solution can be achieved.

Recently, an H_∞ approach to model matching control has been addressed in [5] and [17] where the solution is obtained by minimizing the H_∞-norm of the difference between the model and the closed-loop reference to output transfer matrices. Here, the problem is approached as a minimax LQ stochastic tracking problem, with the reference signal to be tracked at time t supposed to be known up to time $t+\tau$, τ positive or negative. The solution has the form of a two-degrees-of-freedom control law, whose feedback and feedforward actions can be calculated independently. Accordingly, the feedback controller is designed by minimizing a H_∞ mixed-sensitivity problem. In addition, the feedforward path is determined by minimizing a minimum variance performance index costing a suitable model mismatching error.

Under the exact model matching condition no advantage is gained from knowing future samples of the reference. If such a condition is not met, the present approach yields a causal closed-form solution to the inexact model matching problem provided that a stochastic dynamic model of the reference is available. If this is unavailable, a tight approximation to the optimal solution can be obtained by exploiting the reference future.

4.4.1 Problem formulation and results

Consider a *set* of discrete-time multivariable stochastic linear *plants* as in (4.7) where: $y(t) \in \Re^p$ is the output; $u(t) \in \Re^p$ is the input; $\{e(t)\}$ is a white sequence of random vectors in $\Re^p$ with zero-mean and unit covariance; $\mathcal{P}(d) \in \Re_{pp}(d)$ and $\mathcal{Q}(d) \in \Re_{pp}(d)$ are causal *rational* transfer matrices in the unit backward shift operator d. $\Re_{pp}(d)$ denotes the set of square $p \times p$ matrices with elements in $\Re(d)$, the set of polynomial fractions in the indeterminate d. Moreover, let the assumption **(A.3.1.)** be replaced by

(**A.4.1**)
- The plant with transfer matrix $\mathcal{P}(d)$ is strictly causal, square and nonsingular free of unstable hidden modes.
- Every unstable root of the characteristic polynomial (c.p.) of the overall system (4.7) is also a root with the same multiplicity of the c.p. of the subsystem with transfer matrix $\mathcal{P}(d)$.

The full normal-rank assumption in (**A.4.1**) is necessary for complete output controllability of the system. In fact, this is satisfied in general if the normal-rank of the system equals the number of its outputs [15]. A necessary condition for this is that the number of the inputs is greater or equal then the number of the outputs. Consequently, there is no loss of generality in assuming square plants. In fact, if this is not the case, one needs only to select those p-inputs for which the corresponding columns of $\mathcal{P}(d)$ are linearly independent.

Let $\mathcal{P}_m(d) \in \Re_{pp}(d)$ be a stable reference model specified by the designer

$$y_m(t) = \mathcal{P}_m(d)w(t), \tag{4.62}$$

with $w(t) \in \Re^p$ the *reference* signal satisfying (**A.3.3**) and $y_m(t) \in \Re^p$ the output of the reference model.

Let $\bar{u}(t) \in \Re^p$ be a suitable signal standing for the *nominal* input. It is defined by the following relation

$$y_m(t) = \mathcal{P}_m w(t) = \mathcal{P}\bar{u}(t). \tag{4.63}$$

Roughly speaking, $\bar{u}(t)$ represents the input of $\mathcal{P}$ that generates the same output y_m of the reference model $\mathcal{P}_m$ when driven by $w(t)$. From (**A.3.3**), it follows that both $y_m(t)$ and $\bar{u}(t)$ are independent of $e(t)$.

Let (A_2, B_2) be a right coprime matrix fraction description of $\mathcal{P}$,

$$\mathcal{P}(d) = B_2 A_2^{-1} \tag{4.64}$$

with A_2 and B_2 polynomial matrices in $\Re_{pp}[d]$. Further, let (A_{2m}, B_{2m}) be a right coprime matrix fraction description of $\mathcal{P}_m$,

$$\mathcal{P}_m(d) = B_{2m} A_{2m}^{-1}, \tag{4.65}$$

with $A_{2m} \in \Re_{pp}[d]$ and $B_{2m} \in \Re_{pp}[d]$.

From (4.63) follows that $\bar{u}(t)$ is related to $w(t)$ by

$$\bar{u}(t) = \mathcal{P}^{-1}\mathcal{P}_m w(t). \tag{4.66}$$

It results a mean-square (ms) bounded process, viz. $\varepsilon\{\bar{u}(t)'\bar{u}(t)\} < \infty$, if $\mathcal{P}^{-1}\mathcal{P}_m$ is stable transfer matrix. In turn, by taking into account (4.64) and (4.65), this is satisfied if the possible unstable zero structure of $\mathcal{P}$ is cancelled by $\mathcal{P}_m$. In fact, if $\mathcal{P}$ and $\mathcal{P}_m$ do not share the same unstable zero structure, an unbounded ms signal $\bar{u}(t)$ results from (4.66). In order to overcome this severe limitation for non-minimum phase systems one can proceed as follows. Instead of considering $\bar{u}(t)$, introduce the following ms bounded signal

$$\hat{u}(t) := \mathcal{H}(d, d^{-1})w(t), \tag{4.67}$$

defined in terms of a stable but *non-causally defined* (having in general both causal and anticausal components) transfer matrix, to be determined by solving the following ms approximation problem

$$\min_{\mathcal{H}} \varepsilon\{(\hat{y}_m(t)-y_m(t))'(\hat{y}_m(t)-y_m(t))\} = \min_{\mathcal{H}} tr \prec (\mathcal{H}-\mathcal{P}^{-1}\mathcal{P}_m)^*\mathcal{P}^*\Phi_{ww}\mathcal{P}(\mathcal{H}-\mathcal{P}^{-1}\mathcal{P}_m) \succ \tag{4.68}$$

$$\hat{y}_m(t) := \mathcal{P}(d)\mathcal{H}(d, d^{-1})w(t) \tag{4.69}$$

is the approximated model to be matched by the closed-loop system. The solution can be easily found as follows: factor B_2 and B_{2m} as

$$B_2 = B_{2-}B_{2+}, \qquad B_{2m} = B_{2m-}B_{2m+} \tag{4.70}$$

where: B_{2+} and B_{2m+} are strictly Hurwitz monic polynomial matrices in $\Re_{pp}[d]$ ($\det B_{2+} \neq 0$ for $|\, d \,| \leq 1$), while B_{2-} and B_{2m-} are anti-Hurwitz polynomial matrices in $\Re_{pp}[d]$ ($\det B_{2-} \neq 0$ for $|\, d \,| > 1$), that factorize the plant and model numerators in their stable and unstable parts.

Let $E(d) \in \Re_{pp}[d]$ the right-spectral solution of the factorization problem (4.14) that satisfies (**A.3.5**). Next, perform the following coprime matrix fraction decomposition

$$EA_{uo}^{-1}A_{y3}^{-1}B_{2+}^{-1}B_{2-}^{-1}B_{2m-}B_{2m+}A_{2m}^{-1} = B_4^{-1}E_1A_{4m}^{-1} \tag{4.71}$$

with $B_4 \in \Re_{pp}[d]$ anti-Hurwitz and $A_{4m} \in \Re_{pp}[d]$ strictly Hurwitz both in $\Re_{pp}[d]$. A_{uo} and A_{y3} have been defined in (4.13) and actually $\in \Re_{pp}[d]$. Next, solve for the unique matrix polynomial pair (X, Y), minimal w.r.t Y, i.e. $\partial Y < \partial B_4$, the following bilateral Diophantine equation

$$B_4X + YA_{4m} = E_1. \tag{4.72}$$

Then, taking into account (4.71) and (4.72), one obtains

$$\begin{aligned} A_2A_{y3}A_{uo}E^{-1}EA_{uo}^{-1}A_{y3}^{-1}B_2^{-1}\mathcal{P}_m &= A_2A_{y3}A_{uo}E^{-1}B_4^{-1}(B_4X + YA_{4m})A_{4m}^{-1} \\ B_4^{-1}E_1A_{4m}^{-1} &= (XA_{4m}^{-1} + B_4^{-1}Y) \end{aligned} \tag{4.73}$$

XA_{4m}^{-1} is a *stable and causal* transfer matrix. On the opposite, thanks to the degree constraint on Y, $B_4^{-1}Y$ may be regarded as a *strictly anticausal and stable* transfer matrix, i.e.

$$B_4(d)^{-1}Y(d) = (d^{-b}B_4(d))^{-1}(d^{-b}Y(d)) = \hat{B}_4^{-1}(d^{-1})\hat{Y}(d^{-1})$$

with $p = \partial B_4$ and $\partial Y < b$. Then, provided that

$$(\mathbf{A.4.2})\left\{ \quad \bullet \ \det B_4 \text{ has not roots on the unit circle,} \right.$$

the solution is given by

$$\mathcal{H}(d,d^{-1}) = A_2(d)A_{y3}(d)A_u(d)E^{-1}(d)(X(d)A_{4m}^{-1}(d) + \hat{B}_4^{-1}(d^{-1})\hat{Y}(d^{-1})) \tag{4.74}$$

and the following explicit expressions for (4.67) and (4.69) are achieved by taking into account (4.74)

$$\hat{u}(t) = A_2A_{y3}A_{uo}E^{-1}(XA_{4m}^{-1}+\hat{B}_4^{-1}\hat{Y})w(t) \quad \hat{y}_m(t) = B_2A_{y3}A_{uo}E^{-1}(XA_{4m}^{-1}+\hat{B}_4^{-1}\hat{Y})w(t) \tag{4.75}$$

Before continuing, it is pointed out that if $B_4 = I_p$, e.g. when B_2 and B_{2m} do share the same unstable structure, the required solution for (4.72) is given by $Y = 0$ and $X = E_1$ so that $\hat{y}_m(t) = y_m(t)$ and $\hat{u}(t) = \bar{u}(t)$. Thus, no advantages are gained from the knowledge of the future sample of $w(t)$ w.r.t. t in the exact model matching case.

Let $I^t := \{y^t, u^{t-1}, w^{t+\tau}\}$ be the available information at time t. The problem that is addressed hereafter is to find among the *linear non anticipative control laws*

$$u(t) = \mathcal{L}(I^t) \in RL_2^p(I^t) \tag{4.76}$$

the one that, in stochastic steady-state, stabilizes the plant (4.7) and solves the following *minimax LQ stochastic model matching* problem

$$\inf_{\mathcal{L}} \sup_{\bar{Q}}(J) = \inf_{\mathcal{L}} \sup_{\bar{Q}} \varepsilon\{\|\mathcal{W}_y(d)(y(t) - \hat{y}_m(t))\|_{\psi_y}^2 + \|\mathcal{W}_u(d)(u(t) - \hat{u}(t))\|_{\psi_u}^2\}, \tag{4.77}$$

with $\psi_y = \psi_y^T \geq 0, \psi_u = \psi_u^T \geq 0$, $\|v\|_\psi^2 := v'\psi v$ and $\mathcal{W}_y(d)$ and $\mathcal{W}_u(d)$ defined in (4.10) with all the polynomial matrices A_y, B_y, A_u and B_u in $\Re_{pp}[d]$ with full rank over the unit circle and A_y and A_u strictly Hurwitz. The full feedback system structure is shown in Figure 4.3. It is to be pointed out that if $w(t) \equiv 0$ Remark 3.1 applies to the present case. In particular, let

$$u(t) = -\mathcal{K}y(t) = -\mathcal{M}_1^{-1}\mathcal{N}_1 y(t), \tag{4.78}$$

be a solution of the H_∞ mixed-sensitivity problem, viz. (4.7),(4.62)-(4.77) with $w(t) \equiv 0$, with $\mathcal{N}_1(d) \in \Re_{pp}(d)$ and $\mathcal{M}_1(d) \in \Re_{pp}(d)$ stable and causal transfer matrices, satisfying the following identity

$$\mathcal{M}_1 A_2 + \mathcal{N}_1 B_2 = I_p. \tag{4.79}$$

Consider now the class of stabilizing controllers defined in (4.17)

$$\mathcal{M}_1 u(t) = -\mathcal{N}_1 y(t) + v(t) \tag{4.80}$$

As in the previous section, the stated minimax LQ stochastic model matching problem amounts to finding a process $\{v(t)\}$, $v(t) \in RL_2^p(I^t)$, such that (4.80) stabilizes the plant and minimizes (4.77). Further, since according to (**A.3.3**) and (4.80) $RL_2^p(I^t) = RL_2^p(e^t) \bigoplus RL_2^p(w^{t+\tau})$, a convenient orthogonal decomposition for v(t) is

$$v(t) = v_e(t) + v_w(t) \tag{4.81}$$

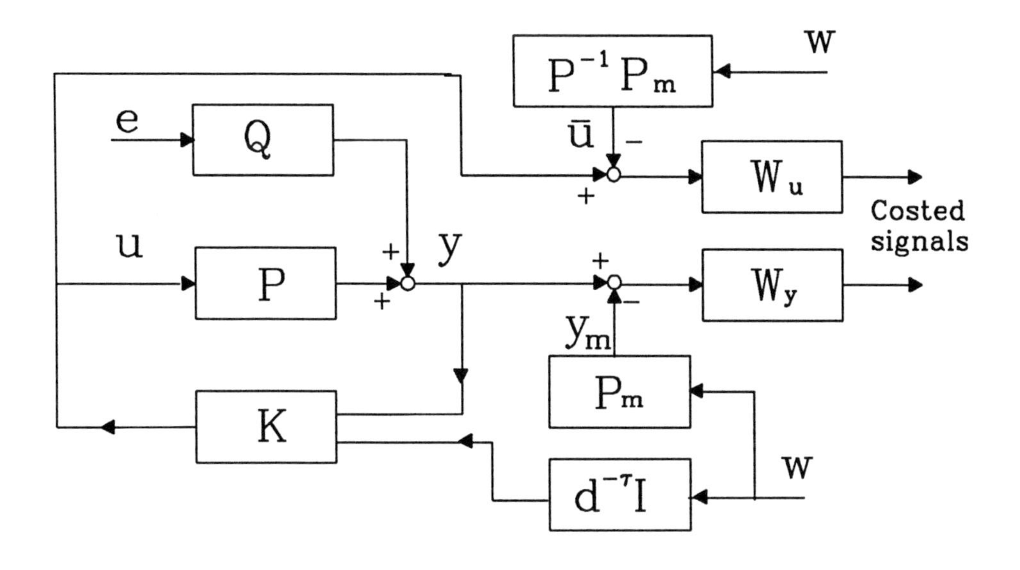

Figure 4.3: Feedback system structure

with $v_e(t) \in RL_2^p(e^t)$ and $v_w(t) \in RL_2^p(w^{t+\tau})$.

With reference to the decomposition (4.19)-(4.20), the quadratic term (4.77) can be split as follows

$$J = J_{ee} + J_{ev} + J_{e\hat{y}_m} + J_{e\hat{u}} + J_{vv} + J_{v\hat{y}_m} + J_{v\hat{u}} + J_{\hat{y}_m\hat{y}_m} + J_{\hat{u}\hat{u}} \tag{4.82}$$

where:

$$\begin{aligned}
J_{ee} &:= \varepsilon\{\|\mathcal{W}_y y_e(t)\|^2_{\psi_y} + \|\mathcal{W}_u u_e(t)\|^2_{\psi_u}\}; \\
J_{ev} &:= \varepsilon\{y_e'(t)\mathcal{W}_y^* \psi_y \mathcal{W}_y y_v(t) + u_e'(t)\mathcal{W}_u^* \psi_u \mathcal{W}_u u_v(t)\}; \\
J_{e\hat{y}_m} &:= -2\varepsilon\{y_e'(t)\mathcal{W}_y^* \psi_y \mathcal{W}_y \hat{y}_m(t)\} = 0 \quad [\mathbf{A.3.3}]; \\
J_{e\hat{u}} &:= -2\varepsilon\{u_e'(t)\mathcal{W}_u^* \psi_u \mathcal{W}_u \hat{u}(t)\} = 0 \quad [\mathbf{A.3.3}]; \\
J_{vv} &:= \varepsilon\{\|\mathcal{W}_y y_v(t)\|^2_{\psi_y} + \|\mathcal{W}_u u_v(t)\|^2_{\psi_u}\}; \\
J_{v\hat{y}_m} &:= -2\varepsilon\{v'(t)B_2^* A_y^{-*} B_y^{-*} \psi_y B_y A_y^{-1} \hat{y}_m(t)\}; \\
J_{v\hat{u}} &:= -2\varepsilon\{v'(t)A_2^* A_u^{-*} B_u^* \psi_u B_u A_u^{-1} \hat{u}(t)\}; \\
J_{\hat{y}_m\hat{y}_m} &:= \varepsilon\{\|\mathcal{W}_y \hat{y}_m(t)\|^2_{\psi_y}\}; \\
J_{\hat{u}\hat{u}} &:= \varepsilon\{\|\mathcal{W}_u \hat{u}(t)\|^2_{\psi_u}\};
\end{aligned} \tag{4.83}$$

Condition (**A.4.2**) ensures $J_{\hat{u}\hat{u}}$ and $J_{\hat{y}_m\hat{y}_m}$ to be finite. In particular, this is always verified in the exact model matching case since $\mathcal{P}_m$ shares the same unstable zero structure of $\mathcal{P}$. On the contrary, (**A.4.2**) imposes some restrictions on the model selection in the inexact model matching case when the plant numerator has some roots on the unit circle.

The main result of the section can now be stated :

Theorem 4.1 - Let (**A.4.1**)-(**A.4.2**),(**A.3.2**)-(**A.3.5**) be fulfilled. Then, the *minimax LQ model matching control law* is given by

$$\mathcal{M}_1(d)u(t) = -\mathcal{N}_1(d)y(t) + A_{y3}A_{uo}E^{-1}u_c(t) \tag{4.84}$$

where : $\mathcal{M}_1$ and $\mathcal{N}_1$ are the transfer matrices in (4.78) and $u_c(t)$ is the *feedforward command* defined by

$$u_c(t) = \varepsilon\{(XA_{4m}^{-1} + \hat{B}_4^{-1}\hat{Y})w(t) \mid w^{t+\tau}\}. \tag{4.85}$$

The residual cost of the control is given by

$$\inf_{v(t)\in L_2^p(I^t)} \sup_{\bar{\mathcal{Q}}}(J) = \inf_{v(t)\in L_2^p(e^t)} \sup_{\bar{\mathcal{Q}}}(J_{ee}) + J_{\hat{y}_m\hat{y}_m} + J_{\hat{u}\hat{u}} - J_{u_cu_c}, \quad \text{with } J_{u_cu_c} = \varepsilon\{u_c(t)'u_c(t)\}. \tag{4.86}$$

Proof - Because of (**A.3.3**) $J_{ev} = J_{ev_e}$, $J_{v\hat{u}} = J_{v_w\hat{u}}$, $J_{v\hat{y}_m} = J_{v_w\hat{y}_m}$ and $J_{e\hat{u}} = J_{e\hat{y}_m} = 0$. Further, $J_{\hat{y}_m\hat{y}_m}$ and $J_{\hat{u}\hat{u}}$ are not affected both by $\mathcal{Q}$ and $v(t)$ and are finite via (**A.4.2**). Consequently,

$$\inf_{v_e,v_w} \sup_{\bar{\mathcal{Q}}}(J) = \inf_{v_e} \sup_{\bar{\mathcal{Q}}}(J_{ee} + J_{ev_e} + J_{v_ev_e}) + \inf_{v_w}(J_{v_wv_w} + J_{v_w\hat{y}_m} + J_{v_w\hat{u}}) + J_{\hat{y}_m\hat{y}_m} + J_{\hat{u}\hat{u}}$$

Last equality follows since J_{ee}, J_{ev_e} and $J_{v_ev_e}$ do not depend on v_w, while $J_{v_rv_r}$, $J_{v_w\hat{y}_m}$ and $J_{v_w\hat{u}}$ are not affected by v_e and $\mathcal{Q}$. Now

$$\inf_{v_e} \sup_{\bar{\mathcal{Q}}}(J_{ee} + J_{ev_e} + J_{v_ev_e})$$

is the same problem as the underlying H_∞ mixed-sensitivity problem (4.11), of which the pair $\mathcal{N}_1$ and $\mathcal{M}_1$ in (4.15) are the solution. Thus $v_e = 0$ and the optimal solution is in the form $\mathcal{M}_1u(t) = -\mathcal{N}_1y(t) + \hat{v}_w(t)$, with $\hat{v}_w(t) \in RL_2^p(w^{t+\tau})$ the one that minimizes $(J_{v_wv_w} + J_{v_w\hat{y}_m} + J_{v_w\hat{u}})$. After direct steps one founds

$$\begin{aligned}
J_{v_wv_w} &= \varepsilon\{v_w'(t)B_2^*A_y^{-*}B_y^*\psi_yB_yA_y^{-1}B_2v_w(t) + v_w'(t)A_2^*A_u^{-*}B_u^*\psi_uB_uA_u^{-1}A_2v_w(t)\} \\
&= \varepsilon\{\|EA_{uo}^{-1}A_{y3}^{-1}v_w(t)\|^2\} \\
J_{v_w\hat{y}_m} &= -\varepsilon\{v_w'(t)B_2^*A_y^{-*}B_y^*\psi_yB_yA_y^{-1}\hat{y}_m(t) + \hat{y}_m'(t)A_y^{-*}B_y^*\psi_yB_yA_y^{-1}B_2v_w(t)\} \\
J_{v_w\hat{u}} &= -\varepsilon\{v_w'(t)A_2^*A_u^{-*}B_u^*\psi_uB_uA_u^{-1}\hat{u}(t) + \hat{u}'(t)A_u^{-*}B_u^*\psi_uB_uA_u^{-1}A_2v_w(t)\},
\end{aligned}$$

and, by applying the usual *completing the square* argument, after direct manipulations, one has

$$\begin{aligned}
&J_{v_wv_w} + J_{v_w\hat{y}_m} + J_{v_w\hat{u}} = \\
= \;&\varepsilon\{\|EA_{uo}^{-1}A_{y3}^{-1}v_w(t) - E^{-*}(A_{uo}^*B_3^*B_y^*\psi_yB_yA_y^{-1}\hat{y}_m(t) + A_{2o}^*B_u^*\psi_uB_uA_u^{-1}\hat{u}(t))\|^2 \\
&-\|E^{-*}(A_{uo}^*B_3^*B_y^*\psi_yB_yA_y^{-1}\hat{y}_m(t) + A_{2o}^*B_u^*\psi_uB_uA_u^{-1}\hat{u}(t))\|^2\},
\end{aligned}$$

Then, the global minimum, among all $v_w \in RL_2^p(w^{t+\tau})$ is attained at

$$\begin{aligned}
\hat{v}_w(t) &= A_{y3}A_{uo}E^{-1}\varepsilon\{E^{-*}(A_{uo}^*B_3^*B_y^*\psi_yB_yA_y^{-1}\hat{y}_m(t) + A_{2o}^*B_u^*\psi_uB_uA_u^{-1}\hat{u}(t)) \mid w^{t+\tau}\} \\
&= A_{y3}A_{uo}E^{-1}\varepsilon\{E^{-*}(A_{uo}^*B_3^*B_y^*\psi_yB_yB_3A_{uo} + A_{2o}^*B_u^*\psi_uB_uA_{2o})A_{uo}^{-1}A_{y3}^{-1}B_2^{-1}\hat{y}_m(t)) \mid w^{t+\tau} \\
&= A_{y3}A_{uo}E^{-1}u_c(t)
\end{aligned}$$

with $u_c(t)$ given by (4.85) via (4.75). □

It is straightforward to verify that $u_c(t)$ is a mean-square (ms) bounded process, being a conditional expectation of an ms bounded process. Hence, it consists in general of a linear combination of conditional expectation of past (notice the difference with the previous section) and future samples of $w(t)$ w.r.t. $w^{t+\tau}$, i.e.

$$u_c(t) = \sum_{i=-\infty}^{\infty} \alpha_i \mathcal{E}\{w(t+i) \mid w^{t+\tau}\}, \tag{4.87}$$

with $\{\alpha_i\}$ a two-sided convergent sequence of constant matrices such that $XA_{4m}^{-1} = \sum_{i=0}^{\infty} \alpha_i d^i$ and $\hat{B}_4^{-1}\hat{Y} = \sum_{i=-\infty}^{-1} \alpha_i d^i$. Further, if $\tau > 0$, (4.87) can be rewritten as

$$u_c(t) = u_c^\tau(t) + \tilde{u}_c(t) = \sum_{i=-\infty}^{\tau} \alpha_{-i} w(t+i) + \sum_{i=\tau+1}^{\infty} \alpha_{-i} \mathcal{E}\{w(t+i) \mid w^{t+\tau}\} \tag{4.88}$$

where only the second term in the right-side of (4.88) depends of the statistical properties of $w(t)$. Further, $u_c^\tau(t)$ can be expressed more conveniently as follows

$$u_c^\tau(t) = (XA_{4m}^{-1} + \sum_{i=-1}^{-\tau} \alpha_i d^i) w(t). \tag{4.89}$$

Corollary 4.1 - Let (**A.4.1**)-(**A.4.2**),(**A.3.2**)-(**A.3.5**) be fulfilled. Let $\hat{u}(t)$ and $\hat{y}_m(t)$ in (4.77) be respectively replaced by $\hat{u}^\tau(t)$ and $\hat{y}_m^\tau(t)$, the following their τ-*steps* approximations

$$\hat{u}^\tau(t) = A_2 A_{y3} A_{uo} E^{-1} u_c^\tau(t), \quad \hat{y}_m^\tau(t) = B_2 A_{y3} A_{uo} E^{-1} u_c^\tau(t) \tag{4.90}$$

Then, the following control law

$$\mathcal{M}_1(d)u(t) = -\mathcal{N}_1(d)y(t) + A_{y3}A_{uo}E^{-1}u_c^\tau(t) \tag{4.91}$$

is the solution of the stated τ-steps approximated minimax LQ model matching problem and provides the exact model matching to $y_v(t) = \hat{y}_m^\tau(t)$ and $u_v(t) = \hat{u}^\tau(t)$.

Proof - To prove the assertion is enough to show that the residual cost of the control becomes

$$\inf_{v(t) \in RL_2^p(I^t)} \sup_{\bar{Q}} J = \inf_{v_e(t) \in RL_2^p(e^t)} \sup_{\bar{Q}} J_{ee}.$$

This follows directly by considering the closed-loop expressions for $y(t)$ and $u(t)$. In fact, by (4.19) and (4.20) one has

$$y(t) - \hat{y}_m^\tau(t) = y_e(t) + y_v(t) - \hat{y}_m^\tau(t) = y_e(t) + B_2 A_{y3} A_{uo} E^{-1} u_c^\tau(t) - \hat{y}_m^\tau(t) = y_e(t)$$
$$u(t) - \hat{u}^\tau(t) = u_e(t) + u_v(t) - \hat{u}^\tau(t) = u_e(t) + A_2 A_{y3} A_{uo} E^{-1} u_c^\tau(t) - \hat{u}^\tau(t) = u_e(t)$$

from which $J_{u_c^\tau u_c^\tau} = J_{\hat{y}_m^\tau \hat{y}_m^\tau} + J_{\hat{u}^\tau \hat{u}^\tau}$ can be argued. □

It is finally to be noticed that the *feedforward command* $u_c(t)$ is determined irrespective of feedback specifications. In fact, it does not depend explicitly on ψ_u,

ψ_y, $\mathcal{W}_y$ and $\mathcal{W}_u$, which, in turn, can be freely chosen as requested by closed-loop specifications.

Example 4.1 - Consider a SISO plant $B_2 = d(1-\alpha d)$ and a H_∞ minimum sensitivity control problem ($\psi_u = 0$ and $\psi_y = \mathcal{W}_y = \mathcal{W}_u = 1$ in (4.77)). Then, for $| \alpha |> 1$ (non minimum-phase plant), we get from (4.14)

$$E(d) = k(1-\alpha^{-1}d), \qquad k := \left(\frac{1+\alpha^2}{1+\alpha^{-2}}\right)^{\frac{1}{2}} \tag{4.92}$$

By selecting the following model $B_{2m} = d$ the optimal feedforward command becomes

$$u_c(t) = k\varepsilon\{\frac{1-\alpha^{-1}d}{1-\alpha d}w(t) \mid w^{t+\tau}\} \tag{4.93}$$

As in (4.88), it can be decomposed as $u_c^\tau(t) + \tilde{u}_c(t)$:

$$u_c^\tau(t) = k\left(\alpha^{-2}r(t) + (\alpha^{-2}-1)\sum_{i=1}^{\tau}\alpha^{-i}w(t+i)\right). \tag{4.94}$$

$$\tilde{u}_c(t) = k(\alpha^{-2}-1)\alpha^{-\tau}\sum_{i=1}^{\infty}\alpha^{-i}\varepsilon\{w(t+\tau+i) \mid w^{t+\tau}\}. \tag{4.95}$$

A τ-steps approximation of the optimal feedforward command can be obtained simply by replacing $u_c(t)$ with $u_c^\tau(t)$. In order to valuate the approximation error, let $w(t) = 1(t)$ be the unitary step function ($w(t) = 1$ for $t \geq 0$, 0 otherwise). The feedforward components of the closed-loop input and output result given by $u_v(t) = \frac{1}{k(1-\alpha^{-1}d)}u_c^\tau(t)$ and $y_v(t) = \frac{d(1-\alpha d)}{k(1-\alpha^{-1}d)}u_c^\tau(t)$ respectively. The steady-state response of the feedforward command (4.94) values $u_c^\tau(\infty) = -\frac{k}{\alpha}(1-\alpha^{-\tau-1}(\alpha+1))$. Consequently, one finds that $u_v(\infty) = \frac{1-\alpha^{-\tau-1}(\alpha+1)}{1-\alpha}$ and $y_v(\infty) = 1-\alpha^{-\tau-1}(\alpha+1)$. As expected, $y_v(\infty) \to 1$ as $\tau \to \infty$ and $u_v(\infty)$ increases with τ becoming unbounded for $\alpha = 1$. The transient of the output step response is reported in Figure 4.4 for $\tau \in [0,3]$ and $\alpha = 3$. It is evident from the plots that the non-minimum phase effect of the unstable zero of the plant is reduced as τ increases. □

Remark 4.2 - A special case arises when (4.71) is a stable transfer matrix. Under this assumption an exact model matching problem results. This is verified if

$$B_{2-}^{-1}B_{2m-} \in \Re_{pp}[d]. \tag{4.96}$$

In fact, since E and A_{2m} are strictly Hurwitz by assumption, the unstable zero structure of B_{2-} can only be cancelled by B_{2m-}. It is to be pointed out that condition (4.96) is the same of the one in [16] for exact model matching achievement. Thus, the *feedforward command* $u_c(t)$ simply reduces to

$$u_c(t) = EA_{uo}^{-1}A_{y3}^{-1}B_{2+}^{-1}UB_{2m+}A_{2m}^{-1}w(t) \tag{4.97}$$

with $U = B_{2-}^{-1}B_{2m-} \in \Re_{pp}[d]$ after common factors simplification. From (4.97) it follows that no advantage is gained from the knowledge of the future samples of

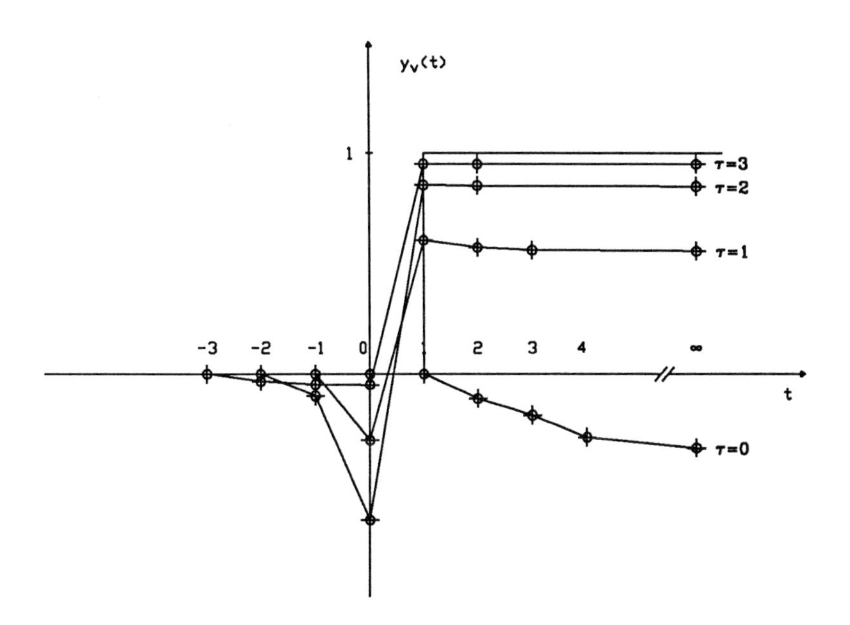

Figure 4.4: Closed-loop step responses for $\alpha = 3$ and $\tau \in \{0, 1, 2, 3\}$.

$w(t)$ w.r.t. t. The feedforward path of the regulated signals in (4.19) and (4.20) are given by

$$y_v(t) = B_2 A_{y3} A_{uo} E^{-1} u_c(t) = \mathcal{P}_m w(t) = y_m(t) \quad (4.98)$$

$$u_v(t) = A_2 A_{y3} A_{uo} E^{-1} u_c(t) = \mathcal{P}^{-1} \mathcal{P}_m w(t) = \bar{u}(t) \quad (4.99)$$

that achieves the desired exact model matching condition with residual cost

$$\inf_{v(t) \in L_2^p(I^t)} \sup_{\bar{Q}}(J) = \inf_{v(t) \in L_2^p(e^t)} \sup_{\bar{Q}}(J_{ee})$$

Some special exact model matching problems, as output decoupled and deadbeat control, can be encompassed into the present framework.

Output decoupled control - Such a problem can be solved as a exact model matching one by imposing a diagonal model $\mathcal{P}_m$. Then, it results that a exact decoupling is possible only if B_{2-} is diagonal. This agrees with the results of [18] where a decoupling procedure is determined in a similar way for stable and stably invertible plants. The present approach extends those results to possibly unstable and non minimum-phase systems.

Deadbeat control -Such a problem can be solved when the *delay* structure of the system, sometimes referred to as the *interactor* [19], is diagonal and the system is minimum-phase. This is satisfied if $B_{2-} = \mathcal{D}iag\{d^{\gamma_i}\}$, $i = 1, 2, ..., p$, where γ_i stands for the minimum i/o delay between any inputs and the i-th output. Thus, a minimum delay deadbeat controller can be obtained by choosing $\mathcal{P}_m = \mathcal{D}iag\{d^{\gamma_i}\}$, any other choice giving rise to an inexact model matching problem. □

4.4.2 ARMA and recurrent reference models

Expression (4.85) can be further elaborated, as done in the previous section, when a stochastic model for the reference realization $w(t)$ is available. In this connection, assume that $w(t)$ be given as in (4.37) with G_2 and F_2 satisfying **(A.3.6)** and $\{\nu(t)\}$, $\nu(t) \in \Re^p$, is a white sequence of random vectors with zero-mean and unitary covariance, $G_2F_2^{-1} \in \Re_{pp}(d)$ is a causal rational transfer matrix and G_2 and F_2 polynomial matrices in $\Re_{pp}[d]$. Next, consider the following coprime matrix fraction decomposition

$$EA_{uo}^{-1}A_{y3}^{-1}B_{2+}^{-1}B_{2-}^{-1}B_{2m-}B_{2m+}A_{2m}^{-1}G_2 = B_5^{-1}E_2A_{5m}^{-1} \tag{4.100}$$

with B_5 anti-Hurwitz and A_{5m} strictly Hurwitz both in $\Re_{pp}[d]$.

Corollary 4.2 - Let **(A.4.1)-(A.4.2)**,**(A.3.2)-(A.3.6)** be fulfilled and the reference be modelled as in (4.37). Then, the optimal *feedforward command* is given by

$$u_c(t) = \Delta A_{5m}^{-1}G_2^{-1}w(t+\tau) \tag{4.101}$$

where $(\Delta(d), \Gamma(d))$ is the unique solution of minimum degree w.r.t. Γ, i.e. $\partial\Gamma < \partial B_5$, of the following bilateral Diophantine equation

$$B_5\Delta + \Gamma F_2 A_{5m} = d^{\max\{\tau,0\}}E_2 \tag{4.102}$$

Proof - First, since B_5 is nonsingular and anti-Hurwitz while F_2A_{4m} is strictly Hurwitz, there exist unique polynomial matrices (Δ, Γ), $\partial\Gamma < \partial B_5$, satisfying (4.102) [12]. The two cases $\tau \geq 0$ and $\tau < 0$ are treated separately. In the former ($\tau \geq 0$), by taking into account (4.100) and (4.101), (4.85) becomes

$$\begin{aligned} u^c(t) &= \varepsilon\{B_5^{-1}E_2A_{5m}^{-1}F_2^{-1}d^\tau\nu(t+\tau) \mid \nu^{t+\tau}\} \\ &= \varepsilon\{B_5^{-1}(B_5\Delta + \Gamma F_2A_{5m})A_{5m}^{-1}F_2^{-1}\nu(t+\tau) \mid \nu^{t+\tau}\} \\ &= \varepsilon\{(\Delta A_{5m}^{-1}F_2^{-1} + B_5^{-1}\Gamma)\nu(t+\tau) \mid \nu^{t+\tau}\} \\ &= \Delta A_{5m}^{-1}F_2^{-1}\nu(t+\tau). \end{aligned} \tag{4.103}$$

The last equality follows from the degree constraint on Γ. In fact, $B_5^{-1}\Gamma = (d^b\hat{B}_5(d^{-1}))^{-1}$ $(d^b\hat{\Gamma}(d^{-1}))$, with $b = \partial B_5$, is a *strictly anticausal* transfer matrix. Next, for $\tau < 0$ case one has

$$\begin{aligned} u_c(t) &= \varepsilon\{B_5^{-1}E_2A_{5m}^{-1}F_2^{-1}\nu(t+\tau) \mid \nu^{t+\tau}\} \\ &= \varepsilon\{(\Delta A_{5m}^{-1}F_2^{-1} + B_5^{-1}\Gamma)\nu(t+\tau) \mid \nu^{t+\tau}\} \\ &= \Delta F_2^{-1}\nu(t+\tau). \ \Box \end{aligned} \tag{4.104}$$

the last equality follows from the same arguments used in the $\tau \geq 0$ case. Finally (4.101) follows from (4.37). □

Last, consider a *recurrent* or *predictable* reference. This is the case when the reference realization is governed by the relation (4.45) where $W(d) \in \Re_{pp}[d]$ is proper, viz. the matrix coefficient of the highest power of d is nonsingular, and marginally stable with all roots of $\det W(d)$ of unit modulus.

Corollary 4.3 - Let (**A.4.1**)-(**A.4.2**),(**A.3.2**)-(**A.3.5**) be fulfilled and the reference predictable as in (4.45). Then, the optimal *feedforward command* is given by

$$u^c(t) = \Delta(d)A_{4m}^{-1}w(t) \tag{4.105}$$

where $(\Delta(d), \Gamma(d)) \in \Re_{pp}[d]$ is the unique solution of minimum degree w.r.t. Δ, i.e. $\partial\Delta < \partial W A_{4m}$ of the following bilateral Diophantine equation

$$B_4\Delta + \Gamma W A_{4m} = E_1 \tag{4.106}$$

Proof - W.l.o.g., one can set $\tau = 0$. Thus

$$\begin{aligned} u^c(t) &= \varepsilon\{B_4^{-1}E_1A_{4m}^{-1}w(t) \mid w^t\} \\ &= \Delta A_{4m}^{-1}w(t) + B_4^{-1}\Gamma W w(t) \\ &= \Delta A_{4m}^{-1}w(t) \end{aligned}$$

That (4.106) can be uniquely solved as stated, follows from the properness of W and from the fact that $\det B_w$ and $\det WA_{4m}$ do not share common roots. □

Example 4.2 - Consider the following non minimum-phase system defined by

$$B_2 = \begin{bmatrix} d(3d+1) & 0 \\ d & d \end{bmatrix}, \qquad A_2 = I_2 \tag{4.107}$$

With respect to the factorization (4.70) one has $B_{2-} = B_2$ and $B_{2+} = I_2$. Next, consider the following model defined by

$$B_{2m} = \begin{bmatrix} d & 0 \\ 0 & d \end{bmatrix}, \qquad A_{2m} = I_2, \qquad B_{2m-} = B_{2m}, \qquad B_{2m+} = I_2 \tag{4.108}$$

It is found that condition (4.96) is not satisfied, so an inexact model matching problem presently results.

By choosing $\psi_u = 0_2$ and $\psi_y = \mathcal{W}_y = \mathcal{W}_u = I_2$ in (4.77), one has $A_{y3} = A_{uo} = I_2$ and E in (4.14) equals $E = \begin{bmatrix} 3+d & 0 \\ 1 & 1 \end{bmatrix}$. The decomposition (4.71) can be easily accomplished and it found: $B_4 = \begin{bmatrix} 3d+1 & 0 \\ 0 & 1 \end{bmatrix}$, $E_1 = \begin{bmatrix} 3+d & 0 \\ 0 & 1 \end{bmatrix}$ and $A_{4m} = I_2$. Thus, the required solution of (4.72) is given by

$$X = \begin{bmatrix} 0.333 & 0 \\ 0 & 1 \end{bmatrix}, \quad Y = \begin{bmatrix} 2.666 & 0 \\ 0 & 0 \end{bmatrix}, \tag{4.109}$$

and the feedforward command $u_c(t) = \begin{bmatrix} u_c^{(1)}(t) & u_c^{(2)}(t) \end{bmatrix}'$ defined in (4.85) results after straightforward steps

$$\begin{aligned} u_c^{(1)}(t) &= \varepsilon\{(0.333 + \frac{2.666d^{-1}}{(3+d^{-1})})w^{(1)}(t) \mid w^{t+\tau}\} \\ &= 0.333w^{(1)}(t) + \frac{2.666}{3}\varepsilon\{\sum_{j=1}^{\infty}(-3)^{-j+1}w^{(1)}(t+j) \mid w^{t+\tau}\} \end{aligned} \tag{4.110}$$

$$u_c^{(2)}(t) = w^{(2)}(t). \tag{4.111}$$

with $w(t) = \left[\begin{array}{cc} w^{(1)}(t) & w^{(2)}(t) \end{array}\right]'$.

Finally, the closed-loop output $y_v(t)$ in (4.20) is given, after simplification steps, by

$$y_v^{(1)}(t) = \frac{d(3d+1)}{0.943(3.370+d)} u_c^{(1)}(t) \tag{4.112}$$

$$y_v^{(2)}(t) = d u_c^{(2)}(t) \tag{4.113}$$

If a stochastic model for the reference is given the results of the Corollary 4.2 may than be applied. To this end, consider the following model

$$G_2 = I_2, \quad F_2 = \left[\begin{array}{cc} 1 & 0 \\ 0 & 3+d \end{array}\right] \tag{4.114}$$

and select $\tau = 2$. The decomposition (4.100) can be accomplished with $B_5 = B_4$, $E_2 = E_1$ and $A_{5m} = I_2$ and the required solution for the Diophantine equation (4.102) is given by

$$\Delta = \left[\begin{array}{cc} \frac{1}{27}(1-9d+3d^2) & 0 \\ 0 & d^2 \end{array}\right], \quad \Gamma = \left[\begin{array}{cc} -\frac{1}{27} & 0 \\ 0 & 0 \end{array}\right] \tag{4.115}$$

and the feedforward command, carried out by (4.101), results

$$u_c^{(1)}(t) = \frac{1}{27}(1-9d+3d^2) w^{(1)}(t+2) \tag{4.116}$$

$$u_c^{(2)}(t) = w^{(2)}(t). \tag{4.117}$$

The closed-loop expressions (4.112) and (4.113) apply with $u^c(t)$ given by (4.116) and (4.117).

If a predictable or recurrent reference sequence is provided the result of Corollary 4.3 may then be applied. To this end, consider the following recurrent model

$$W(d)w(t) = \left[\begin{array}{cc} 1-d+d^2 & 0 \\ 0 & 1+d+d^2 \end{array}\right] w(t) = 0_2. \tag{4.118}$$

The required solution of (4.106) is given by

$$\Delta = \left[\begin{array}{cc} \frac{1}{13}(15-8d) & 0 \\ 0 & 1 \end{array}\right], \quad \Gamma = \left[\begin{array}{cc} \frac{24}{13} & 0 \\ 0 & 0 \end{array}\right], \tag{4.119}$$

and the feedforward command, carried out by (4.105), results

$$u_c^{(1)}(t) = \frac{1}{13}(15-8d) w^{(1)}(t) \tag{4.120}$$

$$u_c^{(2)}(t) = w^{(2)}(t). \tag{4.121}$$

As above, the closed-loop expressions (4.112) and (4.113) apply with $u^c(t)$ given by (4.120) and (4.121).

4.5 Conclusions

In this chapter two special problems of mixed H_2/H_∞ control have been presented. The corresponding two-degrees-of-freedom controller is made up by a feedforward part solving the underlying H_∞ mixed sensitivity problem and by a feedforward part that can be carried out independently by solving an H_2-norm optimization problem.

The solution of the feedback part is standard while the H_2 problem to be solved for feedforward evaluation depends strictly on the available information on the reference signal (ARMA model, recurrent model or future reference samples).

The knowledge of the reference realization with a lead or a delay of τ steps has practical advantages in program control and in some advanced adaptive control schemes. In particular, it has been shown how, when τ is positive and large enough, the feedforward command can be tightly approximated without requiring any stochastic model for the reference.

Bibliography

[1] P. Boekhout, *The H_∞ control design method: A polynomial approach*, Ph.D Thesis, University of Twente, The Netherlands, 1989.

[2] S. Boyd and J. Doyle, "Comparison of peak and RMS gains for discrete-time systems," *Systems and Control Letters*, 9:1-6, 1987.

[3] M.J. Grimble, "Minimization of a combined H_∞ and LQG cost-function for a two-degrees-of-freedom control design," *Automatica*, 25(4):635,638, 1989.

[4] M. Verma and E. Jonckheere, "L_∞ - compensation with mixed sensitivity as broadband matching problem," *Systems and Control Letters*, 4:125-130, 1984.

[5] B.A. Francis, in Systems and optimization. In A. Bagchi and H.Th. Jougen, editors, *Lecture Notes in Control and Information Sciences.* Springer-Verlag, Berlin, 1985.

[6] H. Kwakernaak, "Minimax frequency domain performance and robustness optimization of linear feedback system," *IEEE Transactions on Automatic Control*, AC-30:994–1004, 1985.

[7] E. Jonckheere and M. Verma, "A spectral characterization of H_∞ - optimal feedback performance and its efficient computation," *Systems and Control Letters*, 8:13–22, 1986.

[8] M.J. Grimble, "Optimal H_∞ robustness and its relationship to lqg design problems," *International Journal of Control*, 43:351–372, 1986.

[9] H. Kwakernaak, "A polynomial approach to minimax frequency domain optimization of multivariable feedback system," *International Journal of Control*, 44:117–156, 1986.

[10] A. Casavola, M.J. Grimble, E. Mosca, and P. Nistri, "Continuous-time LQ regulator design by polynomial equations," *Automatica*, 27(3):555-557, 1991.

[11] E. Mosca, A. Casavola and L. Giarré, "Minimax LQ stochastic tracking and servo problems," *IEEE Transaction on Automatic Control*, AC-35(1):95-97, 1990.

[12] V. Kučera, *Discrete Linear Control: the Polynomial Equation Approach.* Chichester:Wiley, 1979.

[13] E. Mosca and G. Zappa, "Matrix fraction solution to the discrete time LQ stochastic tracking and servo problem," *IEEE Transactions on Automatic Control*, AC-34(2):240-242, 1989.

[14] S.H. Wang and C.A. Desoer, "The exact model matching problem of linear multivariable systems," *IEEE Transaction on Automatic Control*, AC-19:347-349, 1972.

[15] W.A. Wolovich, *Linear Multivariable Systems.* Springer-Verlang, New-York, 1974.

[16] H. Elliot and W.A. Wolovich, "Parameterization issues in multivariable systems," *Automatica*, 20(5):533-545, 1984.

[17] B.S. Chen and S.S. Wang, "Optimal model matching control system: Minimax approach," *IEEE Transaction on Automatic Control*, AC-33(6):518-521, 1987.

[18] J.J.Jr Bongiorno, "On the design of optimal decoupled two-degree-of-freedom multivariable feedback control system," In *Proc.* 24th *Conf. Decision and Control*, pages 996,591-592, P.Lauderdale, Florida, 1985.

[19] W.A. Wolovich and P.L. Falb, "Invariants and canonical forms under dynamic compensation," *SIAM J. Control Optimization*, 14:996,1008, 1976.

Chapter 5

Optimal Filtering Problems

A. Ahlén and M. Sternad

5.1 Introduction

In this chapter, we shall demonstrate the power and utility of the polynomial approach in the area of signal processing and communications[1]. By studying specific model structures, considerable engineering insight can be gained.

Minimisation of mean–square error criteria by linear filters will be considered. We shall focus on the optimisation of realisable discrete–time IIR-filters, to be used for prediction, filtering or smoothing of signals. Stochastic models of possibly complex-valued signals are assumed known.

Historically, such problems have been dealt with by applying the classical Wiener–Hopf approach. See e.g. [12], [14], [37], [64]. Even after the Kalman–filter breakthrough [38], many researchers still prefer the frequency domain approach, despite its inferior numerical properties for high order problems. One reason is the relative ease with which an obtained filter can be examined. A quick inspection of the poles and zeros roughly tells us what filter properties could be expected.

While the classical Wiener solution is conceptually elegant it has, until recently, been rather intractable to perform the causal bracket operation $\{\cdot\}_+$ central to the design of realisable filters. In particular, Wiener–smoothing has not been straightforward. With the polynomial approach, pioneered in [40], Diophantine equations now offer an efficient way of automating the causul bracket operation.

Polynomial equations were first used among control engineers. An early result is due to Åström in 1970 [8]. To obtain minimum variance control laws, he derived a Diophantine equation for calculating the d–step prediction of an ARMA process

[1]The chapter includes work partially supported by the Swedish National Board for Technical Development, NUTEK, under contract 87-01573.

$$y(t) = \frac{C(q^{-1})}{D(q^{-1})}e(t) = \frac{(1 + c_1 q^{-1} + \ldots + c_n q^{-n})}{(1 + d_1 q^{-1} + \ldots + d_n q^{-n})}e(t) \tag{5.1}$$

with $C(q^{-1})$ stable and $e(t)$ white and zero mean. The linear d–step predictor, which minimises the mean square estimation error, is given by

$$\hat{y}(t|t-d) = \frac{G(q^{-1})}{C(q^{-1})}y(t-d) \quad . \tag{5.2}$$

Here, the polynomial $G(q^{-1})$ of degree $n-1$, together with a polynomial $F(q^{-1})$ of of degree $d-1$, is the solution to the Diophantine equation

$$C(q^{-1}) = q^{-d}G(q^{-1}) + D(q^{-1})F(q^{-1}) \quad . \tag{5.3}$$

See Theorem 3.1 of [8]. This result was later generalised to multivariable systems by Borison [13]. Compared to earlier Wiener methods, which were based on the manipulation of auto- and cross covariance functions, the polynomial approach offered a considerable simplification. Another contribution (which did not explicitly use Diophantine equations), was the self-tuning smoother of Hagander and Wittenmark [33]. See also [49]. Other early contributions were a filter for a signal vector in white measurement noise [39], and a polynomial method for computing the gain matrix of a Kalman filter [41], both by Kučera.

Fairly recently, the polynomial systems framework has been more systematically utilised to solve signal processing and communications problems. See e.g. [2], [4], [19], [20], [28], [31], [32], [47], [48], [54], [57]. There has been some work related to time–varying filter design [29]. Effort has also been spent on relating Kalman, Wiener, and polynomial methods [9], [28], [44], [56].

When a polynomial approach is used, estimators are calculated from three types of equations: Diophantine equations, polynomial spectral factorisations and coprime factorisations of polynomial matrices. Two dominating approaches have been used for deriving these sets of equations: the *"variational approach"* developed in [4], see also [20], [58], [59] and the *"completing the squares approach"* used e.g. in [28], [40], [41], [44] and [54]. Lately, the *inner–outer factorisation approach*, see e.g. [63], has been utilised in [15]. In [5], that method is interpreted in the polynomial systems framework.

We shall discuss how the classical Wiener approach and the inner–outer factorisation approach relate to the polynomial methods based on variational arguments and completing the squares. The purpose of this discussion is not only to compare advantages and drawbacks, but also to emphasise similarities, to link and increase understanding of the different approaches. To understand how they relate to each other, design equations for a simple filtering problem will be derived using each approach. This is the objective of Section 5.3.

The polynomial approach, based on variational arguments, is then used to study a collection of signal processing and communications problems in Section 5.4–5.6. The selected special problems have features of general interest: multisignal estimation (Section 5.4), discrete time design based on a continuous time problem formulation (Section 5.5), and the approximation of a problem involving a static nonlinearity by a $\mathcal{H}_2$ problem (Section 5.6). Numerical examples are not included, but can be found in the referenced papers. Some concluding remarks, discussing characteristics and suitability of the polynomial approach, are found in Section 5.7.

Remarks on the notation: Let p_j^* denote the complex conjugate of polynomial coefficient p_j. For any complex–valued polynomial $P(q^{-1}) = p_0 + p_1 q^{-1} + \ldots + p_{np} q^{-np}$ in the backward shift operator q^{-1} ($q^{-1}y(t) = y(t-1)$), define

$$P_* \triangleq p_0^* + p_1^* q + \ldots + p_{np}^* q^{np}$$

$$\overline{P} \triangleq q^{-np} P_* = p_{np}^* + p_{np-1}^* q^{-1} + \ldots + p_0^* q^{-np} \quad .$$

Whenever a polynomial in positive powers of q is introduced, it will be denoted with a star, P_*. Rational matrices, or transfer function matrices, are denoted by script letters, for example as $\mathcal{R}(q^{-1})$. For polynomial matrices, P_* means complex conjugate transpose. We denote the trace of P by tr$P$. When appropriate, the complex variable z is substituted for the forward shift operator q. The *degree* of a polynomial matrix is the highest degree of any of its polynomial elements. Polynomial matrices $P(q^{-1})$ are called *stable* if all zeros of $\det P(z^{-1})$ are located in $|z| < 1$. For *marginally stable* polynomial matrices, some zeros of $\det P(z^{-1})$ are located on $|z| = 1$. Arguments of polynomials and rational matrices are often omitted, when there is no risk of misunderstanding.

5.2 A Set of Filtering Problems

A very general linear filtering problem can be formulated in the following way. Based on measurements $z(t)$, up to time $t+m$, a complex–valued vector $f(t) = (f_1(t) \ldots f_\ell(t))^T$ of desired signals is sought. The signals are described by the linear discrete–time stochastic system

$$\begin{pmatrix} z(t) \\ f(t) \end{pmatrix} = \begin{pmatrix} \mathcal{G}_g(q^{-1}) \\ \mathcal{D}_g(q^{-1}) \end{pmatrix} u_g(t) \tag{5.4}$$

and the estimator is

$$\hat{f}(t|t+m) = \mathcal{R}_z(q^{-1}) z(t+m) \quad . \tag{5.5}$$

Here, $\mathcal{G}_g$, $\mathcal{D}_g$, and $\mathcal{R}_z$ are rational matrices of appropriate dimensions and $\{u_g(t)\}$ is a stochastic process, not necessarily white. The weighted estimation error is to be minimised according to some norm, for example $\mathcal{H}_2$ or $\mathcal{H}_\infty$. It is defined by $\mathcal{W}(q^{-1})[f(t) - \hat{f}(t|t+m)]$. See Figure 5.1. A solution to this problem in a $\mathcal{H}_2$ sense will be discussed in Section 5.4.

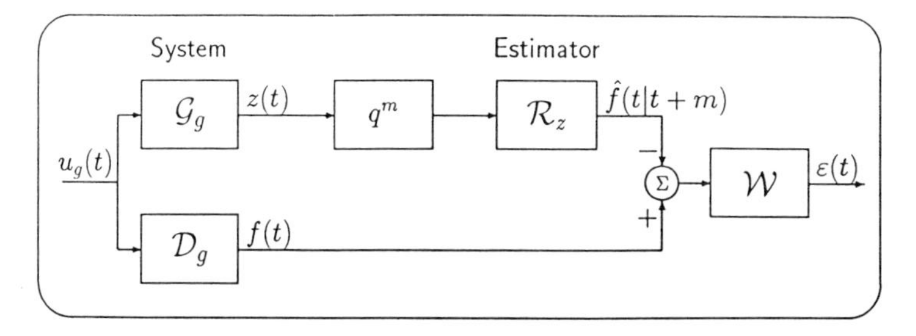

Figure 5.1: A general filtering problem formulation.

While (5.4) is general, it lacks sufficient degree of structure, to obtain solutions which provide useful engineering insight. For the purpose of of this chapter, we will therefore introduce a more detailed structure. It encompasses a number of special cases, to be separately studied in Sections 5.3–5.6. We split the vector $u_g(t)$ into two parts

$$u_g(t) = \begin{pmatrix} u(t) \\ w(t) \end{pmatrix}$$

where $w(t)$ represents additive measurement noise, uncorrelated to the desired signal $f(t)$. We also introduce explicit stochastic models $u(t) = \mathcal{F}(q^{-1})e(t)$ and $w(t) = \mathcal{H}(q^{-1})v(t)$, with $\mathcal{F}$ and $\mathcal{H}$ not necessarily stable. The noises $\{e(t)\}$ and $\{v(t)\}$ are mutually uncorrelated and stationary vector sequences. They have zero means[2] and covariance matrices $\phi \geq 0$ and $\psi \geq 0$, of dimensions $k|k$ and $r|r$, respectively. Furthermore, define the measurement vector as

$$z(t) \triangleq \begin{pmatrix} y(t) \\ a(t) \end{pmatrix} \tag{5.6}$$

where $y(t) = (y_1(t) \ldots y_p(t))^T$ are noisy measurements and $a(t) = (a_1(t) \ldots a_h(t))^T$ is an auxilliary measurement vector, uncorrupted by the noise $w(t)$. (One example could be directly measurable inputs to the system.) The model structure (5.4) is thus specialised to

$$\begin{aligned} \begin{pmatrix} y(t) \\ a(t) \\ f(t) \end{pmatrix} &= \begin{pmatrix} \mathcal{G}(q^{-1}) & I \\ \mathcal{G}_a(q^{-1}) & 0 \\ \mathcal{D}(q^{-1}) & 0 \end{pmatrix} \begin{pmatrix} u(t) \\ w(t) \end{pmatrix} \\ \begin{pmatrix} u(t) \\ w(t) \end{pmatrix} &= \begin{pmatrix} \mathcal{F}(q^{-1}) & 0 \\ 0 & \mathcal{H}(q^{-1}) \end{pmatrix} \begin{pmatrix} e(t) \\ v(t) \end{pmatrix} \end{aligned} \tag{5.7}$$

Compare to (5.4), with $\mathcal{D}_g = [\mathcal{D}\ 0]$. Above, $\mathcal{G}$, $\mathcal{G}_a$, $\mathcal{F}$, $\mathcal{H}$, and $\mathcal{D}$ are transfer function matrices of dimensions $p|s$, $h|s$, $s|k$, $p|r$, and $\ell|s$ respectively. See Figure 5.2. From

[2]One way of handling nonzero means in on–line applications is outlined in subsection 5.3.9.

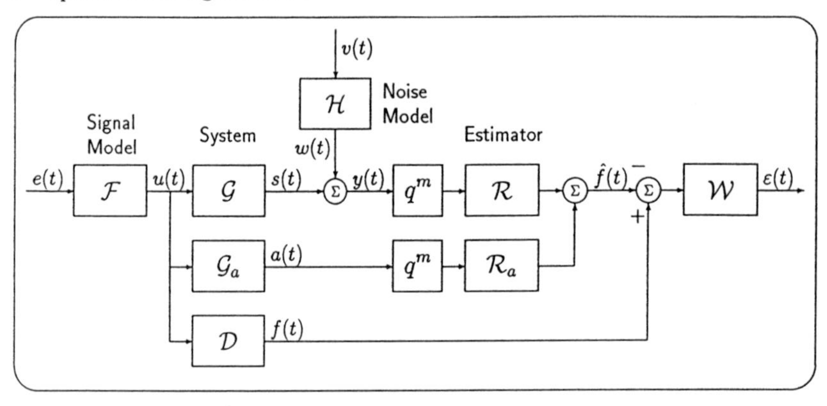

Figure 5.2: Unifying structure for a collection of filtering problems. The signal $f(t)$ is to be estimated from data up to time $t+m$.

the measurements $z(t)$, up to time $t+m$, our aim is to optimise a linear estimator of $f(t)$

$$\hat{f}(t|t+m) = \mathcal{R}_z(q^{-1})z(t+m) \quad ; \quad \mathcal{R}_z(q^{-1}) = [\mathcal{R}(q^{-1}) \; \mathcal{R}_a(q^{-1})] \tag{5.8}$$

with $\mathcal{R}$ and $\mathcal{R}_a$ being stable and causal transfer function matrices of dimensions $\ell|p$ and $\ell|h$, respectively. Depending on m, the estimator constitutes a predictor ($m < 0$), a filter ($m = 0$) or a fixed lag smoother ($m > 0$).

We will consider minimisation of the mean square estimation error (MSE). Introduce the quadratic criterion

$$J \; = \; \mathrm{tr}E(\varepsilon(t)\varepsilon^*(t)) \; = \; E(\varepsilon^*(t)\varepsilon(t)) \; = \; \sum_{i=1}^{\ell} E|\varepsilon_i(t)|^2 \tag{5.9}$$

where

$$\varepsilon(t) \; = \; (\varepsilon_1(t)\ldots\varepsilon_\ell(t))^T \; \triangleq \; \mathcal{W}(q^{-1})(f(t) - \hat{f}(t|t+m)) \; . \tag{5.10}$$

Above, $\mathcal{W}(q^{-1})$ is a stable and causal transfer function weighting matrix, of dimension $\ell|\ell$. It may be used to emphasise filtering performance in certain frequency ranges. The criterion (5.9) is to be minimised, under the constraint of realisability (internal stability and causality) of the filter $\mathcal{R}_z(q^{-1})$.

The structure depicted in Figure 5.2 covers a large set of different problems. We shall in this chapter discuss the following collection:

- Scalar prediction, filtering or smoothing: $\mathcal{G} = \mathcal{D} = \mathcal{W} = 1$, $\mathcal{G}_a = 0$. (Section 5.3.)
- Multivariable deconvolution and linear equalisation: $\mathcal{G}_a = 0$. (Section 5.4.)
- Numerical differentiation of scalar signals and state estimation: $\mathcal{W} = 1$, $\mathcal{G}_a = 0$, $u(t)$ state vector, $\mathcal{G}$ and $\mathcal{D}$ constant vectors. (Section 5.5.)

- Decision feedback equalisation of a scalar symbol sequence: $\mathcal{W} = \mathcal{F} = \mathcal{D} = 1$, $\mathcal{G}_a = q^{-m-1}$. (Section 5.6.)

5.3 Solution Methods

We have recently investigated and developed a *variational approach* for solving filtering problems as well as LQG control problems. We open this section by a presentation of the underlying general ideas, before going into details and comparisons with other approaches. (Application to $\mathcal{H}_2$-optimal control is discussed in [58] and in Chapter 3 of this volume [59].)

5.3.1 Optimisation by variational arguments

Consider the criterion (5.9), (5.10) and the estimator (5.8). Introduce an *alternative weighted estimate*

$$\hat{d}(t|t+m) \;=\; \mathcal{W}(q^{-1})\hat{f}(t|t+m) + \nu(t) \;=\; \mathcal{W}(q^{-1})\mathcal{R}_z(q^{-1})z(t+m) + \nu(t) \quad (5.11)$$

where a stationary signal $\nu(t)$ represents a modification of the (weighted) estimate (5.8). See Figure 5.3. The optimal estimate $\hat{f}(t)$ must be such that no admissible variation can improve the criterion value.

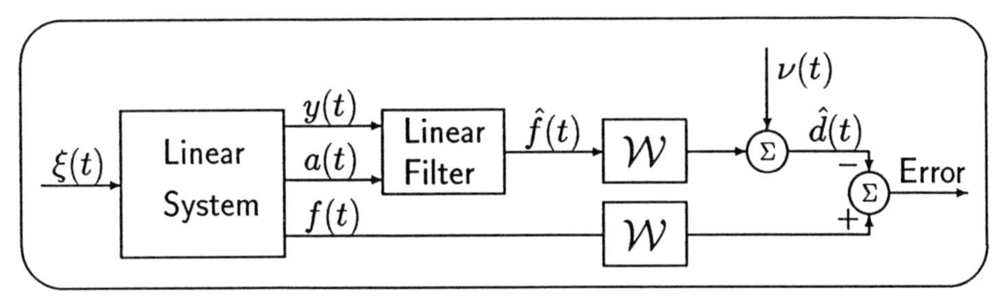

Figure 5.3: A variational approach to the general estimation problem (with driving noise $\xi(t) = (e(t)^T \; v(t)^T)^T$), discussed in Section 5.2. The estimate $\hat{f}(t|t+m)$ is perturbed by a variation $\nu(t)$.

All admissible variations can be represented by $\nu(t) = \mathcal{T}(q^{-1})z(t+m)$, where $\mathcal{T}(q^{-1})$ is some stable and causal rational matrix. Any nonstationary modes of $z(t)$ must be cancelled by zeros of $\mathcal{T}(q^{-1})$. Except for these requirements, $\mathcal{T}(q^{-1})$ is arbitrary. Since $\mathcal{W}(q^{-1})f(t) - \hat{d}(t) = \varepsilon(t) + \nu(t)$, the use of the modified estimator (5.11) results in the criterion value

$$\begin{aligned}\overline{J} &= \mathrm{tr}E\{\mathcal{W}(q^{-1})f(t) - \hat{d}(t|t+m)\}\{\mathcal{W}(q^{-1})f(t) - \hat{d}(t|t+m)\}^* \\ &= \mathrm{tr}\{E\varepsilon(t)\varepsilon^*(t) - E\varepsilon(t)\nu^*(t) - E\nu(t)\varepsilon^*(t) + E\nu(t)\nu^*(t)\} \qquad (5.12)\end{aligned}$$

where $\varepsilon(t)$ is the error (5.10), obtained with the estimator (5.8). If the *cross–terms* in (5.12) are zero, $\nu(t) \equiv 0$ evidently minimises $\bar{J}$, since $\text{tr}E\nu(t)\nu^*(t) > 0$ if any component of $\nu(t)$ has nonzero variance. Then, the estimator (5.8) is optimal. Orthogonality between the error and any admissible linear function of the measurements, $E\nu^*(t)\varepsilon(t) = \text{tr}E[\varepsilon(t)\nu^*(t)] = 0$, guarantees optimality.

Of the two cross terms, it is sufficient to consider only $\text{tr}E\varepsilon(t)\nu^*(t)$, for symmetry reasons. Now, assume $\varepsilon(t)$ in (5.10) to be stationary. (This is evidently true if $z(t)$ and $f(t)$ are stationary, since $\mathcal{W}(q^{-1})$ and $\mathcal{R}_z(q^{-1})$ are required to be stable. However, if $z(t)$ or $f(t)$ are generated by unstable models, stationarity will have to be verified separately, *after* the derivation. This will be exemplified in subsection 5.3.9.) Parsevals formula can then be used to convert the orthogonality requirement $\text{tr}E\varepsilon(t)\nu^*(t) = 0$ into the frequency–domain relation

$$\text{tr}[E\varepsilon(t)\nu^*(t)] = \text{tr}\frac{1}{2\pi j}\oint_{|z|=1} \phi_{\varepsilon\nu^*}\frac{dz}{z} = 0 \qquad (5.13)$$

The rational $\ell|\ell$–matrix $\phi_{\varepsilon\nu^*}$ is the cross spectral density. It can be simplified by using a spectral factorisation, derived by expressing the measurement vector $z(t)$ in innovations form.

When $\ell > 1$, it seems very hard to determine the estimator from the scalar condition (5.13). An important insight is that the derivation becomes easy if we instead require $E\varepsilon(t)\nu^*(t) = 0$. This corresponds to the *elementwise* conditions

$$E\varepsilon_m(t)\nu_n^*(t) = \frac{1}{2\pi j}\oint_{|z|=1} \phi_{\varepsilon\nu^*}^{mn}\frac{dz}{z} = 0 \qquad m = 1\ldots\ell\ ,\ n = 1\ldots\ell \qquad (5.14)$$

which, of course, imply (5.13). These ℓ^2 conditions determine the estimator $\mathcal{R}_z(q^{-1})$. They are fulfilled if *the integrands are made analytic inside the integration path* $|z| = 1$. All poles inside the unit circle should be cancelled by zeros.

A rational matrix $\mathcal{G}(z^{-1})$ can be represented by polynomial matrices as a matrix fraction description (MFD), either left or right: $\mathcal{G} = A_1^{-1}B_1 = B_2A_2^{-1}$. See [36]. Using the *left* polynomial matrix fraction description, the relations (5.14) can be evaluated collectively, rather than individually, when $\ell > 1$. They then reduce to a linear polynomial (matrix) equation, a (bilateral) *Diophantine equation.*

The variational approach can be summarised as a step by step procedure.

1. Parametrise the system by rational transfer functions, represented by polynomial fractions or left MFD's. Define a polynomial spectral factorisation from the spectral density of $z(t)$.

2. Define the estimation error $\varepsilon(t)$ and introduce an admissible variation $\nu(t)$ of the estimate. Express $E\varepsilon(t)\nu^*(t)$ in the frequency domain using Parseval's formula and simplify, by inserting the spectral factorisation.

3. Fulfill the orthogonality requirement $E\varepsilon(t)\nu^*(t) = 0$ by cancelling all poles in $|z| < 1$, in every element of the integrand, by zeros. This leads to linear polynomial (matrix) equation(s), which determines the estimator. For *stable* systems, the derivation ends here.

4. For *marginally stable* and *unstable* signal–generating systems, verify stationarity of $\varepsilon(t)$.[3]

For some problems, the solution is simplified if the variational term is set to a sum of several terms, for example, $\nu(t) = \mathcal{T}_1 y(t+m) + \mathcal{T}_2 a(t+m)$. Orthogonality with respect to each of them, separately, may be achieved. Instead of one large design equation, several smaller ones can then be obtained in Step 3. This situation occurs, when more than one filter is to be optimised and at least one of the filters can be attached to one of the driving noises only. The decision feedback equalisation problem, discussed in Section 5.6, and the feedforward–feedback control problem, see the proof of Theorem 3.1 in Chapter 3 of this volume, are such cases.

More details on the above technique can be found in [4]. See also [58] and [59] for applications to control problems. It will now be exemplified and compared to other methods by solving a simple filtering problem. Another detailed illustration of the approach, for a multisignal estimation problem, is found in Section 5.4.

5.3.2 A scalar filtering problem

Consider the expressions (5.7)–(5.10) and set $\mathcal{G} = 1$, $\mathcal{G}_a = 0$ ($\Rightarrow \mathcal{R}_a = 0$), $\mathcal{F} = C/D$, $\mathcal{H} = M/N$, $\mathcal{D} = 1$, $\mathcal{W} = 1$, $\mathcal{R}_z = [\mathcal{R} \;\; 0]$. Let $\phi = \lambda_e$, $\psi = \lambda_v$. Thus, the signal $f(t) = s(t) = [C(q^{-1})/D(q^{-1})]e(t)$ is to be estimated from noisy measurements

$$y(t) = s(t) + \frac{M(q^{-1})}{N(q^{-1})}v(t) \qquad (5.15)$$

up to time $t+m$, using an estimator $\hat{s}(t|t+m) = \mathcal{R}(q^{-1})y(t+m)$, with $\mathcal{R}$ stable and causal. See Figure 5.4. All model polynomials, with degree nc, nd, etc, are monic. Signal and noise models are assumed stable. Discussion of unstable models is deferred to subsection 5.3.9. The following assumption guarantees problem solvability.

Assumption A. The signal and noise ARMA–models $s(t) = (C/D)e(t)$ and $w(t) = (M/N)v(t)$ are stable and causal, and are assumed to have no common zeros on the unit circle.

---

[3]For strictly unstable systems (poles in $|z| > 1$), the solution to the Diophantine equation(s) obtained in Step 3 may be non–unique, in very rare cases. An additional equation, obtained by requiring stationarity of $\varepsilon(t)$, must then be solved in conjunction with the other equations. See Subsection 5.3.9.

The measurements $\{y(t)\}$ can also be described by the innovations model

$$y(t) = \frac{\beta(q^{-1})}{D(q^{-1})N(q^{-1})}(\sqrt{\lambda_\epsilon}\epsilon(t)) \tag{5.16}$$

where the innovations sequence $\sqrt{\lambda_\epsilon}\epsilon(t)$ is white and has variance λ_ϵ. The monic polynomial $\beta(q^{-1}) = 1 + \beta_1 q^{-1} + \ldots + \beta_{n\beta} q^{-n\beta}$ is stable, under Assumption A. It is the (polynomial) *spectral factor*. We shall now see how the solution to this problem can be derived in four different ways.

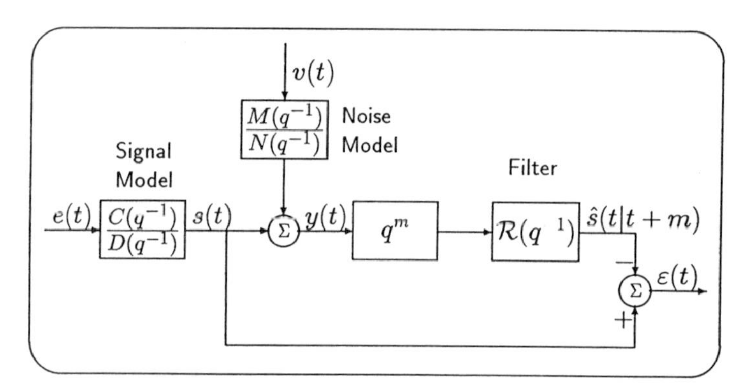

Figure 5.4: A scalar output filtering, prediction or smoothing problem. The signal $\{s(t)\}$ is to be estimated from $\{y(t+m)\}$. In the noise–free case, this problem reduces to prediction of an ARMA–process $s(t) = (C/D)e(t)$.

5.3.3 The variational approach

Let us optimise $\mathcal{R}$ above, following the procedure for the variational approach.

1. Obtain spectral densities of $y(t)$, $\Phi_y(e^{j\omega})$, from both (5.15) and (5.16), and set them equal. This gives the spectral factorisation equation

$$r\beta\beta_* = CC_*NN_* + \rho MM_*DD_* \tag{5.17}$$

   where $r = \lambda_\epsilon/\lambda_e$, $\rho \triangleq \lambda_v/\lambda_e$ and $\beta(z^{-1})$ is stable and monic.

2. Use the error expression

$$\varepsilon(t) = (1 - q^m\mathcal{R})\frac{C}{D}e(t) - q^m\mathcal{R}\frac{M}{N}v(t) \tag{5.18}$$

   and the estimator variation $\nu(t) = \mathcal{T}(q^{-1})y(t+m)$, where $\mathcal{T}(q^{-1})$ is any stable and causal transfer function. The first cross term in (5.12) is then

$$E\varepsilon(t)\nu^*(t) =$$

$$= E\left((1-q^m\mathcal{R})\frac{C}{D}e(t)\right)\left(\mathcal{T}q^m\frac{C}{D}e(t)\right)^* - E\left(q^m\mathcal{R}\frac{M}{N}v(t)\right)\left(\mathcal{T}q^m\frac{M}{N}v(t)\right)^*$$

$$= \frac{\lambda_e}{2\pi j}\oint_{|z|=1}\frac{(1-z^m\mathcal{R})z^{-m}CC_*NN_* - \rho\mathcal{R}MM_*DD_*}{DD_*NN_*}\mathcal{T}_*\frac{dz}{z}$$

$$= \frac{\lambda_e}{2\pi j}\oint_{|z|=1}\frac{(z^{-m}CC_*NN_* - \mathcal{R}r\beta\beta_*)}{DD_*NN_*}\mathcal{T}_*\frac{dz}{z} \qquad (5.19)$$

where Parseval's formula and (5.17) were utilised.

3. In the integrand of (5.19), the stable polynomials D and N contribute poles in $|z| < 1$, while the poles of $\mathcal{T}_*$ and the zeros of D_*N_* are in $|z| > 1$. Furthermore, $\beta(z^{-1}) = \bar{\beta}(z)/z^{n\beta}$ may contribute poles at the origin. We see that N and β can be cancelled by $\mathcal{R}$, if $\mathcal{R} = \mathcal{R}_1 N/\beta$ in (5.19). Thus, while β is cancelled directly, N may be factored out of the numerator, to cancel N in the denominator. The remaining poles of the integrand inside $|z| = 1$, are eliminated if (and only if)

$$\frac{(z^{-m}CC_*N_* - \mathcal{R}_1 r\beta_*)}{D}\frac{1}{z} = L_*$$

for some polynomial $L_*(z)$. It suffices for $\mathcal{R}_1$ to be a polynomial, in order to obtain a (polynomial) Diophantine equation. With $\mathcal{R}_1 = Q_1$ and q exchanged for z, we obtain polynomials $Q_1(q^{-1})$, and $L_*(q)$, as the *unique*[4] solution to the linear Diophantine equation

$$q^{-m}CC_*N_* = r\beta_*Q_1 + qDL_* \ . \qquad (5.20)$$

Thus, the optimal estimator

$$\hat{s}(t|t+m) = \frac{Q_1(q^{-1})N(q^{-1})}{\beta(q^{-1})}y(t+m) \qquad (5.21)$$

is obtained by solving (5.17) for β (and r) and (5.20) for Q_1 (and L_*). It can be noted that the d–step predictor (5.2) for ARMA processes (5.1) is a special case of the solution above[5]. Note also that the transfer function $\mathcal{T}_*$ in the variational term does not influence the solution at all.

[4] See subsection 5.3.8.

[5] With a stable C and no measurement noise ($\rho = 0, M = N = 1$), we have $\beta = C, r = 1$. Then $\beta_* = C_*$ is a factor of two terms in (5.20), so it must also be a factor of qDL_*. Set $L_* = C_*L_{1*}$ in (5.20), cancel C_* and multiply by q^m, to obtain $C = q^mQ_1 + D(q^{m+1}L_{1*})$. Now, with $m = -d$, $G(q^{-1}) = Q_1(q^{-1})$ and $F(q^{-1}) = \bar{L}_1(q^{-1}) \triangleq q^{-d+1}L_{1*}(q)$, we obtain (5.3).

5.3.4 Completing the squares

While the variational method is based on manipulation of the orthogonality relation $E\varepsilon\nu^* = 0$, the "completing the squares" approach is a way of deriving the filter by manipulating the criterion $\mathrm{tr}E(\varepsilon\varepsilon^*)$ itself. The goal is to express the criterion as a sum of several terms, of which some can be minimised in a straightforward way. The other terms are either zero or are unaffected by the filter.

The completing the squares approach has been used in the time domain by Kučera in e.g. [40], [44]. The frequency domain variant discussed below has been used, for example, by Roberts and Newmann in [54] and by Grimble in [28]. In the example of subsection 5.3.2, the criterion (5.9) can be expressed by

$$\begin{aligned} J &= E|s(t) - \mathcal{R}y(t+m)|^2 \\ &= E\left|(1 - q^m\mathcal{R})\frac{C}{D}e(t)\right|^2 + E\left|q^m\mathcal{R}\frac{M}{N}v(t)\right|^2 \\ &= \frac{\lambda_e}{2\pi j}\oint_{|z|=1}\left((1 - z^m\mathcal{R})(1 - z^{-m}\mathcal{R}_*)\frac{CC_*}{DD_*} + \rho\mathcal{R}\mathcal{R}_*\frac{MM_*}{NN_*}\right)\frac{dz}{z} \\ &= \frac{\lambda_e}{2\pi j}\oint_{|z|=1}\left(\frac{CC_*}{DD_*} - z^m\mathcal{R}\frac{CC_*}{DD_*} - \frac{CC_*}{DD_*}z^{-m}\mathcal{R}_* + \mathcal{R}\mathcal{R}_*\frac{r\beta\beta_*}{DD_*NN_*}\right)\frac{dz}{z} \ . \end{aligned}$$

In the third equality, we used Parseval's formula and in the last, the spectral factorisation (5.17) was inserted. Completing the square gives

$$\begin{aligned} J &= \frac{\lambda_e}{2\pi j}\oint_{|z|=1} r\left(\frac{\beta}{DN}\mathcal{R} - \frac{z^{-m}CC_*N_*}{r\beta_*D}\right)\left(\frac{\beta_*}{D_*N_*}\mathcal{R}_* - \frac{z^mC_*CN}{r\beta D_*}\right)\frac{dz}{z} \\ &+ \frac{\lambda_e}{2\pi j}\oint_{|z|=1}\left(\frac{CC_*}{DD_*} - \frac{CC_*CC_*NN_*}{r\beta\beta_*DD_*}\right)\frac{dz}{z} \triangleq J_1 + J_2 \ . \end{aligned} \tag{5.22}$$

The first term in (5.22), J_1, depends on $\mathcal{R}$ while the second term, J_2, does not. If $\mathcal{R}$ were not restricted to be realisable (internally stable and causal), the problem could have been solved by choosing $\mathcal{R}$ such that $J_1 = 0$. (This would constitute the so–called non–realisable Wiener filter.)

A realisable $\mathcal{R}$ can only eliminate the *causal* parts of the integrand of J_1. Since $(\beta/DN)\mathcal{R}$ is causal, it remains to partition $(z^{-m}CC_*N_*)/(r\beta_*D)$. Let

$$\frac{z^{-m}CC_*N_*}{r\beta_*D} = \frac{Q_1}{D} + \frac{zL_*}{r\beta_*} \tag{5.23}$$

for some polynomials Q_1 and L_*. The term $Q_1(z^{-1})/D(z^{-1})$ represents the causal part and $(zL_*(z)/(r\beta_*(z))$ the (strictly) noncausal part. By setting the right hand side of (5.23) on common denominator form, we obtain

$$z^{-m}CC_*N_* = r\beta_*Q_1 + zDL_* \tag{5.24}$$

which is (5.20), with z exchanged for q. Using (5.23) to express J_1 gives

$$J_1 = \frac{\lambda_e}{2\pi j}\oint_{|z|=1} r\left(\frac{\beta}{DN}\mathcal{R} - \frac{Q_1}{D} - \frac{zL_*}{r\beta_*}\right)\left(\frac{\beta}{DN}\mathcal{R} - \frac{Q_1}{D} - \frac{zL_*}{r\beta_*}\right)_* \frac{dz}{z} \quad .$$

By expanding the integrand, J_1 may be written as a sum of four terms:

$$\begin{aligned}
V_1 &= \frac{\lambda_e}{2\pi j}\oint_{|z|=1} r\left(\frac{\beta}{DN}\mathcal{R} - \frac{Q_1}{D}\right)\left(\frac{\beta_*}{D_*N_*}\mathcal{R}_* - \frac{Q_{1*}}{D_*}\right)\frac{dz}{z} \\
V_2 &= -\frac{\lambda_e}{2\pi j}\oint_{|z|=1}\left(\frac{\beta}{DN}\mathcal{R} - \frac{Q_1}{D}\right)\frac{z^{-1}L}{\beta}\frac{dz}{z} \\
V_3 &= -\frac{\lambda_e}{2\pi j}\oint_{|z|=1} z\frac{L_*}{\beta_*}\left(\frac{\beta_*}{D_*N_*}\mathcal{R}_* - \frac{Q_{1*}}{D_*}\right)\frac{dz}{z} \\
V_4 &= \frac{\lambda_e}{2\pi j}\oint_{|z|=1}\frac{LL_*}{r\beta\beta_*}\frac{dz}{z} \quad .
\end{aligned}$$

For any causal and stable choice of the rational filter $\mathcal{R}$, all poles of the integrand of V_3 will be located outside the unit circle, since β, D and N are all stable. Hence, $V_3 = 0$. (Note that it is crucial that zL_*/β_* is strictly noncausal, starting with a free z. This z cancels the pole at the origin of V_3.) For symmetry reasons, $V_2 = 0$. The term V_4 does not depend on $\mathcal{R}$. Thus, the criterion J_1 is minimised by minimising V_1. We readily obtain $V_1 = 0$ by choosing

$$\frac{\beta}{DN}\mathcal{R} - \frac{Q_1}{D} = 0 \quad .$$

This gives

$$\mathcal{R} = \frac{Q_1 N}{\beta}$$

where Q_1, together with L_*, is the solution to (5.24) and β is the stable polynomial spectral factor obtained from (5.17). The minimal criterion value is $J_{\min} = J_2 + V_4$. This derivation should be compared to steps 2. and 3. in the derivation in subsection 5.3.3.

5.3.5 The classical Wiener solution

Wiener filters are traditionally designed by first whitening the measurements and then multiplying them by the cross spectral density, $\phi_{f\epsilon}$, between desired signal and whitened mesurement. See, for example, [12], [14]. For the example of subsection 5.3.2, with $f(t) = s(t)$, the causal Wiener filter is $\hat{s}(t|t+m) = \{\phi_{s\epsilon}\}_+\epsilon(t+m)$, where $\epsilon(t+m) = \mathcal{V}(q^{-1})y(t+m)$ is the whitened measurement. Thus,

$$\mathcal{R} = \{\phi_{s\epsilon}\}_+\mathcal{V} = \{\phi_{sy}\mathcal{V}_*\}_+\mathcal{V} \quad . \tag{5.25}$$

Above, ϕ_{sy} is the cross–spectral density of desired signal and measurement. The notation $\{\cdot\}_+$ represents the use of only the causal part of the expression $\{\cdot\}$. The whitening filter is denoted $\mathcal{V}(q^{-1})$ and its conjugate $\mathcal{V}_*(q)$. Apart from a scale factor $1/\sqrt{\lambda_\epsilon}$, it is the inverse of the innovations model (5.16):

$$\mathcal{V} = \frac{DN}{\sqrt{\lambda_\epsilon}\beta} \ . \tag{5.26}$$

The expression (5.25) is elegant. However, ϕ_{sy} is not explicit, in terms of polynomial coefficients of rational transfer functions of the signal and noise models. The polynomial systems framework is of help here. It can be used to evaluate the term $\{\cdot\}_+$ in an efficient way.

Since $e(t)$ and the noise $v(t)$ are mutually uncorrelated and the measurement is $y(t+m) = s(t+m) + (M/N)v(t+m)$, we readily obtain

$$\phi_{sy} = \phi_{s(t)y(t+m)} = \phi_{s(t)s(t+m)} = \frac{C}{D}z^{-m}\frac{C_*}{D_*}\lambda_e \ .$$

Thus, (5.25) becomes, with $r \triangleq \lambda_\epsilon/\lambda_e$,

$$\mathcal{R}(q^{-1}) = \left\{\frac{C}{D}q^{-m}\frac{C_*}{D_*}\lambda_e\frac{D_*N_*}{\sqrt{\lambda_\epsilon}\beta_*}\right\}_+\frac{DN}{\sqrt{\lambda_\epsilon}\beta} = \left\{q^{-m}\frac{CC_*N_*}{Dr\beta_*}\right\}_+\frac{DN}{\beta} \ . \tag{5.27}$$

Extraction of the causal part $\{\cdot\}_+$ of the double–sided function, corresponds to performing a partial fraction expansion. Let

$$q^{-m}\frac{C(q^{-1})C_*(q)N_*(q)}{D(q^{-1})r\beta_*(q)} = \frac{Q_1(q^{-1})}{D(q^{-1})} + \frac{\tilde{L}_*(q)}{r\beta_*(q)} \tag{5.28}$$

for some polynomials $Q_1(q^{-1})$ and $\tilde{L}_*(q)$. Terms without delay should appear exclusively in the causal part, so the noncausal part starts with a free q–term. Thus, let $\tilde{L}_*(q) \triangleq qL_*(q)$. (This avoids the occurrence of an error pointed out by Chen [16].) Multiplying both sides of (5.28) by $Dr\beta_*$ then gives

$$q^{-m}CC_*N_* = r\beta_*Q_1 + qDL_* \ .$$

Once again, this is precisely the linear Diophantine equation (5.20). Thus, the causal Wiener filter is

$$\mathcal{R}(q^{-1}) = \left\{\frac{Q_1}{D} + \frac{qL_*}{r\beta_*}\right\}_+\frac{DN}{\beta} = \frac{Q_1DN}{D\beta} \tag{5.29}$$

which, of course, coincides with (5.21), if the stable factor D is cancelled. (Unstable systems are not allowed in the classical Wiener formulation.) The link between partial fraction expansion and Diophantine equations was noted by Grimble [28], and has also been independently noted by us and by others. This link also plays a key role in the "completing the squares"–reasoning, cf (5.23).

5.3.6 The inner–outer factorisation approach

Vidyasagar [63] has discussed a factorisation approach to optimal filtering. This subsection is based on that approach. To explain it, we need a brief recapitulation of inner and outer matrices and their properties. Consider rational matrices with n rows and m columns, having stable discrete–time transfer functions as elements. Let such matrices be denoted $\mathcal{P}^{n|m}(z^{-1})$, or just $\mathcal{P}$, and their conjugate transpose $\mathcal{P}_*^{m|n}(z)$ (or $\mathcal{P}_*$). We need the following definitions (see [27] and [63]).

- A stable rational matrix $\mathcal{P}^{n|m}(z^{-1})$, $n \geq m$, is *inner* if $\mathcal{P}_*\mathcal{P} = I_m$ for almost all $|z| = 1$. It is *co–inner* if $n \leq m$ and $\mathcal{P}\mathcal{P}_* = I_n$ for almost all $|z| = 1$.

- A stable rational matrix $\mathcal{P}^{n|m}(z^{-1})$, $n \leq m$, is *outer* if and only if it has full row rank n, $\forall\, |z| \geq 1$. In other words, it has no zeros in $|z| \geq 1$. It is *co–outer* when $n \geq m$ if and only if it has full column rank m, $\forall\, |z| \geq 1$.

- A stable rational matrix $\mathcal{P}^{n|m}(z^{-1})$, with full rank $p \triangleq \min\{m, n\}$ for all $z = e^{j\omega}$ (no zeros on the unit circle), has an *inner–outer factorisation*

$$\mathcal{P}^{n|m} = \mathcal{P}_i^{n|p}\mathcal{P}_o^{p|m} \tag{5.30}$$

with the outer factor $\mathcal{P}_o$ having a stable right inverse. It also has a *co–inner–outer factorisation*

$$\mathcal{P}^{n|m} = \mathcal{P}_{co}^{n|p}\mathcal{P}_{ci}^{p|m} \tag{5.31}$$

with the co–outer factor $\mathcal{P}_{co}$ having a stable left inverse. If $n \leq m$, the co–outer matrix is square, and its inverse is unique.

Inner and co–inner matrices are generalisations of scalar all–pass links. Multiplication by a (co)inner matrix does not affect the spectral density or power of a signal vector. The important property of outer and co–outer matrices is that they are *stably* invertible. Additionally, the inverses are *causal* if the instantaneous gain matrices $\mathcal{P}_o(0)$ and $\mathcal{P}_{co}(0)$ have full rank p.

Now, minimising (5.9) is, for the filtering example of subsection 5.3.2, equivalent to minimising

$$J = \left\|\left[\frac{C}{D}\lambda_e^{1/2} \quad 0\right] - \mathcal{R}\left[z^m\frac{C}{D}\lambda_e^{1/2} \quad z^m\frac{M}{N}\lambda_v^{1/2}\right]\right\|_2^2 \tag{5.32}$$

where $\| x(z^{-1}) \|_2^2 = (1/2\pi j)\mathrm{tr} \oint_{|z|=1} x x_* dz/z$.

The idea is now to factor the second term of (5.32) as

$$U \triangleq \left[z^m\frac{C}{D}\lambda_e^{1/2} \quad z^m\frac{M}{N}\lambda_v^{1/2}\right] = U_{co}U_{ci} \tag{5.33}$$

where U_{co} is co–outer of dimension $1|1$ and U_{ci} is co–inner of dimension $1|2$. The scalar co–outer will have a stable inverse, if the left hand side of (5.33) has full rank 1 for all $|z| = 1$. [6] The inverse $U_{co}^{-1}(z^{-1})$ is causal if and only if $U_{co}(0) \neq 0$.

[6] In other words, C and $\lambda_v^{1/2}M$ should have no common factors with zeros on $|z| = 1$. This corresponds to the condition for existence of a stable spectral factor in (5.17).

By invoking (5.33), the criterion (5.32) can be written as

$$J = \left\| \left[\frac{C}{D} \lambda_e^{1/2} \;\; 0 \right] - \mathcal{R} U_{co} U_{ci} \right\|_2^2 . \tag{5.34}$$

Now, multiplying the interior of the norm in (5.34) from the right by U_{ci*}, which is normpreserving, and using the co–inner property, $U_{ci}U_{ci*} = 1$ on $|z| = 1$, gives

$$J = \left\| \left[\frac{C}{D} \lambda_e^{1/2} \;\; 0 \right] U_{ci*} - \mathcal{R} U_{co} \right\|_2^2 .$$

By decomposing into *causal* and *noncausal* parts, the causal and stable filter $\mathcal{R}$, which minimises J, is readily found from the requirement that $\mathcal{R}$ should eliminate the whole causal part. Thus,

$$\mathcal{R} U_{co} = \left\{ \left[\frac{C}{D} \lambda_e^{1/2} \;\; 0 \right] U_{ci*} \right\}_+ \tag{5.35}$$

where $\{\cdot\}_+$, as before, represents the causal part. The optimal filter thus becomes

$$\mathcal{R} = \left\{ \left[\frac{C}{D} \lambda_e^{1/2} \;\; 0 \right] U_{ci*} \right\}_+ U_{co}^{-1} . \tag{5.36}$$

The inverse U_{co}^{-1} is stable by definition.

The factorisation–based solution thus consists of first performing a co–inner–outer factorisation (5.33) and then the causal–noncausal factorisation required in (5.35). We will now emphasise the correspondence of these two steps to the previous solutions.

If the spectral factorisation (5.17) has been solved, the co–inner and co–outer factors can be obtained as

$$U_{co} = \frac{\lambda_\epsilon^{1/2} \beta}{DN} \qquad U_{ci} = \left[\frac{\lambda_e^{1/2} z^m CN}{\lambda_\epsilon^{1/2} \beta} \quad \frac{\lambda_v^{1/2} z^m MD}{\lambda_\epsilon^{1/2} \beta} \right] . \tag{5.37}$$

It is easily verified that $U = U_{co}U_{ci}$ and that, with $r = \lambda_\epsilon / \lambda_e$, $\rho = \lambda_v / \lambda_e$, cf. (5.17), we obtain

$$U_{ci} U_{ci*} = \frac{\lambda_e CN(CN)_* + \lambda_v MD(MD)_*}{\lambda_\epsilon \beta \beta_*} = 1 . \tag{5.38}$$

Furthermore, U_{co} given by (5.37) has no zero in $|z| \geq 1$, and is therefore stably invertible, whenever a stable spectral factor β exists. The construction above is an application of the standard way of performing inner–outer factorisations: by means of spectral factorisation, see e.g. [27].

Using (5.37), the optimal filter (5.36) can be expressed as

$$\mathcal{R} = \left\{ \left[\lambda_e^{1/2} \frac{C}{D} \;\; 0 \right] U_{ci*} \right\}_+ U_{co}^{-1} = \left\{ \frac{\lambda_e z^{-m} CC_* N_*}{\lambda_\epsilon D \beta_*} \right\}_+ \frac{DN}{\beta} \tag{5.39}$$

where the scalar $\lambda_\epsilon^{-1/2}$ from U_{co}^{-1} has been absorbed into the $\{\cdot\}_+$–factor.

The causal bracket operation is the same as in the classical Wiener–solution. Thus, exchange q for z and introduce polynomials $Q_1(q^{-1})$ and $L_*(q)$, such that the impulse response of the rational function inside the brackets of (5.39) can be expressed as the sum of a causal and a noncausal term

$$\frac{\lambda_e q^{-m} C(q^{-1}) C_*(q) N_*(q)}{\lambda_\epsilon D(q^{-1}) \beta_*(q)} = \frac{Q_1(q^{-1})}{D(q^{-1})} + \frac{q L_*(q)}{\lambda_\epsilon \beta_*(q)} \quad . \tag{5.40}$$

Thus, the $\{\ \cdot\ \}_+$–factor equals Q_1/D. By setting the expression (5.40) on a common denominator, we obtain the Diophantine equation (5.20). The estimator (5.39) equals (5.21):

$$\mathcal{R} = \frac{Q_1}{D} \frac{DN}{\beta} = \frac{Q_1 N}{\beta} \quad . \tag{5.41}$$

Observe that the inverse of the co–outer, $\mathcal{U}_{co}$, is nothing but the well–known whitening filter $\mathcal{V}$ in (5.26), from the classical Wiener solution. As in that case, unstable D–polynomials are not allowed.

5.3.7 A comparative discussion

Above, we have presented four different routes to the MSE–optimal solution. Which route to be preferred is more or less a matter of taste and background. There are, however, some aspects a problem solver should be aware of. We shall briefly summarise them below.

Evidently, the four approaches arrive, one way or another, at a polynomial spectral factorisation and a Diophantine equation. In the variational approach, spectral factorisation arises as an obvious simplification of the cross spectral density $\phi_{\varepsilon\nu^*}$. See (5.19). The same is true for the "completing the squares"-approach, but there it simplifies the criterion expression. In the classical Wiener solution, the spectral factorisation determines the whitening filter while in the inner–outer factorisation approach, it is part of the inner outer–factorisation. In particular, it is defined by the inner property (5.38). The inverse of the outer matrix is the whitening filter in the classical solution, while the bracket term in (5.39) is just another way of writing $\{\phi_{sy}\mathcal{V}_*\}_+$, cf. (5.25).

Spectral factorisation can be avoided in noise–free situations, with stably invertible models, such as the prediction problem (5.1)–(5.3). In problems with noise, it can be avoided only in very special cases, such as the optimisation of decision feedback equalisers. That problem is discussed in Section 5.6.

It is interesting to note how the Diophantine equation arises. In all formulations except in the variational approach, it originates from a causal–noncausal partitioning, where the causal factor $\{\cdot\}_+$ is sought. In the variational approach, it arises from the requirement that the variational term should be orthogonal to the error.

Problems with unstable models can be handled by the variational approach (see subsection 5.3.9 below) and by the completing the squares–method. They cannot be handled by the classical Wiener approach or by inner–outer factorisation.

A disadvantage with the "completing the squares" approach is that it will , in difficult problems, be hard to complete the square: the solution has to be known (or suspected) in order to find it. On the other hand it requires, in essense, the simplest mathematics: just quadratic forms are needed. (This is more apparent in a time domain formulation.) In the classical solution, it might be difficult to find the right way from the expression (5.25) to an explicit solution. In particular, this is not straightforward for the problem discussed in Section 5.6. The same is true for the inner–outer factorisation approach.

A main advantage with the variational approach is that it leads to the solution along a constructive and systematic route. This is of considerable importance in more difficult problem formulations. See e.g. Section 5.4.

Another advantage is that the free q–factor in e.g. (5.20) emerges automatically from the cancellation of the free z in (5.19). In the other methods, one has to, somewhat arbitrarily, include direct terms only in the causal part of the causal–noncausal partitioning, to avoid a suboptimal solution [16]. A disadvantage with the variational approach is that some extra calculations are required to obtain the criterion value. In the "completing the squares" approach, this comes as a bonus.

5.3.8 The scalar Diophantine equation

While Diophantine equations in general have an infinite number of solutions, equations arising from linear quadratic design problems mostly have a unique solution. This is a consequence of two requirements, which are easily seen for the scalar Diophantine equations obtained in this chapter, see e.g. (5.20):

1. Filter causality requires Q_1 to be a polynomial only in q^{-1}.

2. Optimality restricts L_* to be a polynomial only in q.

For equations with these properties, the following result can be established.

Theorem 5.3.1

Consider the scalar Diophantine equation

$$C(q,q^{-1}) = A(q,q^{-1})X(q^{-1}) + B(q,q^{-1})Y(q) \tag{5.42}$$

where

$$\begin{aligned}
C(q,q^{-1}) &\triangleq c_{nc1}q^{nc1} + \ldots + c_0 + \ldots + c_{-nc2}q^{-nc2} \\
A(q,q^{-1}) &\triangleq a_{na1}q^{na1} + \ldots + a_0 + \ldots + a_{-na2}q^{-na2} \neq 0 \\
B(q,q^{-1}) &\triangleq b_{nb1}q^{nb1} + \ldots + b_0 + \ldots + b_{-nb2}q^{-nb2} \neq 0 \ .
\end{aligned}$$

Let d be the the number of linearly dependent equations in the corresponding system of linear equations. Then, (5.42) has a *unique* solution

$$X(q^{-1}) = x_o + x_1 q^{-1} + \ldots + x_{nx} q^{-nx} \quad ; \quad Y(q) = y_o + y_1 q + \ldots + y_{ny} q^{ny}$$

with degrees[7]

$$nx = \max\{nc2, nb2\} - na2 \quad ; \quad ny = \max\{nc1, na1\} - nb1 \tag{5.43}$$

if and only if common factors of A and B are also factors of C and

$$nb1 + na2 - d = 1 \quad . \tag{5.44}$$

□

Proof: See [4].

The equation (5.20) fulfills (5.44). There we have $nb1 = 1$ (because of the free q–factor) and $na2 = 0$. Since β_* (unstable) and D (stable) cannot have common factors, the corresponding system of equations has full rank. Consequently, $d = 0$ in (5.44). From (5.43), we obtain the degrees $nQ_1 = \max(nc+m, nd-1)$, $nL = \max(nc+nn-m, n\beta) - 1$.

5.3.9 Unstable signal models

Let us remove the assumption of stability of D and N in the problem described in subsection 5.3.2. The complete solution then turns out to include a second Diophantine equation. However, we will argue that the original equation is sufficient in filtering problems of practical interest. Assumption A is now exchanged for

Assumption B.
The signal and noise models $s(t) = (C/D)e(t)$ and $w(t) = (M/N)v(t)$ are causal and have no unstable hidden modes. They have no common zeros on the unit circle and no common poles on or outside the unit circle.

The requirement of no common unstable modes corresponds to detectability of a state space model.

For unstable systems, an innovations model (5.16) can still be defined. It should, more properly, be called a generalised innovations model, with $\beta(q^{-1})$ being a generalised (polynomial) spectral factor [56]. Under Assumption B, the spectral factorisation equation (5.17) will have a unique stable solution.

In the variational approach, the stationarity of the variation $\nu(t) = \mathcal{T}y(t+m)$ has to be guaranteed. (The modified estimation error is $\varepsilon(t) + \nu(t)$. Assuming $\varepsilon(t)$ to be zero mean stationary, we could never obtain a lower MSE by adding to it a

[7]The degrees (5.43) are obtained from the requirement that the variables $X(q^{-1})$ and $Y(q)$ should cover the maximal powers of q^{-1} and q, respectively, in the other terms of (5.42).

nonstationary signal $\nu(t)$, with variance tending to infinity.) Stationarity of $\nu(t)$ is guaranteed by requiring $\mathcal{T}$ to contain all unstable poles as zeros. For example, set $\mathcal{T} = DN\mathcal{T}_1$ with $\mathcal{T}_1$ stable and causal. Then, the factor $(1/D_*N_*)\mathcal{T}_*$ in (5.19) has poles only outside $|z| = 1$. The rest of the reasoning remains unchanged. The optimal linear estimator still satisfies (5.20) and (5.21).

For unstable signal (and noise) models, two different situations are now possible.

Case 1: β_* and D have no common factors. Under Assumption B, the equation (5.20) remains *uniquely solvable*. Since β_* has zeros only in $|z| > 1$, this holds for *marginally stable* models, where D (or N) has zeros on $|z| = 1$. These are the unstable models of most interest in filtering problems. They are used for describing signals and noise with drifting or sinusoid behaviour. The use of signal models with poles at $z = 1$ is also a trick to avoid bias when estimating stationary signals with nonzero mean.

Stationarity of the error $\varepsilon(t)$ in subsection 5.3.3 is verified in the following way. (See also Appendix A and B to Chapter 3 of this volume [59].) The use of (5.21) in (5.18) gives

$$\varepsilon(t) = \left(1 - q^m \frac{Q_1 N}{\beta}\right) \frac{C}{D} e(t) - q^m \frac{Q_1 N}{\beta} \frac{M}{N} v(t) \quad .$$

Cancellation of N in the last term is assumed to be exact. (If D is stable and N unstable, $\varepsilon(t)$ is therefore stationary.) Let us evaluate the first term at the zeros of D in $|z| \geq 1$, denoted $\{z_j\}$. When (5.17) and (5.20) are evaluated at $\{z_j\}$, their most right–hand terms (but no other terms) vanish. Use of this fact gives

$$1 - q^m \frac{Q_1 N}{\beta}\bigg|_{z=z_j} = 1 - q^m \left(q^{-m} \frac{CC_*N_*}{r\beta_*}\right) \frac{N}{\beta}\bigg|_{z=z_j} = 0 \quad .$$

Thus, the transfer function from $e(t)$ to the error $\varepsilon(t)$ remains finite for all $z \geq 1$, including $\{z_j\}$. Unstable poles are cancelled by zeros. In SISO problems, the reasoning above is straightforward. In multi–signal estimation problems, additional conditions will often have to be imposed, to avoid "impossible" problem formulations, for which no finite minimal criterion value exists.

Note that for signals with nonzero mean, the presence of a zero at $z = 1$ in the transfer function from $s(t)$ to $\varepsilon(t)$ precludes biased estimates. The presence of such a zero is assured by including a pole at $z = 1$ in the signal model C/D.

Case 2: β_* and D have common factors. Under assumption B, those factors must also be factors of the left–hand side of (5.20)[8]. Thus, the Diophantine equation remains solvable, but the solution becomes *non–unique*. We obtain a linear dependence in the equations, represented by $d > 0$ in (5.44). Only one of these solutions

[8] From the spectral factorisation (5.17), it is evident that factors common to D and $r\beta\beta_*$ must also appear in CC_*NN_*. Since (D, N) and (D, C) are not allowed to have unstable common factors, these factors must be present in C_*N_*.

corresponds to a stationary error[9]. The correct solution is obtained by *requiring* that D is cancelled in the transfer function from $e(t)$ to $\varepsilon(t)$. Thus, we require that

$$\beta - q^m Q_1 N = XD \tag{5.45}$$

for some polynomial $X(q^{-1})$. This is the second Diophantine equation. An alternative variant is obtained by multiplying this equation by $r\beta_*$. This gives

$$r\beta\beta_* = q^m(r\beta_* Q_1)N + r\beta_* XD \ .$$

The use of (5.17) and (5.20), and cancellation of D, gives the equation

$$\rho M M_* D_* = -q^{m+1} L_* N + r\beta_* X \ . \tag{5.46}$$

Any one of the equations (5.45) or (5.46) can be solved in conjunction with (5.20), in the same way as in the feedback design of Example 1 in Chapter 3. Then, the unique optimal $Q_1(q^{-1})$ is obtained, together with $L_*(q)$ and $X(q^{-1})$.

The need for a second Diophantine equation in certain situations has been emphasised by Kučera [42] for feedback control problems and by Grimble [30] and Chisci and Mosca [17] for filtering problems.

The additional equation complicates the solution, but it is required only in the exceptional Case 2. In the open loop filtering problems considered here, that situation is furthermore of little practical interest. It corresponds to estimation of exponentially increasing, "exploding", time series[10]. There would be severe problems with variable overflow, except for short data series. Furthermore, the stationarity of the error depends on exact cancellation. Arbitrarily small modelling errors or roundoff errors would ruin the result completely in the long run. When signals are nonstationary, the problem of model errors is furthermore larger than for stationary signals. In a nonlinear world, linear time–invariant models are good (but not perfect) descriptions of time series only around *stationary* operating points. The sensitivity problem is still serious, but more acceptable, in the important case of poles on $|z| = 1$.

For these reasons, estimation problems for strictly unstable models, with a theoretical need for an additional Diophantine equation, will not be considered in the following.

5.4 Multisignal Deconvolution

Let us now consider the problem of deconvolution or input estimation, as presented in [4]. The formulation includes all problems described by the general structure of

[9] The demonstration of a finite transfer function utilised in Case 1 cannot be used for the zeros of common factors of β_* and D. Both (5.17) and (5.20) vanish completely at those zeros.

[10] This claim does not hold for estimation within a stabilised closed loop. One example is a state estimator used in conjunction with a state feedback, which stabilises the unstable mode.

Figure 5.1. The solution illustrates the application of the variational approach to multi–signal filtering problems.

In many areas, it is of interest to estimate the input to a linear system. One interesting application is the reconstruction of stereophonic sound, described by Nelson et.al. [51]. Others are described in [2], [21], [22], [46], and the references therein.

Let the noise–corrupted measurement $y(t)$ and the input $u(t)$ be described by

$$\begin{aligned} y(t) &= A^{-1}Bu(t) + N^{-1}Mv(t) \\ u(t) &= D^{-1}Ce(t) \ . \end{aligned} \tag{5.47}$$

Here, (A, B, N, M, D, C) are polynomial matrices of dimensions $p|p$, $p|s$, $p|p$, $p|r$, $s|s$ and $s|k$, respectively. As before, $\{e(t)\}$ and $\{v(t)\}$ are mutually uncorrelated zero mean stochastic processes. They have covariance matrices $\phi \geq 0$ and $\psi \geq 0$, of dimensions $k|k$ and $r|r$, respectively. The matrix B need not be stably invertible. It may not even be square. From data $y(t)$ up to time $t+m$, an estimator

$$\hat{f}(t|t+m) = \mathcal{R}(q^{-1})y(t+m) \tag{5.48}$$

of a filtered version $f(t)$ of the input $u(t)$

$$f(t) = T^{-1}Su(t)$$

is sought. The quadratic estimation error (5.9) is to be minimised with dynamic weighting $\mathcal{W} = U^{-1}V$. This corresponds to the choice $\mathcal{G} = A^{-1}B$, $\mathcal{G}_a = 0$ ($\Rightarrow \mathcal{R}_a = 0$), $\mathcal{F} = D^{-1}C$, $\mathcal{H} = N^{-1}M$, $\mathcal{D} = T^{-1}S$, $\mathcal{W} = U^{-1}V$, and $\mathcal{R}_z = [\mathcal{R}\ 0]$ in (5.7)–(5.10). See Figure 5.5. The filter $T^{-1}S$, with T and S of dimensions $\ell|\ell$ and $\ell|s$, may represent additional dynamics in the problem description (cf [19],[20]), a frequency shaping weighting filter (cf [2]), or a selection of certain states.

When $\{u(t)\}$ represents a sequence of transmitted symbols in a communication network, (5.48) represents a *linear equaliser*. Its output is then fed into a decision device in order to recover the the transmitted symbols. See e.g. [26], [57].

Introduce the following assumptions.

Assumption 1: The polynomial matrices $A(q^{-1})$, $N(q^{-1})$, $D(q^{-1})$, $T(q^{-1})$, $U(q^{-1})$ and $V(q^{-1})$ all have stable determinants and non–singular leading coefficient matrices. (Thus, they have stable and causal inverses.)

Assumption 2: The spectral density of $y(t)$, $\Phi_y(e^{j\omega})$, is nonsingular for all ω.

The spectral density matrix, Φ_y, will be expressed using a *polynomial* matrix spectral factorisation. (It is preferable to avoid the numerically difficult task of performing spectral factorisation of rational matrices, and instead use factorisation of polynomial matrices. For this, there exist efficient numerical algorithms [35], [43].) In

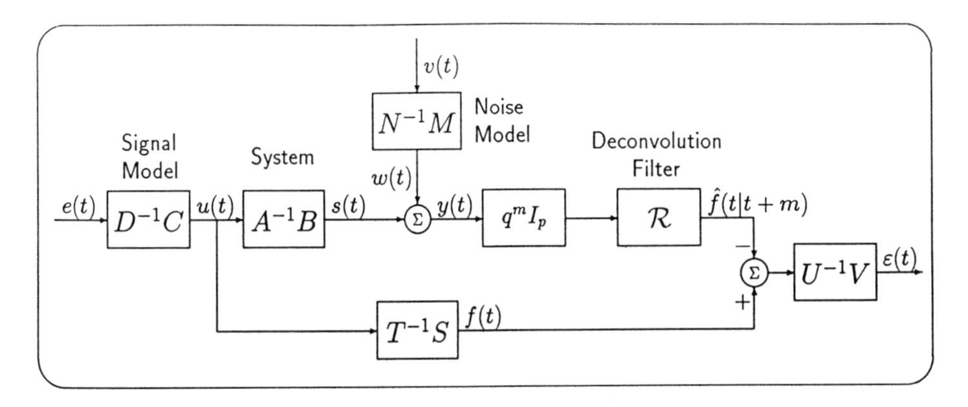

Figure 5.5: A generalised multi–signal deconvolution problem. The vector sequence $\{f(t)\}$ is to be estimated from the measurements $\{y(t)\}$, up to time $t+m$.

order to achieve this, two *coprime factorisations* have to be introduced:

$$\begin{aligned} \tilde{D}^{-1}\tilde{B} &= BD^{-1} \\ \tilde{N}^{-1}\tilde{A} &= \tilde{D}AN^{-1} \ . \end{aligned} \tag{5.49}$$

Here, the polynomial matrices $\tilde{D}$, $\tilde{N}$ and $\tilde{A}$ have dimension $p|p$, while $\tilde{B}$ has dimension $p|s$. The factorisations constitute the calculation of irreducible left MFD's ($\tilde{D}, \tilde{B}$ and $\tilde{N}, \tilde{A}$ are left coprime) from right MFD's. Thus, no unstable common factors are introduced. Since D and N are stable, $\tilde{D}$ and $\tilde{N}$ will be stable. Using (5.49), inverse matrices in the expression for Φ_y can be factored out to the left and right, leaving a polynomial matrix in the middle. We obtain

$$\Phi_y \;=\; A^{-1}BD^{-1}C\phi C_*D_*^{-1}B_*A_*^{-1} + N^{-1}M\psi M_*N_*^{-1} \;=\; \alpha^{-1}\beta\beta_*\alpha_*^{-1} \tag{5.50}$$

where

$$\beta\beta_* = \tilde{N}\tilde{B}C\phi C_*\tilde{B}_*\tilde{N}_* + \tilde{A}M\psi M_*\tilde{A}_* \tag{5.51}$$

and

$$\alpha \triangleq \tilde{N}\tilde{D}A \ .$$

Under Assumption 2, a stable $p|p$ spectral factor β, with $\det\beta(z^{-1}) \neq 0$ in $|z| \geq 1$ and nonsingular leading matrix β_0, can always be found.

Now, the optimal estimator can be derived as outlined in subsection 5.3.1. Let $\varepsilon(t) = U^{-1}V(f(t) - \hat{f}(t|t+m))$ be the filtered error and $\nu(t) = \mathcal{T}(q^{-1})y(t+m)$ the variation. Since $e(t)$ and $v(t)$ are assumed uncorrelated, we obtain

$$\begin{aligned} E\varepsilon(t)\nu^*(t) = EU^{-1}V[(T^{-1}S - q^m\mathcal{R}A^{-1}B)D^{-1}Ce(t) - q^m\mathcal{R}N^{-1}Mv(t)] \\ [\mathcal{T}q^m(A^{-1}BD^{-1}Ce(t) + N^{-1}Mv(t))]^* \end{aligned}$$

$$= \frac{1}{2\pi j} \oint_{|z|=1} U^{-1} V \{ z^{-m} T^{-1} S D^{-1} C \phi C_* D_*^{-1} B_* A_*^{-1}$$

$$- \mathcal{R}[A^{-1} B D^{-1} C \phi C_* D_*^{-1} B_* A_*^{-1} + N^{-1} M \psi M_* N_*^{-1}] \} \mathcal{T}_* \frac{dz}{z} \ . \qquad (5.52)$$

The use of (5.49) and (5.50) in (5.52) gives, with $\alpha_*^{-1} = \tilde{N}_*^{-1} \tilde{D}_*^{-1} A_*^{-1}$,

$$E\varepsilon(t)\nu^*(t) = \frac{1}{2\pi j} \oint_{|z|=1} U^{-1} \{ z^{-m} V T^{-1} S D^{-1} C \phi C_* \tilde{B}_* \tilde{N}_* - V \mathcal{R} \alpha^{-1} \beta \beta_* \} \alpha_*^{-1} \mathcal{T}_* \frac{dz}{z} \ . \qquad (5.53)$$

Since A, $\tilde{D}$ and $\tilde{N}$ are stable, all elements of $\alpha^{-1} = A^{-1} \tilde{D}^{-1} \tilde{N}^{-1}$ have poles only in $|z| < 1$. Elements of β may contribute poles at the origin, since they are polynomials in z^{-1}. These factors to the right of $\mathcal{R}$ can be cancelled directly by $\mathcal{R}$. Moreover, introduce the additional coprime left MFD

$$\tilde{T}^{-1} \tilde{S} = V T^{-1} S D^{-1} \qquad (5.54)$$

with a stable $\tilde{T}$ of dimension $\ell|\ell$ and $\tilde{S}$ of dimension $\ell|s$. Apply it in the first term of the integrand of (5.53). If $\mathcal{R}$ contains $V^{-1}\tilde{T}^{-1}$ as a left factor, $\tilde{T}^{-1}$ can then be factored out to the left. Thus,

$$\mathcal{R} = V^{-1} \tilde{T}^{-1} Q_1 \beta^{-1} \alpha \qquad (5.55)$$

where $Q_1(z^{-1})$, of dimension $\ell|p$, is undetermined. With (5.55) inserted, (5.53) becomes

$$E\varepsilon(t)\nu^*(t) = \frac{1}{2\pi j} \oint_{|z|=1} U^{-1} \tilde{T}^{-1} \{ z^{-m} \tilde{S} C \phi C_* \tilde{B}_* \tilde{N}_* - Q_1 \beta_* \} \alpha_*^{-1} \mathcal{T}_* \frac{dz}{z} \ .$$

All poles, of every element, of $\alpha_*^{-1} \mathcal{T}_*$ are located outside $|z| = 1$, since α is stable and $\mathcal{T}$ is causal and stable. In order to fulfill (5.14) collectively, we require

$$z^{-m} \tilde{S} C \phi C_* \tilde{B}_* \tilde{N}_* = Q_1 \beta_* + z \tilde{T} U L_* \ . \qquad (5.56)$$

This is a linear polynomial matrix equation, a bilateral Diophantine equation[11]. Here, $Q_1(z^{-1})$ and $L_*(z)$ are polynomial matrices, of dimension $\ell|p$, with degrees

$$nQ_1 \leq \max(nc + n\tilde{s} + m, n\tilde{t} + nu - 1) \ ; \ nL \leq \max(nc + n\tilde{b} + n\tilde{n} - m, n\beta) - 1 \ . \qquad (5.57)$$

[11]A general rule is that the stable inverses which can not be cancelled by $\mathcal{R}$ directly, must be factored out to the left. When cancelled later, this will define a Diophantine equation, such as (5.56). (Note that if $\tilde{S}C$ had been factored out as well, (5.56) could not have become a *polynomial* matrix Diophantine equation.)

With β and $\tilde{T}U$ stable, $\det \beta_*$ and $\det \tilde{T}U$ will have no common factors. Thus, a unique solution to (5.56) exists. See [4] or the reasoning in Section 3.3 [59].

The design equations thus consist of the coprime factorisations (5.49), (5.54), the left spectral factorisation (5.51), the Diophantine equation (5.56) and the filter expression (5.55). The derivation above constitutes a slight generalisation of the derivation in [4], to the case of frequency dependent weighting $\mathcal{W} = U^{-1}V \neq I_\ell$. For scalar systems, the solution reduces to the one presented in [2], [20]. An alternative derivation, based on the inner–outer approach, can be found in [5], which is a comment on [15].

The minimal criterion value is obtained by inserting (5.55), (5.50), (5.54), and (5.56), in this order, into the criterion J in (5.9). When $\mathcal{W} = I_\ell$, we obtain, with $H \triangleq \tilde{N}\tilde{B}C$,

$$\frac{1}{2\pi j}\oint \operatorname{tr}\{L_*\beta_*^{-1}\beta^{-1}L + T^{-1}SD^{-1}C(\phi - \phi H_*\beta_*^{-1}\beta^{-1}H\phi)C_*D_*^{-1}S_*T_*^{-1}\}\frac{dz}{z} \quad . \tag{5.58}$$

The minimal criterion value consists of two terms. The first term involves the sometimes so called "dummy"-polynomial L_*. In deconvolution problems, it can be given a nice interpretation: it represents the unavoidable error caused by incomplete inversion of the system $A^{-1}B$. Only the use of an infinite smoothing lag can make the first term vanish, unless the system is minimum phase and there is no noise. One can show that $L \to 0$ when $m \to \infty$ [19].

The rule in the derivation technique is to cancel what can be cancelled directly, by means of $\mathcal{R}$. The rest of the terms contributing poles in $|z| < 1$ must be factored out to the left, to be taken care of by L_* and Q_1. It is instructive to note how the Diophantine equation interacts with the cross–term (5.53). It has to absorb contributing parts of the integrand which cannot be cancelled directly by $\mathcal{R}$. L_* represents the remainder. There exists a very special case in which perfect input estimation is possible. It is the case of minimum–phase systems without noise, with $q^m B$ square and stably and causally invertible. Consider this situation and let $S = T = I_\ell$. Then, $\mathcal{R} = D^{-1}\tilde{B}^{-1}\tilde{D}Aq^{-m}$ makes the integrand of (5.53) zero directly. Consequently, there is nothing left for L_* to take care of, so L_* must be zero. By utilising (5.49) we obtain $\mathcal{R} = q^{-m}B^{-1}A$, that is, the inverse system.

For scalar systems, the deconvolution problem has also been studied in an adaptive setting, see [3]. Multivariable adaptive deconvolution, for the special case of white input and noise, has been discussed in [22] and [47]. Crucial for an adaptive algorithm to work, is that the model polynomials can be estimated from the output only. In [1], the identifiability properties of the scalar deconvolution problem are investigated and conditions for parameter identifiability are given. If similar conditions exist for the multivariable problem, is still an open question.

The considered deconvolution problem turns out to be dual to the *feedforward* control problem (with rational weights) discussed in Chapter 3, Section 3.3, of this volume [59]. See [10]. It is very simple to demonstrate this duality. Reverse all arrows, interchange summation points and node points and transpose all rational matrices. Then, the block diagram for the other problem is obtained. The transposition explains why the system is described by *left* MFD's in the filtering problems, while *right* MFD's are used in the control problem.

Note that the problem set–up contains the *general filtering problem* described by (5.4), (5.5) as the special case $v(t) = 0$. See Figure 5.1. The solution derived here thus solves *all* problems discussed in this chapter. (By duality, it also solves all $\mathcal{H}_2$ feedforward control problems.) However, it does not provide the same degree of expliciteness as do the solutions in Section 5.3 and Sections 5.5–5.6. One can simply not "see through" all the generality[12]. One of our convictions is that *structure gives insight.* Therefore, we have, in each specific problem, abandoned generality for structure, in order to gain insight, and also to simplify the solution.

5.5 Differentiation and State Estimation

The problem of estimating derivatives from measured data can be treated as an application of state estimation. It is an important engineering problem, which has been extensively studied over the years, see for example, [7], [18], [25], [53], [62], and the references therein. In radar applications, velocity estimation from position data is of interest [24]. Other applications are, for example, estimation of heating rates from temperature data or net flow rates in tanks from level data. The estimation of derivatives is challenging because of its sensitivity to measurement noise. We will here describe a design method developed by our collegue Bengt Carlsson. For more details, see [18] and [19].

Since the derivative is a continuous–time concept, it is appropriate to base the discrete–time filter design on a continuous–time problem formulation. Let a continuous-time scalar signal $s_c(t_c)$ be characterised as a linear stochastic process

$$s(t_c) = G(p)e_c(t_c) \tag{5.59}$$

where $e_c(t_c)$ is zero mean white noise, with spectral density $\lambda_c/2\pi$. The argument t_c denotes continuous time, and $G(p)$ is a rational function of the derivative operator $p \triangleq d/dt$,

$$G(p) = \frac{b_o p^{\delta-n-1} + b_1 p^{\delta-n-2} + \ldots + b_{\delta-n-1}}{p^\delta + a_1 p^{\delta-1} + \ldots + a_\delta} \quad . \tag{5.60}$$

The transfer function has order $\delta \geq n+1$ and pole excess (relative degree) $\geq n+1$. Here, we think of the expression (5.59) as a model describing the spectral properties of the signal. We assume λ_c and $G(p)$ to be time–invariant. The signal $s(t_c)$ is sampled with sampling period h. The objective is to seek the n'th order derivative of the signal $s(t_c)$

$$f(t_c) \triangleq \frac{d^n s(t_c)}{dt_c^n} = p^n G(p) e_c(t_c)$$

[12]For example, the solution to the decision feedback equalisation problem does not involve any (polynomial) spectral factorization. To see this from the general solution would be very hard.

$$= \frac{b_o p^{\delta-1} + b_1 p^{\delta-2} + \ldots + b_{\delta-n-1} p^n}{p^\delta + a_1 p^{\delta-1} + \ldots + a_\delta} e_c(t_c) \tag{5.61}$$

at the time instants $t_c = th$; $t = 0, 1, \ldots$.

Let us outline a solution, which is derived and discussed in more detail in [19]. The stochastic model (5.59)–(5.61) can be represented in state space form, denoting the state vector $u(t)$, as

$$\begin{aligned} du(t_c) &= Au(t_c)dt + BdW(t_c) \\ s(t_c) &= H_1 u(t_c) \\ f(t_c) &= H_2 u(t_c) \ . \end{aligned} \tag{5.62}$$

Here, $dW(t_c) = e_c(t_c)dt$ represents Wiener increments and H_1, H_2 are vectors. The internal structure of the matrices in (5.62) depends on how (5.60) and (5.61) is represented. See [19].

Stochastic sampling (see e.g. [8]) of (5.62), results in the discrete-time representation

$$\begin{aligned} u(t+1) &= Fu(t) + e_v(t) \\ s(t) &= H_1 u(t) \\ f(t) &\triangleq \left. \frac{d^n s(t_c)}{dt_c^n} \right|_{t_c = th} = H_2 u(t) \end{aligned} \tag{5.63}$$

where $F = e^{Ah}$. Note that $f(t)$ is *exactly* the derivative at the sampling instants th. We assume the system to have poles in $|z| \leq 1$ and the pair (F, H_1) to be detectable. (Possible unobservable modes must be stable.) The column vector $e_v(t)$ consists of discrete-time stationary white noise elements with zero mean. Their covariance matrix equals

$$Ee_v(t)e_v(t)^T \triangleq \lambda_c R_e = \lambda_c \int_0^h e^{A\tau} BB^T e^{A^T \tau} d\tau \ . \tag{5.64}$$

Note that while the continuous-time noise process $e_c(t_c)$ is scalar, $e_v(t)$ will be a vector of dimension $s = \dim A$. In general, R_e has full rank. The effect of all components of $e_v(t)$ on $f(t) = H_2 u(t)$ can *not*, in general, be calculated exactly from their effect on $s(t) = H_1 u(t)$, *unless* the covariance matrix R_e has rank 1. (When the sampling frequency increases, R_e approaches a rank 1-matrix.)

Measurements of the signal $s(t)$ are assumed to be corrupted by a discrete-time noise $w(t)$, described below by a discrete–time ARMA model

$$y(t) = s(t) + w(t) \ . \tag{5.65}$$

In order to fit this problem into the parametrisation (5.7), we will convert the state space model (5.63) into a transfer-function based model. For this reason, introduce the characteristic polynomial $D(q^{-1})$, of degree nd equal to the number of states s, and the polynomial matrix $\mathbf{C}(q^{-1})$ as

$$D(q^{-1}) \triangleq \det(I - q^{-1}F) \quad ; \quad \mathbf{C}(q^{-1}) \triangleq \operatorname{adj}(I - q^{-1}F)q^{-1} \ . \tag{5.66}$$

Note that $\mathbf{C}$ has dimension $s|s$. Also, note that we use a bold face $\mathbf{C}$ here, to distinguish between a polynomial matrix, a polynomial, such as e.g. $D(q^{-1})$, and constant matrices like H_1 and R_e. Hence, the sampled system can be expressed as

$$\begin{aligned} u(t) &= \frac{\mathbf{C}(q^{-1})}{D(q^{-1})} e_v(t) \qquad\qquad Ee_v(t)e_v(t)^T = \lambda_c R_e \\ w(t) &= \frac{M(q^{-1})}{N(q^{-1})} v(t) \qquad\qquad\qquad Ev(t)^2 = \lambda_v \ . \\ y(t) &= H_1 u(t) + w(t) \quad ; \quad f(t) = H_2 u(t) \end{aligned} \tag{5.67}$$

Assume the parameters of the continuous-time model (5.59)–(5.61), and those of the noise description, to be known a priori or correctly estimated in some way. The discrete-time model (5.68) is then obtained by stochastic sampling. We seek the stable time-invariant linear estimator of the n'th derivative

$$\hat{f}(t|t+m) = \frac{Q^c(q^{-1})}{R^c(q^{-1})} y(t+m) \tag{5.68}$$

which minimises the mean square estimation error $\varepsilon(t) = f(t) - \hat{f}(t|t+m)$. This corresponds to the choices $\mathcal{G} = H_1$, $\mathcal{G}_a = 0$ ($\Rightarrow \mathcal{R}_a = 0$), $\mathcal{F} = (1/D)\mathbf{C}$, $\mathcal{H} = M/N$, $\mathcal{D} = H_2$, $\mathcal{W} = 1$ and $\mathcal{R}_z = [Q^c/R^c \ \ 0]$ in (5.7)–(5.10). See Figure 5.6.

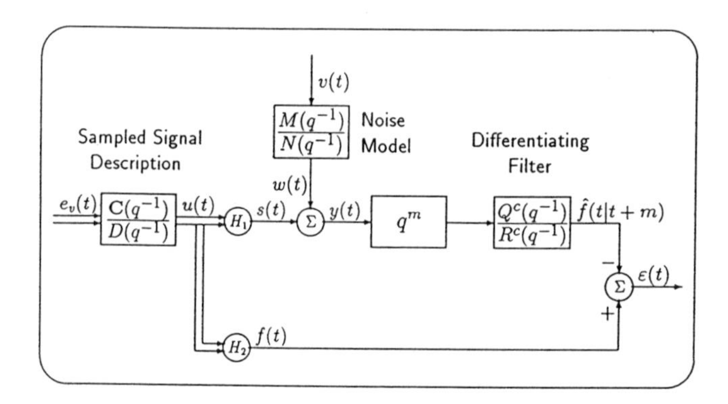

Figure 5.6: A state estimation problem, which here represents a differentiation problem based on a continuous time model. The state $f(t)$ is the derivative of $s(t)$. It is to be estimated from measurements $y(t+m)$.

Introduce the following polynomials, obtained from the model (5.68)

$$P_{ij} = p_{nc}^{ij} q^{nc} + \ldots p_o^{ij} + \ldots p_{-nc}^{ij} q^{-nc} \triangleq H_i \mathbf{C}(q^{-1}) R_e \mathbf{C}_*(q) H_j^T \ , \ i,j = 1,2 \tag{5.69}$$

Also, with $\eta \triangleq \lambda_v / \lambda_c$, introduce the polynomial spectral factorisation

$$\tau \beta \beta_* = P_{11} N N_* + \eta D D_* M M_* \tag{5.70}$$

defining the stable and monic spectral factor $\beta(q^{-1}) = 1 + \beta_1 q^{-1} + \ldots + \beta_{n\beta} q^{-n\beta}$ of degree $n\beta = \max\{nc + nn, nd + nm\}$ and a scalar τ. As mentioned before, a stable spectral factor β exists, if and only if the two terms on the right hand side of (5.70) have no common factors with zeros on the unit circle. (If $\eta = 0$, the first term should have no zeros on the unit circle.)

The polynomials P_{ij} and β have specific interpretations. Note, from (5.68) and (5.69), that for stationary signals (stable D and N), the spectral densities of $\{s(t)\}$ and $\{f(t)\}$ are given by

$$\phi_s(\omega) = \frac{\lambda_c}{2\pi} \frac{P_{11}}{DD_*} \qquad \phi_f(\omega) = \frac{\lambda_c}{2\pi} \frac{P_{22}}{DD_*} \qquad \phi_{fs}(\omega) = \frac{\lambda_c}{2\pi} \frac{P_{21}}{DD_*} \tag{5.71}$$

where $e^{-i\omega T}$ and $e^{i\omega T}$ have been substituted for q^{-1} and q in all polynomials.

If D and N are stable, the spectral density of the measurement sequence $\{y(t)\}$ is given by

$$\begin{aligned} \phi_y(\omega) &= \phi_s(\omega) + \phi_w(\omega) = \frac{\lambda_c}{2\pi} \frac{P_{11}}{DD_*} + \frac{\lambda_v}{2\pi} \frac{MM_*}{NN_*} \\ &= \frac{\lambda_c}{2\pi} \frac{\tau \beta \beta_*}{DD_* NN_*} . \end{aligned} \tag{5.72}$$

As usual, the spectral factor β thus represents the numerator of an innovations model.

Theorem 5.5.1

Consider the sampled signal model described by (5.68), with all zeros of D and N in $|z| \leq 1$. Assume that a stable spectral factor β, defined by (5.70), exists. A stable linear estimator (5.68) of the derivative then attains the minimum of the criterium J in (5.9) if and only if it has the same coprime factors as

$$\frac{Q^c}{R^c} = \frac{Q_1^c N}{\beta} . \tag{5.73}$$

Here $Q_1^c(q^{-1})$, together with a polynomial $L_*^c(q)$, is the unique solution to the linear polynomial equation

$$q^{-m} P_{21} N_* = \tau \beta_* Q_1^c + q D L_*^c \tag{5.74}$$

with polynomial degrees

$$\begin{aligned} nQ_1^c &= \max\{nc + m, nd - 1\} \\ nL^c &= \max\{nc + nn - m, n\beta\} - 1 \ . \end{aligned} \tag{5.75}$$

The minimal variance of the estimation error is given by

$$E\varepsilon(t)^2_{\min} = \frac{\lambda_c}{2\pi i} \oint_{|z|=1} \{ \underbrace{\frac{L^c L_*^c}{\tau\beta\beta_*}}_{I} + \eta \underbrace{\frac{MM_* P_{22}}{\tau\beta\beta_*}}_{II} + \underbrace{\frac{NN_*[P_{11}P_{22} - P_{12}P_{21}]}{\tau\beta\beta_* DD_*}}_{III} \} \frac{dz}{z} \tag{5.76}$$

□

Proof: See [19], where the optimality is verified using a non–constructive variant of the variational approach.

This solution considers estimation of one state variable only: the derivative of order n. It is straightforward to estimate several state variables, or even all of them, with different smoothing lags for each one. If $f(t)$ is a vector, the estimation of component i of $f(t)$ does not affect the estimation of component j. The total estimator of $f(t)$ can then be obtained as ℓ parallel scalar estimators. We thus obtain a set of independent Diophantine equations of the type (5.74), one for each estimated state variable. The scalar spectral factorisation remains unaltered.

This way of expressing a state estimator can be seen as an alternative way of computing a stationary Kalman filter/predictor/smoother, for systems with scalar measurements $y(t)$. For systems with multiple measurements, a matrix spectral factorisation would be required. It is then doubtful if a polynomial solution offers any computational advantage.

If the characteristic polynomial D is marginally stable, both $f(t)$ and the estimate $\hat{f}(t)$ will, in general, be nonstationary sequences. The estimation error $\varepsilon(t) = f(t) - \hat{f}(t)$ will, however, be a stationary zero mean sequence, with a finite minimal variance given by (5.76). (This implies that marginally stable factors of D, in the denominator of term III in (5.76) are cancelled by numerator factors.)

The three terms in (5.76) can be interpreted as follows.

Term I represents the effect of a *finite smoothing lag* m. As is shown in [19], $L^c \to 0$ when $m \to \infty$. Term 1 then vanishes.

Term II depends on the noise $w(t)$. It represents the unavoidable performance degradation due to noise, which cannot be eliminated, even with an arbitrarily large smoothing lag m. The term vanishes in the noise-free case ($\eta = 0$).

Term III remains even when $m \to \infty$ and $\eta = 0$. It represents aliasing effects. Asymptotically, when $h \to 0$ and the covariance matrix $\lambda_c R_e$ (defined by (5.64)) approaches a rank 1-matrix, the term vanishes. See [19].

The differentiation problem can also be posed in a discrete time setting, without assuming an explicit underlying continuous–time model. The problem becomes a scalar variant of the general problem of Section 5.4, with $u(t)$ being the signal of interest and $\mathcal{D} = T^{-1}S$ representing a discrete–time approximation of the derivative operator $(i\omega)^n$. Based on such an approximation, which can be designed by well–known means [18], [53], the polynomial approach provides an estimator which optimally takes noise and transducer dynamics into account. See [19], [20]. The use of a discrete–time approximation of the derivative operator mostly results in an additional perfomance loss, compared to the formulation outlined above. (However, if the continuous time system is known, this loss can be eliminated by using an optimal derivative approximation. See [19].)

5.6 Decision Feedback Equalisation

We finally turn our interest to an important problem in digital communications, and present a polynomial solution derived in [57]. When digital data are transmitted over a communication channel, intersymbol interference and noise prevent a receiver from always detecting the symbols correctly. Consider a received sampled data sequence $y(t)$. It is described as a sum of channel output $s(t)$ and noise $w(t)$ by the following linear stochastic discrete time model

$$y(t) \quad = \quad s(t) + w(t) \quad = \quad q^{-k}\frac{B(q^{-1})}{A(q^{-1})}u(t) + \frac{M(q^{-1})}{N(q^{-1})}v(t) \qquad (5.77)$$

The first right–hand term of (5.77) represents a dispersive linear communication channel, with $\{u(t)\}$ being the transmitted data sequence. The channel model includes pulse shaping, receiver filter and a transmission delay of k samples. Baseband operation on a complex channel is assumed. The second term describes a coloured noise, where the colour may be caused, for example, by effects of receiver filters or leakage from other channels. The sequence $\{v(t)\}$ is a discrete-time white noise. It is zero mean, stationary and independent of $u(t)$.

The polynomials in (5.77), having degrees δa, δb etc, are assumed known a priori or correctly estimated. Except for the $B(q^{-1})$-polynomial, which has an arbitrary nonzero leading coefficient b_o, all polynomials are monic. It is realistic to assume $A(q^{-1})$ and $M(q^{-1})$ to be stable polynomials, while $B(q^{-1})$ can have zeros anywhere and $N(q^{-1})$ may have zeros in $|z| \leq 1$.

The sequence $\{u(t)\}$ is here assumed to be white. It may be real or complex. One example is the use of p-ary symmetric Pulse Amplitude Modulated (PAM) signals. Then, $u(t)$ is a real, white, zero mean sequence which attains values $\{-p+1, \ldots, -1, +1, \ldots, p-1\}$ with some probability distribution. In other modulation schemes, such as Quadrature Amplitude Modulation (QAM), the model coefficients

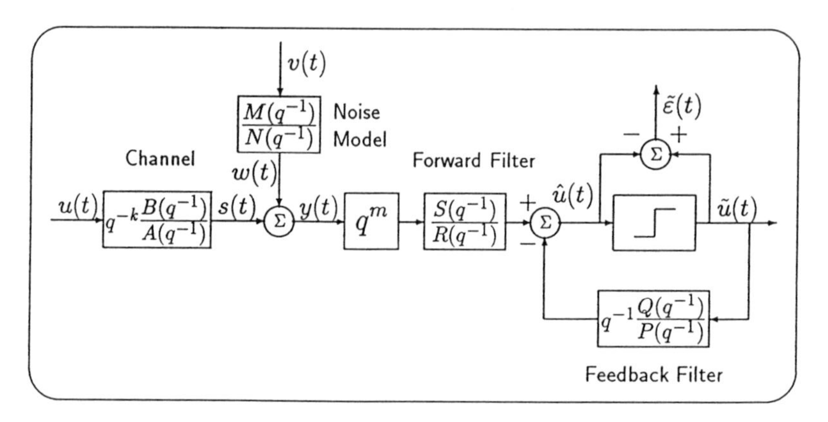

Figure 5.7: The Decision feedback equalisation problem. The estimate $\hat{u}(t|t+m)$ is obtained by subtracting decisioned data $\tilde{u}(t)$, fed back through a filter. The estimate $\hat{u}(t|t+m)$ is fed into a decision device to recover the transmitted sequence.

and signals in (5.77) are complex-valued. For the source coding scheme of interest, define

$$\lambda_u \triangleq E|u(t)|^2 \quad ; \quad \rho \triangleq E|v(t)|^2/E|u(t)|^2 \ . \tag{5.78}$$

The data sequence $u(t)$ is to be reconstructed from measurements of $y(t)$. As has been mentioned in Section 5.4, this can be accomplished by a linear equaliser. Superior performance is, however, achieved with a Decision Feedback Equaliser (DFE) for moderate and high signal to noise ratios. A DFE is a nonlinear filter, which involves a decision circuit. Decisioned data are fed back through a linear filter to improve the estimate. See e.g. [11], [23], [50], [52], [55], and the references therein. The bit error rate of a DFE is in many cases several orders of magnitude lower than for a linear equaliser.

Previously available design principles for the linear filters of a DFE have either provided optimal filters that are not realisable, or realisable filters with a suboptimal transversal (FIR) structure. Here, we will introduce a *general IIR decision feedback equaliser* (GDFE), see Figure 5.7,

$$\hat{u}(t|t+m) = \underbrace{\frac{S(q^{-1})}{R(q^{-1})}}_{\text{forward filter}} y(t+m) - \underbrace{q^{-1}\frac{Q(q^{-1})}{P(q^{-1})}}_{\text{feedback filter}} \tilde{u}(t) \tag{5.79}$$

Above, m is the number of lags (smoothing lag) and $\tilde{u}(t)$ is decisioned data, for example $\text{sign}(\hat{u}(t))$, when PAM is used with $p = 2$. The denominator polynomials $R(q^{-1})$ and $P(q^{-1})$ are assumed to be monic, and required to be stable. The sampling rate is assumed to equal the symbol rate. If $u(t)$ is complex-valued, the coefficients of the filters must be complex.

Given a received sequence $y(t)$, a model (5.77), (5.78) and a smoothing lag $m \geq k$, the problem is to find polynomials (S, R, Q, P) which minimise the MSE criterion

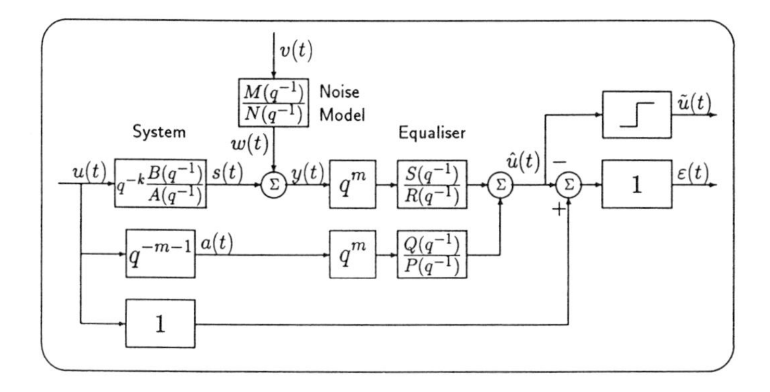

Figure 5.8: The decision feedback equalisation problem, when correct past decisions are assumed. The signal $u(t)$ is to be estimated from measurements $z(t+m)$.

$E|\varepsilon(t)|^2 = E|u(t) - \hat{u}(t|t+m)|^2$.[13] Because of the presence of a nonlinear decision circuit, it is impossible to obtain closed-form expressions for an optimal estimator. As in most previous treatments of the DFE–problem, we will simplify the problem by assuming *correct past decisions.*

If previous decisions are correct, they can be used to completely eliminate the interference, caused by past symbols, at the current received signal. In contrast to linear equalisers, this can be achieved without any noise amplification. This is more easily understood if Figure 5.7 is redrawn as in Figure 5.8. No inversion has to be done. Instead, a feedforward from $u(t-1)$ is used. The nonlinearity is now outside the signal path from $u(t)$ and $v(t)$ to $\varepsilon(t)$. Thus, by assuming correct past decisions we can transform the problem into a LQ optimisation problem[14].

This problem formulation corresponds to the choices $\mathcal{G} = q^{-k}B/A$, $\mathcal{G}_a = q^{-m-1}$, $\mathcal{F} = \mathcal{D} = 1$, $\mathcal{H} = M/N$, $\mathcal{W} = 1$, and $\mathcal{R}_z = [S/R \;\; Q/P]$ in (5.7)-(5.10).

Introduce the following polynomials:

$$\begin{aligned} \tau(q^{-1}) &\triangleq BN = \tau_o + \tau_1 q^{-1} + \ldots + \tau_{\delta\tau} q^{-\delta\tau} \\ \gamma(q^{-1}) &\triangleq AM = 1 + \gamma_1 q^{-1} + \ldots + \gamma_{\delta\gamma} q^{-\delta\gamma} \\ \alpha(q^{-1}) &\triangleq \gamma + q^{-1}Q = 1 + \alpha_1 q^{-1} + \ldots + \alpha_{\delta\alpha} q^{-\delta\alpha} \; . \end{aligned} \quad (5.80)$$

We are now able to state the following result.

[13]It could be argued that a more relevant criterion is minimum probability of decision errors (MPE), which leads to a nonlinear optimisation problem. However, Monsen [50] has concluded that consideration of MPE and MSE lead to essentially the same error probability.

[14]For low signal to noise ratios, the assumption of correct past decisions is not appropriate. Because of the high noise level, incorrect decisions will occur. They may even start a burst of errors. This phenomenon is known as "*error propagation*". If too many error bursts occur, they will deteriorate the performance considerably and could in fact make the equaliser useless. For a discussion, see [23] and [57].

Theorem 5.6.1

Assume the received data to be accurately described by (5.77),(5.78). The general DFE (5.79) then attains the global minimum of $J = E|u(t) - \hat{u}(t|t+m)|^2$, if and only if the filters S/R and Q/P have the same coprime factors as

$$\frac{S}{R} = \frac{S_1 N}{M} \qquad \frac{Q}{P} = \frac{Q}{AM} \tag{5.81}$$

where $S_1(q^{-1})$ and $Q(q^{-1})$, together with polynomials $L_{1*}(q)$ and $L_{2*}(q)$, satisfy the two coupled polynomial equations

$$q^{m-k}\tau S_1 + \gamma L_{1*} = \alpha \tag{5.82}$$
$$-\rho\gamma_* S_1 + q^{-m+k}\tau_* L_{1*} = qL_{2*} \tag{5.83}$$

with polynomial degrees

$$\begin{aligned} \delta S_1 &= \delta L_1 = m-k \\ \delta Q &= \delta L_2 = \max(\delta\gamma, \delta\tau) - 1 \ . \end{aligned} \tag{5.84}$$

The minimal mean square estimation error is

$$E|\varepsilon(t)|^2_{\min} = \frac{\lambda_u}{2\pi j}\oint_{|z|=1} L_1 L_{1*} + \rho S_1 S_{1*}\frac{dz}{z} = \lambda_u\left(\sum_{j=0}^{m-k} |\ell_j|^2 + \rho|s_j|^2\right) \tag{5.85}$$

□

Proof: See [57], where optimality is verified using a non–constructive variant of the variational approach. A variation $\nu(t) = \mathcal{T}_1 y(t+m) + \mathcal{T}_2 u(t-1)$ with two terms is introduced. Orthogonality with respect to each of these terms is verified. The Diophantine equations arise from the two orthogonality requirements.

Remark: Note that (5.82) and (5.83) represent two coupled polynomial equations, containing *four* unknown polynomials ($S_1(q^{-1})$, $Q(q^{-1})$, $L_{1*}(q)$, $L_{2*}(q)$). Also, note that *no spectral factorisation is required.* The solution obtained here is, of course, a special case of the one obtained in Section 5.4. It is, however, difficult to derive it from that solution. Instead, by formulating a scalar problem which utilises both $y(t)$ and $a(t)$ in Figure 5.2, the solution is readily obtained.

An explicit solution to (5.82) and (5.83) is given by the following result.

Theorem 5.6.2

The polynomials S_1, $\overline{L}_1$ and Q, calculated in the following way, provide the unique solution to the polynomial equations (5.82) and (5.83).

1. Solve for the coefficients of the polynomials $\dot{S}_1(q^{-1})$ and $\overline{L}_1(q^{-1})$ in

$$\left[\begin{array}{ccc|ccc} \tau_o & & 0 & 1 & & 0 \\ \vdots & \ddots & & \gamma_1 & \ddots & \\ \tau_{m-k} & \cdots & \tau_o & \gamma_{m-k} & \cdots\gamma_1 & 1 \\ & & & & & \\ \rho & \rho\gamma_1^* \cdots & \rho\gamma_{m-k}^* & -\tau_0^* & \cdots & -\tau_{m-k}^* \\ & \ddots & \rho\gamma_1^* & & \ddots & \vdots \\ 0 & & \rho & 0 & & -\tau_0^* \end{array}\right] \left[\begin{array}{c} s_o \\ \vdots \\ s_{m-k} \\ \\ l_{m-k}^* \\ \vdots \\ l_0^* \end{array}\right] = \left[\begin{array}{c} 0 \\ \vdots \\ 1 \\ \\ 0 \\ \vdots \\ 0 \end{array}\right] \tag{5.86}$$

2. With $\{s_i\}$ and $\{\ell_j^*\}$ obtained from step 1, perform the multiplication

$$\left[\begin{array}{ccc|ccc} \tau_o & & 0 & 1 & & 0 \\ \vdots & \ddots & & \gamma_1 & \ddots & \\ \vdots & & \tau_o & \vdots & & 1 \\ \tau_{\delta\tau} & & \vdots & \gamma_{\delta\gamma} & & \gamma_1 \\ & \ddots & \vdots & & \ddots & \vdots \\ 0 & & \tau_{\delta\tau} & 0 & & \gamma_{\delta\gamma} \end{array}\right] \left[\begin{array}{c} s_o \\ \vdots \\ s_{m-k} \\ l_{m-k}^* \\ \vdots \\ l_o^* \end{array}\right] = \left[\begin{array}{c} 0 \\ \vdots \\ 0 \\ 1 \\ \alpha_1 \\ \vdots \\ \alpha_{\delta\alpha} \end{array}\right] \tag{5.87}$$

   yielding the coefficients of the polynomial $\alpha(q^{-1})$.

3. Finally, calculate the polynomial $Q(q^{-1})$ from (5.80)

$$Q(q^{-1}) = q(\alpha(q^{-1}) - \gamma(q^{-1})) \quad . \tag{5.88}$$

□

The equivalent equalised channel (from $u(t+m)$ to $\hat{u}(t|t+m)$) will be

$$C_{eq} = q^{-k}\frac{BNS_1}{AM} - q^{-m-1}\frac{Q}{AM} = q^{-m} - q^{-k}\overline{L}_1(q^{-1}) \tag{5.89}$$

Equation (5.86) is obtained in the following way. Write (5.82) and (5.83) in matrix form. Select all rows with *known* right hand sides and combine them into a new system of equations, in the coefficients of S_1 and L_{1*} only. For details, see Appendix B of [57]. (Observe that the polynomial α is defined monic in (5.80). With the leading coefficients of α fixed and nonzero, we avoid the trivial solution $S_1 = L_{1*} = 0$.) The matrix blocks in (5.86) are quadratic. If $\tau(q^{-1})$ or $\gamma(q^{-1})$ are of order $< m-k$, zeros are used to fill up the corners of the blocks. The second step represents calculation of α from equation (5.82), with known S_1 and $\overline{L}_1$: $\tau S_1 + \gamma\overline{L}_1 = q^{-m+k}\alpha$. By further substitutions, a linear system for determining S_1 of only half the size of (5.86) can be derived. See [57].

An important question is if a unique solution to (5.86) can always be found without any restrictions on, for example, the coprimeness of $\tau(q^{-1})$ and $\gamma(q^{-1})$.

Theorem 5.6.3

If ρ and the leading coefficient of B, b_o, are not both zero, then (5.86) will always have a unique solution, $(S_1, \overline{L}_1)$. □

Proof: See [57].

Remark: When both $|b_o|$ $(= |\tau_o|)$ and the noise variance ratio ρ are small, the system (5.86) may be badly conditioned.

Summing up, one can conclude that an equaliser can be calculated using (5.81) and (5.86)-(5.88) (Theorem 5.6.2). This procedure works under very general conditions (Theorem 5.6.3). The resulting equaliser is MSE-optimal (Theorem 5.6.1). The minimal criterion value is given by (5.85), assuming correct past decisions. The properties of the optimal DFE are emphasised in some more detail below:

1. It is efficient to whiten the noise. The forward filter S/R contains the inverse noise description in cascade with a transversal filter S_1 of order $m-k$. After noise inversion, we have to equalise a channel $q^{-k}\tau/\gamma = q^{-k}BN/AM$. Therefore, the polynomials S_1, Q and P are determined exclusively by the polynomials τ and γ, *not* by their separate factors A, B, M and N.

2. A conventional DFE-structure (transversal filters both in the forward and feedback loops in (5.79)) is optimal if and only if $M = 1$ and $A = 1$. In other words, the channel must be adequately described by a transversal filter, and the noise statistics by an autoregressive process.

3. Theorem 5.6.1 provides us with an optimal *filter structure* and optimal *polynomial degrees*. Hence, unnecessary overparametrisation is avoided. It also gives guidelines on how to choose filter degrees in a conventional DFE.

4. In the criterion (5.85), the second term $\rho S_1 S_{1*}$ represents noise transmission. The first term $L_1 L_{1*}$ is caused by residual intersymbol interference from the first $m-k$ taps of the equalised channel $(\lambda_u \sum_{j=1}^{m-k} |\ell_j|^2)$. It is also caused by the deviation of the reference tap (at time index $m-k$) from 1 $(\lambda_u |\ell_o|^2)$. See (5.89). We thus get a nice interpretation of one of the extra "dummy"-polynomials. As in all DFE's, the equalised channel impulse response beyond time index $m-k$ is cancelled completely by the feedback filter. See (5.89). Past symbols thus do not affect the present decision.

5. In the noise-free case $(\rho = 0)$, $L_{1*} = L_{2*} = 0$, see (5.83). For any $m \geq k$, this gives $|\varepsilon(t)|^2 = 0$, *even when B is unstable.* The reason for this remarkable property is that, instead of inverting the system, the estimator uses feedforward from $u(t-1)$. See Figure 5.8.

6. The denominator polynomials R and P are stable by construction, since A and M are stable. In adaptive algorithms, stability of the estimates $\hat{A}$ and $\hat{M}$, or of $\hat{R}$ and $\hat{P}$, would be required.

The use of the algorithm above in an adaptive equaliser for the American digital mobile radio standard is investigated in [45]. Combined with a novel and efficient channel estimator, it has achieved very good performance.

5.7 Concluding Remarks

Why and when should a filter designer use the polynomial approach? What advantages does it offer from an engineering point of view, compared to e.g. a state space approach [6] or Wiener design of FIR filters [34]? We shall in this section give some answers.

• Many properties of the resulting filter can be disclosed by inspection only. See, for example, the remarks to the solutions obtained for the problems in Sections 5.4–5.6. Such information is hard to obtain from a corresponding state space approach. The obtained filters can also be examined directly using classical concepts, such as frequency response, poles and zeros.

• The solution is often explicit, in terms of the model polynomials. (Note, for example, the presence of the noise model denominator N as numerator factor of the filters (5.20) and (5.73).) This not only helps a designer to gain engineering insight, but also to build in design requirements. An example is the suggestion in subsection 5.3.9 to use integrating signal models to avoid bias for non–zero mean signals. The minimal criterion value can often be interpreted in terms of effects of different design constraints. For example, in the differentiation problem of Section 5.5, performance of the estimator is limited by the effects of aliasing, noise and finite smoothing lag. A designer will not only be able to calculate the limits of performance, but also to understand them.

• If an incorrect filter structure, with insufficient degrees of freedom is assumed, a solution will not exist. In the polynomial derivation techniques, the warning signals for this are inconsistencies or degenerated polynomial degrees. A polynomial solution thus leads to the optimal structure and degrees of filters and their polynomials. In contrast to the Wiener design of FIR filters [34], unnecessary overparametrisation is avoided. This is of considerable importance, if the solution is to be used in an indirect adaptive algorithm.

• Fixed lag smoothing does not complicate the solution or decrease insight, nor do singular situations (where white noise is not present in all measurements), or the introduction of frequency–dependent weighting matrices in the criteria.

Of course, the approach has limitations as well as strengths. Compared to Kalman filtering, polynomial methods seem less well suited to some off–line problems such as fixed point smoothing or fixed interval smoothing [6].

The derivation of solutions to multi–signal estimation problems is achieved with almost the same number of algebraic steps as for scalar problems. Compare Section 5.4 to subsection 5.3.3. However, the design equations themselves become considerably more complex when matrix spectral factorisations and coprime factorisations are required. The structure of a Kalman estimator, and the numerical routines required for obtaining it, remain unchanged regardless of the dimensions of $y(t)$ and $f(t)$. In contrast, there is a considerable step in complexity between polynomial solutions to scalar problems and to multivariable problems.

It is well known that the zeros of polynomials of high order are sensitive to variations in the coefficients. Therfore, solutions based on the polynomial approach will often have inferior numerical properties, as compared to a corresponding state space approach, in particular for high order problems. There exist algorithms for solutions of Riccati–equations that are very well–behaved numerically [43]. Therefore, we suggest that for high order problems, a designer uses the polynomial approach in order to derive optimal filters and to gain engineering insight, but uses a state space approach for performing spectral factorisations.

Performance robustness is another important issue related to the discussed approach, as well as to any other filter design method. How well does a designed filter perform under non–ideal conditions and in presence of modelling errors? The performance of the estimators designed in this chapter can be sensitive to model errors, if the filters have poles or zeros close to the unit circle. A methodology which is flexible enough to encompass a variety of design requirements, and which allows the designer to build performance robustness into the design, is presented in [60], [61].

Bibliography

[1] A. Ahlén, "Identifiability of the deconvolution problem," *Automatica*, vol. 26, pp. 177–181, 1990.

[2] A. Ahlén and M Sternad, "Optimal deconvolution based on polynomial methods," *IEEE Transactions on Acoustics, Speech and Signal Processing*, vol. 37, pp. 217-226, 1989.

[3] A. Ahlén and M. Sternad, "Adaptive input estimation," *IFAC Symposium ACASP-89, Adaptive Systems in Control and Signal Processing*, pp. 631-636, Glasgow, UK, 1989.

[4] A. Ahlén and M. Sternad, "Wiener filter design using polynomial equations," *IEEE Transactions on Signal Processing*, vol. 39, no 11, pp. 2387–2399, 1991.

[5] A. Ahlén and M. Sternad, "Filter design via inner–outer factorization: Comments on 'Optimal deconvolution filter design based on orthogonal principle'," submitted to Signal Processing, 1992.

[6] B. D. O. Anderson and J. B. Moore, *Optimal Filtering.* Prentice–Hall, Englewood Cliffs, NJ, 1979.

[7] R. S. Anderssen and P. Bloomfield, "Numerical differentiation procedures for non-exact data," *Numerische Mathematik*, vol. 22, pp. 157-182, 1974.

[8] K. J. Åström, *Introduction to Stochastic Control Theory.* Academic Press, New York, 1970.

[9] J. F. Barret and T. J. Moir, "A unified approach to multivariable discrete–time filtering based on the Wiener theory," *Kybernetika (Prague)*, vol. 23, pp 177–197, 1987.

[10] B. Bernhardsson and M. Sternad, "Feedforward control is dual to deconvolution." To appear in the *International Journal of Control*, 1992.

[11] C. A. Belfiore and J. H. Park, "Decision feedback equalization," *Proc IEEE*, vol. 67, pp. 1143-1156, 1979.

[12] H. W. Bode and C. E. Shannon, "A simplified derivation of linear least square smoothing and prediction theory," *Proceedings of the I.R.E.*, vol. 38, pp. 417–425, April 1950.

[13] U. Borison, "Self–tuning regulators for a class of multivariable systems," *Automatica*, vol 15, pp. 209–215, 1979.

[14] J. A. Cadzow, *Foundations of Digital Signal Processing and Data Analysis.* Macmillan, New York, 1987.

[15] B–S. Chen and S–C. Peng, "Optimal deconvolution filter design based on orthogonal principle," *Signal Processing*, vol. 25, pp. 361–372, 1991.

[16] C.–T. Chen, "On digital Wiener filters," *Proc IEEE*, vol. 64, pp. 1736–1737, 1976.

[17] L. Chisci and E. Mosca, "A general polynomial solution to the MMSE deconvolution problem," *IEEE Transactions on Signal Processing*, vol 39, pp. 962–965, 1991.

[18] B Carlsson, Digital differentiating filters and model–based fault detection. *Ph.D. thesis* (Uppsala Dissertations from the Faculty of Science, no 28), Uppsala University, Sweden, 1989.

[19] B. Carlsson, A. Ahlén and M. Sternad, "Optimal differentiation based on stochastic signal models," *IEEE Transactions on Acoustics, Speech and Signal Processing*, vol. 39, pp. 341–353, 1991.

[20] B. Carlsson, M. Sternad and A. Ahlén, "Digital differentiation of noisy data measured through a dynamic system," *IEEE Transactions on Signal Processing*, vol. 40, pp. 218–221, 1992.

[21] G. Demoment and R. Reynaud, "Fast minimum variance deconvolution," *IEEE Transactions on Acoustics, Speech and Signal Processing,* vol. 33, pp. 1324–1326, 1985.

[22] Z. L. Deng, "White–noise filter and smoother with application to seismic data deconvolution," in *Preprints 7th IFAC/IFORS Symp. Identification Syst. Parameter Estimation*, York, U.K., pp. 621–624, 1985.

[23] D. L. Duttweiler, J. E. Mazo and D. G. Messerschmitt, "An upper bound on the error probability in decision-feedback equalization," *IEEE Transactions on Information Theory*, vol. 20, pp. 490-497, 1974.

[24] F. Farina and F. A. Studer, *Radar Data Processing. Volume I.* Research Student Press and Wiley, New York, 1985.

[25] S. Fioretti, L. Jetto and T. Leo, A discrete-time delay system for efficient simultaneous derivative estimation. In *Signal Processing III: Theories and applications*, eds I T Young et al, Elsevier Science Publishers B V (North-Holland), pp. 89-92, 1986.

[26] S. M. Fitch and L. Kurz, "Recursive equalization in data transmission - A design procedure and performance evaluation," *IEEE Transactions on Communication*, vol. 23, pp. 546-550, 1975.

[27] B. A. Francis, *A Course in H_∞ Control Theory.* Springer–Verlag, Berlin, 1987.

[28] M. J. Grimble, "Polynomial systems approach to optimal linear filtering and prediction," *International Journal of Control,* vol. 41, pp. 1545–1564, 1985.

[29] M. J. Grimble, "Time–varying polynomial systems approach to multichannel optimal linear filtering," *Proc. American Control Conference*, Boston, WA6-10:45, pp. 168-174, 1985.

[30] M. J. Grimble, "Single versus double Diophantine equation debate: comments on 'A polynomial approach to Wiener filtering'," *International Journal of Control,* vol 48, pp. 2161–2165, 1988.

[31] M. J. Grimble and M. A. Johnsson, *Optimal Control and Stochastic Estimation.* Wiley, Chichester, 1988.

[32] M. J. Grimble and A. ElSayed, "Solution of the H_∞ optimal linear filtering problem for discrete–time systems," *IEEE Transactions on Acoustics, Speech and Signal Processing,* vol. 38, pp. 1092–1104, 1990.

[33] P. Hagander and B. Wittenmark, "A self–tuning filter for fixed–lag smoothing," *IEEE Transactions on Information Theory,* vol. 23 pp. 377–384, 1977.

[34] S. Haykin, *Adaptive Filter Theory.* Prentice–Hall, Englewood Cliffs, NJ, 1986.

[35] J. Ježek and V. Kučera, "Efficient algorithm for matrix spectral factorization," *Automatica,* vol. 21, pp. 663–669, November 1985.

[36] T. Kailath, *Linear Systems.* Prentice–Hall, Englewood Cliffs, New Jersey, 1980.

[37] T. Kailath, *Lectures on Wiener and Kalman Filtering.* Springer, Wien, 1981.

[38] R. E. Kalman, "A new approach to linear filtering and prediction problems," Trans. AMSE, Journal of Basic Engineering, vol. 82D, pp. 35-45, 1960.

[39] V. Kučera, "Transfer–function solution of the Kalman–Bucy filtering problem," *Kybernetika (Prague),* vol 14, 110–122, 1978.

[40] V. Kučera, *Discrete Linear Control. The Polynomial Equation Approach.* Wiley, Chichester, 1979.

[41] V. Kučera, "New results in state estimation and regulation," *Automatica,* vol. 17, pp. 745–748, 1981.

[42] V. Kučera, "The LQG control problem: a study of common factors," *Problems of Control and Information Theory (Hungary),* vol 13, pp. 239–251, 1984.

[43] V. Kučera, "Factorization of rational spectral matrices: a survey," Preprints, *IEE Control'91,* Edinburgh, pp. 1074–1078, March 1991.

[44] V. Kučera, *Analysis and Design of Discrete Linear Control Systems.* Academia, Prague and Prentice Hall International, London, 1991.

[45] L. Lindbom, *Adaptive equalizers for fading mobile radio channels.* Licenciate Thesis, Department of Technology, Uppsala University, Sweden, September 1992.

[46] J. M. Mendel, *Optimal Seismic Deconvolution. An Estimation–Based Approach.* Academic Press, New York, 1983.

[47] T. J. Moir, "Optimal deconvolution smoother," *IEE proceedings*, vol. 133, pp. 13–18, 1986.

[48] T. J. Moir, "A polynomial approach to optimal and adaptive filtering with application to speech enhancement," *IEEE Transactions on Signal Processing*, vol 39, pp. 1221–1224, 1991.

[49] T. M. Moir and M.J. Grimble, "Optimal self–tuning filtering, smoothing and prediction for discrete time multivariable processes," *IEEE Transactions on Automatic Control*, vol. 19, pp. 128–137, 1984.

[50] P. Monsen, "Feedback equalization for fading dispersive channels," *IEEE Transactions on Information Theory*, vol. 17, pp. 56-64, 1971.

[51] P. A. Nelson, H. Hamada and S. J. Elliot, "Adaptive inverse filters for stereophonic sound reproduction," *IEEE Transactions on Signal Processing*, vol 40, pp. 1621–1632, 1992.

[52] S. U. H. Qureshi, "Adaptive equalization," *Proc IEEE*, vol. 73, pp. 1349-1387, 1985.

[53] L. R. Rabiner and K. Steiglitz, "The design of wide-band recursive and non-recursive digital differentiators," *IEEE Transactions on Audio Electronics*, vol. 18, pp. 204-209, 1970.

[54] A. P. Roberts and M. M. Newmann, "Polynomial approach to Wiener filtering," *International Journal of Control*, vol. 47, pp. 681–696, 1988.

[55] J. Salz, "Optimum mean-square decision feedback equalization," *Bell System Technical Journal*, vol. 52, pp. 1341-1374, 1973.

[56] U. Shaked, "A generalized transfer function approach to linear stationary filtering and steady–state optimal control problems," *International Journal of Control*, vol. 29, pp. 741–770, 1976.

[57] M. Sternad and A. Ahlén, "The structure and design of realizable decision feedback equalizers for IIR–channels with coloured noise," *IEEE Transactions on Information Theory,* vol. 36, pp. 848–858, 1990.

[58] M. Sternad and A. Ahlén, "A novel derivation methodology for polynomial–LQ controller design," *IEEE Transactions on Automatic Control*, vol. 37, Oct. 1992.

[59] M. Sternad and A. Ahlén, "LQ controller design and self–tuning control," Chapter 3 of K. Hunt. ed. *Polynomial Methods in Optimal Control and Filtering.* Control Engineering Series, Peter Peregrinus, London, 1992.

[60] M. Sternad and A. Ahlén, "Robust filtering based on probabilistic descriptions of model errors," *2nd IFAC Workshop on System Structure and Control*, Prague, Sept. 3–5, 1992.

[61] M. Sternad and A. Ahlén, "Robust filtering and feedforward control based on probabilistic descriptions of model errors," To appear in *Automatica*, vol 29, 1993.

[62] S. Usui and I. Amidror, "Digital low-pass differentiation for biological signal processing," *IEEE Transactions on Biomedical Engineering*, vol. 29, pp. 686-693, October 1982.

[63] M. Vidyasagar, *Control Systems Synthesis. A Factorization Approach.* MIT Press, Cambridge, MA, 1985.

[64] N. Wiener, *Extrapolation, Interpolation and Smoothing of Stationary Time Series.* The Technology Press and Wiley, New York, 1950.

Chapter 6

H_∞ Filtering

M. J. Grimble

6.1 Introduction

The range of signal processing problems is increasing rapidly as new communications techniques emerge and as new application problems arise. Novel filtering techniques must be introduced for new problems in accoustics, speech and fault monitoring. The H_∞ filter has very different properties and has different frequency and time domain characteristics to the more familiar least-square based filters. For example, the magnitude of the power spectrum of the estimation error can be reduced to a lower value with the H_∞ filter than with any other filter. It does not of course minimize the variance of the estimation error but there are applications where the H_∞ technique is more appropriate. The H_∞ filter can also be shown to provide a solution to the robust estimation problem where the signal or noise sources are uncertain (Xie [1]).

The full potential of the H_∞ the filter has still to be explored but Shaked [2] has reported useful results for seismic signal deconvolution and there are indications that the filter may have applications in fault detection and radar systems.

The H_∞ filtering problem was first considered by Grimble [3] using a polynomial systems approach. Smoothing, prediction and filtering problems were solved, for both scalar and multivariable systems, by Elsayed and Grimble ([4], [5], [6]). However, the filtering problem considered in the previous polynomial based work was not in standard model form and is not therefore as generally applicable as the results which follow. That is, most of the previous work on polynomial based H_∞ filtering problems assumed a particular plant and noise structure, whereas the standard system model used here allows a wide class of system models to be used.

6.1.1 *State Space Approach and Relationship to Polynomial Solution*

The Riccati equation approach to solving the H_∞ optimal control problem (Doyle *et al* [7]) has recently emerged as a very practical means of calculating H_∞ controllers. The same state equation based technique may be applied to solving the H_∞ filtering problem (Shaked, [2]; Basar, [8]; Nagpol and Khargonekar, [9]). Stoorvogel [22] has summarized the main results on the state-space approach to solving H_∞ problems.

The solution of the H_∞ filtering problem which follows is based upon Grimble (1992,[24]) that provides the polynomial matrix equivalent of the Doyle *et. al* ([7]) state-space results. The polynomial solution is the direct equivalent of the very successful Doyle, Glover, Khargonekar and Francis (DGKF) state-space solution. The filter derived is sub-optimal, since the filter is obtained for a chosen value of γ (the value of the H_∞ norm) and iteration must be performed to find the minimum value of γ. The original polynomial solution to the H_∞ filtering problem provided the optimal solution but these algorithms are more complicated in the multivariable case. The

of γ. The original polynomial solution to the H_∞ filtering problem provided the optimal solution but these algorithms are more complicated in the multivariable case. The equations for this type of solution do however hold at the optimum, whereas the Riccati motivated equations break down as the optimum is approached. This does not cause a problem in the practical application of either the following results or those developed by Doyle *et al*, since the equations do seem well conditioned when γ is increased to a level slightly above the optimum.

A further difference between the optimal polynomial solution and the approach considered below is that the filter is assumed to be in a model based (Kalman filtering) form, so that the solution provides the constant filter gain matrix, rather than the transfer function of the total filter.

6.1.2 *Game Theory Solution*

The sub-optimal H_∞ filtering problem is solved here via a game theory approach (Bryson and Ho, [23]). That is, a minimum variance optimal filtering problem is solved by minimizing the cost with respect to the feedback gain K_{f2} and maximizing the cost with respect to a special ***worst case*** error signal. This error signal is found by introducing a fictitious loop with gain K_{f1} which generates this worst case signal. This latter loop is not to be implemented but is only a device to find the worst-case generalized error.

This method enables a straightforward Kučera type of optimization argument to be followed to obtain the optimal H_∞ filter. The cost-function is expanded in terms of the filter gain $K_f = [K_{f1} \;\; K_{f2}]$ and a completing the squares argument is introduced to derive the causal filter. This enables the cost-function to be separated into terms depending upon K_{f1} and K_{f2}. The two sets of terms have opposite signs and the minimax solution is that which sets these terms to zero. The derivation is conceptually simple and links are established, to the state equation philosophy.

The synthesis theory is presented in a stochastic setting. The H_∞ control problem is usually concerned with a mapping between deterministic $L_2[0, \infty)$ ***energy*** signal spaces. The stochastic setting may, however, be more appropriate in filtering problems, where the H_∞ norm also provides a measure of system gain.

6.2 Stochastic System and Signals

A very general problem setting is to be employed. The filtering problem is posed in terms of the standard system model shown in Fig. 6.1, where G(s) is the generalized plant model and K_{f2} denotes the filter gain. The generalized plant G(s) includes the actual plant W(s) and absorbs the weighting functions included in the cost-function. The signal w incorporates all external inputs to the system including measurement noise and disturbances. This signal is assumed to be stochastic and of bounded power. A stochastic setting for the H_∞ problem was employed in references [3]-[6]. The generalised error signal z denotes the estimation error to be costed and the signal y denotes the filter innovations signals.

6.2.1 *Standard system model : state equation description*

The standard system model which enables a wide range of H_∞ filtering problems to be considered has a generalized error output z which represents the estimation error. The estimator structure will be assumed to be of the usual model based Kalman filtering form, even though the solution will be determined later using polynomial matrices. The computed gain is of course different to that in the Kalman filtering problem but the feedback structure is chosen to be the same. The system and estimator equations may be listed as:

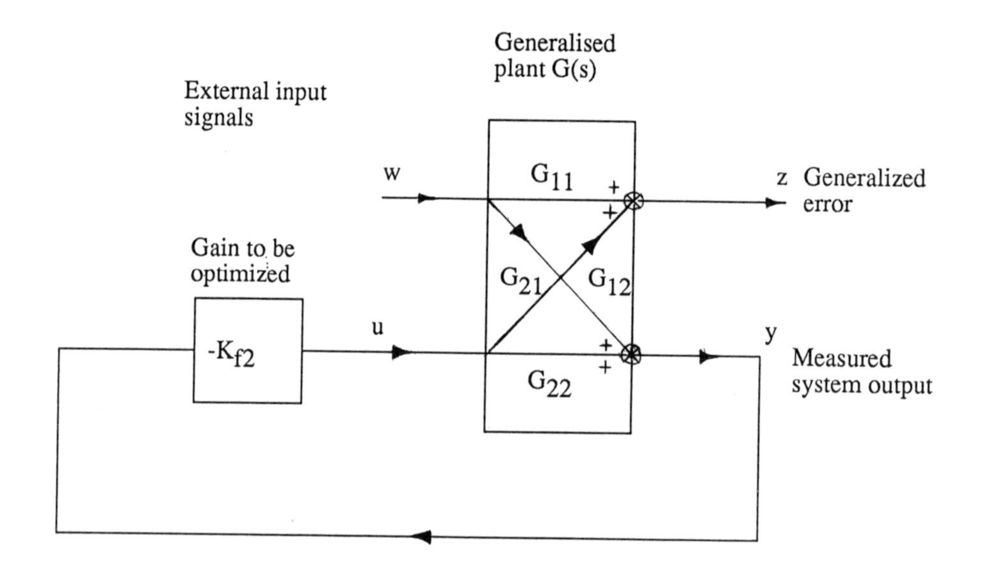

Fig. 1 : ***Standard System Model Drawn in Classical Feedback Form***

Signal generation model:

$$\dot{x}(t) = \overline{A}x(t) + \overline{B}_1 w(t) + \overline{B}_2 u_o(t) \tag{6.1}$$

$$z_o(t) = \overline{C}_1 x(t) + \overline{D}_{12} u_o(t) \tag{6.2}$$

$$y_o(t) = \overline{C}_2 x(t) + \overline{D}_{21} w(t) \tag{6.3}$$

where $x(t) \in R^n$, $z_o(t) \in R^{p_o}$, $w(t) \in R^q$, $u_o(t) \in R^m$ and $y_o(t) \in R^p$ denote the state, output to be estimated, disturbance and noise, control and observations signals respectively. The output matrix for the signal to be estimated and the observations can be defined as:

$$\overline{C} = \begin{bmatrix} \overline{C}_1 \\ \overline{C}_2 \end{bmatrix} \in R^{(p_o+p)xn} \tag{6.4}$$

The following assumptions are required:

A1. $\overline{D}_{21}\overline{D}_{21}^T > 0$ (6.5)

A2. $(\overline{C}_1, \overline{A})$ is detectable and $(\overline{A}, \overline{B}_1)$ is stabilizable.

Filter structure:

$$\dot{\hat{x}}(t) = \overline{A}\,\hat{x}(t) + K_{f2}(y_o(t) - \overline{C}_2\hat{x}(t)) + \overline{B}_2 u_o(t) \tag{6.6}$$

$$\hat{z}_o(t) = \overline{C}_1\hat{x}(t) + \overline{D}_{12} u_o(t) \tag{6.7}$$

where $\hat{x}(t) \in R^n$, and $\hat{z}_o(t) \in R^{p_o}$, represent the estimated state and output, respectively. Notice that the matrices related directly to the underlying state equation model are given overbars to distinguish them from the polynomial system models which are used later.

6.2.2 ***Combined plant and filter model state-equations***

The ***generalized error*** which represents the first output from the system transfer to be optimized is given as (using (6.2) and (6.7)):

$$z(t) = z_o(t) - \hat{z}_o(t) = \overline{C}_1(x(t) - \hat{x}(t)) \tag{6.8}$$

The second output from the standard system model corresponds with the innovations signal (using (6.3)):

$$y(t) = y_o(t) - \overline{C}_2\hat{x}(t) = \overline{C}_2(x(t) - \hat{x}(t)) + \overline{D}_{21} w(t) \tag{6.9}$$

and the second input corresponds with the filter gain output (employing negative-feedback convention):

$$u(t) = -K_{f2}y(t). \tag{6.10}$$

The state $x_s(t) \triangleq x(t) - \hat{x}(t)$ can describe the dynamics of the resulting process, since from (6.1) and (6.6):

$$\dot{x}(t) - \dot{\hat{x}}(t) = \overline{A}(x(t) - \hat{x}(t)) + \overline{B}_1 w(t) - K_{f2}y(t). \tag{6.11}$$

The standard system model state equation which describes the dynamics of the estimation error but not the complete dynamics of the system (6.1) - (6.3) and the filter (6.6) - (6.7), may be written as:

Estimation model:

$$\dot{x}_s(t) = \overline{A}x_s(t) + \overline{B}_1 w(t) + u(t) \tag{6.12}$$

$$z(t) = \overline{C}_1 x_s(t) \tag{6.13}$$

$$y(t) = \overline{C}_2 x_s(t) + \overline{D}_{21} w(t) \tag{6.14}$$

6.2.3 ***Standard system model transfer-function description***

The standard system model transfer-function, shown in Fig. 6.1, can be expressed:

$$G(s) = \begin{bmatrix} G_{11}(s) & G_{12}(s) \\ G_{21}(s) & G_{22}(s) \end{bmatrix} = \left[\begin{array}{c|cc} \overline{A} & \overline{B}_1 & I \\ \hline \overline{C}_1 & 0 & 0 \\ \overline{C}_2 & \overline{D}_{21} & 0 \end{array}\right] \tag{6.15}$$

The two system outputs can be expressed in terms of these transfer functions as:

$$z = \overline{C}_1(sI-\overline{A})^{-1}\overline{B}_1 w \quad + \quad \overline{C}_1(sI-\overline{A})^{-1}u \tag{6.16}$$

$$y = (\overline{C}_2(sI-\overline{A})^{-1}\overline{B}_1 + \overline{D}_{21})w \quad + \quad \overline{C}_2(sI-\overline{A})^{-1}u \tag{6.17}$$

and thus define the following transfers:

$$G_{11}(s) = \overline{C}_1(sI-\overline{A})^{-1}\overline{B}_1 \quad , \quad G_{12}(s) = \overline{C}_1 (sI-\overline{A})^{-1} \tag{6.18}$$

$$G_{21}(s) = \overline{C}_2(sI-\overline{A})^{-1}\overline{B}_1 + \overline{D}_{21} \quad , \quad G_{22}(s) = \overline{C}_2 (sI-\overline{A})^{-1} \tag{6.19}$$

Also define the resolvent matrix as : $\Phi(s) = (sI-\overline{A})^{-1}$ (6.20)

and the matrix:

$$\begin{bmatrix} Q & S \\ S^T & R \end{bmatrix} = \begin{bmatrix} \overline{B}_1 \\ \overline{D}_{21} \end{bmatrix} \begin{bmatrix} \overline{B}_1^{\ T} & \overline{D}_{21}^{\ T} \end{bmatrix} = \begin{bmatrix} \overline{B}_1 \overline{B}_1^{\ T} & \overline{B}_1 \overline{D}_{21}^{\ T} \\ \overline{D}_{21} \overline{B}_1^{\ T} & \overline{D}_{21} \overline{D}_{21}^{\ T} \end{bmatrix} \tag{6.21}$$

Substituting for $u = -K_{f2}y$ (from (6.10)) into (6.16) obtain:

$$z = G_{11}w + G_{12}u \tag{6.22}$$

$$u = -K_{f2}(G_{21}w + G_{22}u) = -(I + K_{f2}G_{22})^{-1}K_{f2}G_{21}w \tag{6.23}$$

$$z = (G_{11} - G_{12}(I + K_{f2}G_{22})^{-1}K_{f2}G_{21})w \tag{6.24}$$

The linear fractional transformation between the exogenous input w and the generalized error z, whose H_∞ norm is to be minimized, can now be defined as:

$$T_{zw} = G_{11} - G_{12}(I + K_{f2}G_{22})^{-1}K_{f2}G_{21}. \tag{6.25}$$

The linear fractional transformation T_{zw} is of course to be minimized in an H_∞ sense. The H_∞ norm of T_{zw} can be interpreted as the maximal RMS gain, for any frequency and any type of input, of the system. The H_∞ filtering problem involves the minimization of

$$J_{f\infty} = \|T_{zw}\|_\infty.$$

The H_∞ problem involves limiting the gain of the system but it can also be given an equivalent power spectral density interpretation. Hence,

$$\| T_{zw}(j\omega)\|_\infty^2 = \lambda_{max}(T_{zw}^*(j\omega)T_{zw}(j\omega)) = \lambda_{max}(T_{zw}(j\omega)T_{zw}^*(j\omega))$$

where λ_{max} is the maximum over all eigenvalues and over all frequencies ω. The term

$\Phi_{zz}(j\omega) = T_{zw}(j\omega)T_{zw}^*(j\omega)$ denotes the power spectral density of the generalized error system output z given an input w of zero-mean white-noise with identity covariance matrix. The equivalent stochastic problem is therefore to minimize the power spectral density of the output of the linear system T_{zw} driven by the white noise input signal w.

6.3 Riccati Equation For J Spectral Factorization

The state-space based game Riccati equation results of Doyle *et. al* [7] which underpins the polynomial analysis to follow will now be introduced. These results will be used to generate the J spectral-factor equation needed in the polynomial solution and they also provide the return-difference relationships which determine the filter stability properties. The relationship between the H_∞ Riccati equation and the solution of the J spectral-factorization problem was first established by Grimble ([10]) and it holds for both the control and filtering problems.

The following Hamiltonian matrix is similar to that which occurs in H_2 optimal filtering theory but it has an additional term depending upon the scalar γ^2. The scalar γ^2 represents an upper bound on the H_∞ norm of the filtering error. The minimum value of γ^2 corresponds to the optimum value of the H_∞ norm and is found by iteration. Let the H_∞ Hamiltonian matrix be defined as:

$$J_\infty = \begin{bmatrix} (\bar{A} - \bar{B}_1\bar{D}_{21}^T R^{-1}\bar{C}_2)^T & \gamma^{-2}C_1^T\bar{C}_1 - C_2^T R^{-1}\bar{C}_2 \\ -\bar{B}_1(I - D_{21}^T R^{-1}\bar{D}_{21})B_1^T & -(\bar{A} - \bar{B}_1\bar{D}_{21}^T R^{-1}\bar{C}_2) \end{bmatrix} \tag{6.26}$$

Let $Y = \text{Ric}(J_\infty)$. That is, let Y denote the unique solution of the Algebraic Riccati Equation (ARE) for which the eigenvalues of $\bar{A}_c = A - (YC^T + S_f)R_f^{-1}\bar{C}$ lie strictly in the left-half of the complex plane. The constant matrices R_f and S_f are defined as:

$$R_f \triangleq \begin{bmatrix} -\gamma^2 I & 0 \\ 0 & R \end{bmatrix} \quad \text{and} \quad S_f \triangleq [0 \quad S] \tag{6.27}$$

6.3.1 ***H_∞ Riccati equation***

The game or H_∞ Riccati equation which defines the stabilizing solution $Y \geq 0$ is defined by (6.26) as:

$$(\bar{A} - \bar{B}_1\bar{D}_{21}^T R^{-1}\bar{C}_2)Y + Y(\bar{A} - \overline{B_1 D}_{21}^T \bar{R}^{-1}\bar{C}_2)^T + Y(\gamma^{-2}\bar{C}_1^T C_1 - \bar{C}_2^T R^{-1}\bar{C}_2)Y$$

$$+ \bar{B}_1(I - \bar{D}_{21}^T R^{-1}\bar{D}_{21})\bar{B}_1^T = 0 \tag{6.28}$$

This equation may be written, in terms of the Kalman gain, as:

$$K_{f2} = (Y\bar{C}_2^T + \bar{B}_1\bar{D}_{21}^T)R^{-1} = (Y\bar{C}_2^T + S)R^{-1} \tag{6.29}$$

giving an alternative form of the Riccati equation as:

$$(\bar{A} - K_{f2}\bar{C}_2)Y + Y(\bar{A} - K_{f2}\bar{C}_2)^T + Y(\gamma^{-2}\bar{C}_1^T C_1)Y$$

$$+ (\bar{B}_1 - K_{f2}\bar{D}_{21})(\bar{B}_1^T - \bar{D}_{21}^T K_{f2}^T) = 0 \tag{6.30}$$

6.3.2 ***Return-difference equation***

Recall the definition of the resolvent matrix $\Phi(s)$, and introduce its adjoint:

$$\Phi(s) \triangleq (sI - \bar{A})^{-1} \quad \text{and} \quad \Phi^*(s) = (-sI - \bar{A}^T)^{-1} \tag{6.31}$$

then after substituting from (6.29) in equation (6.30):

$$-\Phi Y - Y\Phi^* + \gamma^{-2}\Phi Y\bar{C}_1^T C_1 Y\Phi^* + \Phi(Q - K_{f2}RK_{f2}^T)\Phi^* = 0 \tag{6.32}$$

Define the signal model transfer-function matrix $W_{c2} = C_2\Phi$, then premultiplying by C_2 and postmultiplying by $\bar{C}_2^T$, obtain;

$$-\overline{W}_{c2}Y\overline{C}_2^T - \overline{C}_2 Y\overline{W}_{c2}^* + \gamma^{-2}\overline{W}_{c2}Y\overline{C}_1^T C_1 Y\overline{W}_{c2}^* + \overline{W}_{c2}Q\overline{W}_{c2}^*$$

$$= \overline{W}_{c2}K_{f2}RK_{f2}^T\overline{W}_{c2}^* \tag{6.33}$$

but from (6.19): $G_{21} \triangleq \overline{W}_{c2}\overline{B}_1 + \overline{D}_{21}$ and hence,

$$G_{21}G_{21}^* = \overline{W}_{c2}Q\overline{W}_{c2}^* + R + S^T\overline{W}_{c2}^* + \overline{W}_{c2}S. \tag{6.34}$$

Substituting in (6.33) the desired return-difference relationship becomes:

$$G_{21}G_{21}^* + \gamma^{-2}\overline{W}_{c2}Y\overline{C}_1^T C_1 Y\overline{W}_{c2}^* = F_{f2}RF_{f2}^* \tag{6.35}$$

where $F_{f2} \triangleq (I + \overline{W}_{c2}K_{f2}) = I + \overline{C}_2\Phi K_{f2}$ (6.36)

and $F_{f2}^{-1} \varepsilon H_\infty$. This return-difference equation determines the stability characteristics of the optimal estimator to be implemented.

6.3.3 *Fictitious return-difference equation*

The return-difference relationships derived below is needed for obtaining the J spectral-factorization equation. The return-difference relationship is referred to as being fictitious since the output z cannot normally be measured. The feedback loop in the Kalman filter will not therefore include the output map $\overline{C}_1$. However, in the game problem solution it is assumed that the worst case scenario may be found assuming this output is available. The resulting additional loop involves the fictitious gain K_{f1}, as shown by the broken line in Fig. 6.2. Thus, a second return-difference relationship emerges during the solution and this is important, since it motivates the definition of the J spectral factors.

To derive the *fictitious* return-difference relationship first define:

$$\overline{W}_c = \overline{C}\,\Phi = \begin{bmatrix} \overline{C}_1 \\ \overline{C}_2 \end{bmatrix} \Phi \tag{6.37}$$

$$K_f = \begin{bmatrix} K_{f1} & K_{f2} \end{bmatrix} \triangleq (Y\overline{C}^T + S_f)R_f^{-1}$$

$$= [-Y\overline{C}_1^T\gamma^{-2},\ (Y\overline{C}_2^T + S)R^{-1}]. \tag{6.38}$$

Equation (6.32) may now be pre and post-multiplied by $\overline{C}$ and $\overline{C}^T$, respectively:

$$-\overline{W}_c Y\overline{C}^T - \overline{C}\,Y\overline{W}_c^* + \overline{W}_c Q\overline{W}_c^* = \overline{W}_c(-K_{f1}\gamma^2K_{f1}^T + K_{f2}RK_{f2}^T)\overline{W}_c^*$$

but from (6.38) $K_fR_f - S_f = Y\overline{C}^T$ and hence substituting obtain:

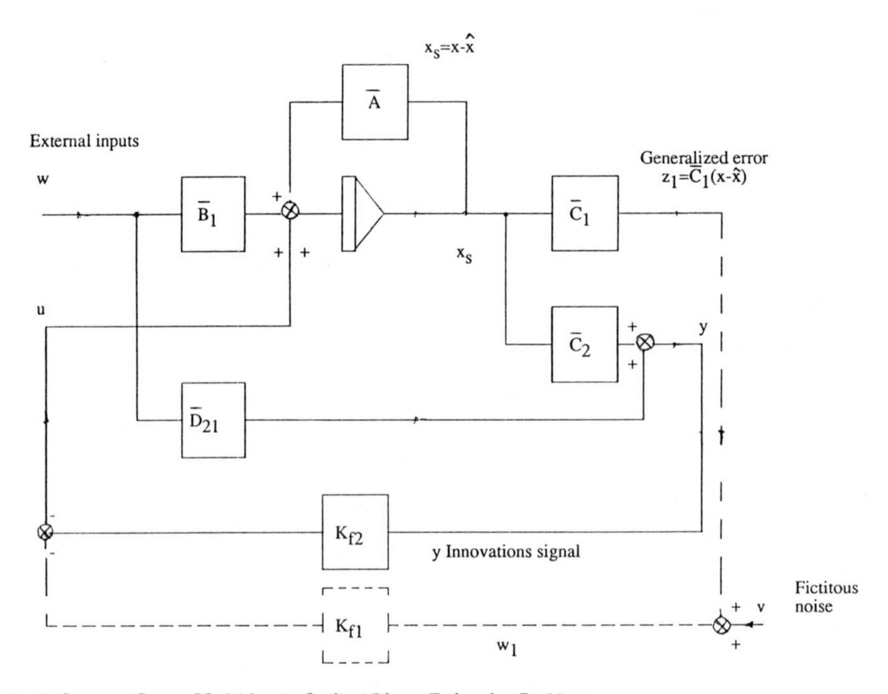

Fig. 2 : **Standard System Model for the Optimal Linear Estimation Problem**

$$\overline{W}_c Q \overline{W}_c^* + R_f + \overline{W}_c S_f + S_f^T \overline{W}_c^* = (I + \overline{W}_c K_f) R_f (I + K_f^T \overline{W}_c^*). \tag{6.39}$$

The return-difference relationship (6.39) may be written in terms of the *standard system model* transfer-functions. Expand the quadratic term and obtain:

$$\overline{W}_c Q \overline{W}_c^* + R_f + \overline{W}_c S_f + S_f^T W_c^* = \begin{bmatrix} G_{11}G_{11}^* - \gamma^2 I & G_{11}G_{21}^* \\ G_{21}G_{11}^* & G_{21}G_{21}^* \end{bmatrix}$$

$$= G_{.1}G_{.1}^* + \Gamma \tag{6.40}$$

$$\text{where} \quad \Gamma \triangleq \begin{bmatrix} -\gamma^2 I & 0 \\ 0 & 0 \end{bmatrix} \quad \text{and } G_{.1} \triangleq \begin{bmatrix} G_{11} \\ G_{21} \end{bmatrix}. \tag{6.41}$$

6.3.4 ***J spectral-factorization and relationship to Riccati equation results***

The filter J spectral-factor Y_f which is required in the optimization argument may be defined, using (6.39) and writing $R_f = H_r J H_r^T$, as:

$$Y_f J Y_f^* = \overline{W}_c Q \overline{W}_c^* + R_f + \overline{W}_c S_f + S_f^T \overline{W}_c^*$$

$$= (I + W_c K_f) H_r J H_r^T (I + K_f^T \overline{W}_c^*) = F_f R_f F_f^* \tag{6.42}$$

where H_r, $J \in R^{(p_o+p)x(p_o+p)}$ and $F_f(s) \in R(s)^{(p_o+p)x(p_o+p)}$ are defined as:

$$H_r \triangleq \begin{bmatrix} \gamma I & 0 \\ 0 & R^{1/2} \end{bmatrix}, \quad J \triangleq \begin{bmatrix} -I & 0 \\ 0 & I \end{bmatrix} \quad \text{and } F_f \triangleq (I + \overline{W}_c K_f). \tag{6.43}$$

The J spectral factor may be defined in terms of the *fictitious* return-difference matrix F_f, using (6.42), as:

$$Y_f = F_f H_r \quad \text{where } F_f \triangleq I + \overline{W}_c K_f \ . \tag{6.44}$$

Comparison of (6.40) and (6.42) reveals that the J spectral-factorization relationship may also be written as:

$$Y_f J Y_f^* = G_{.1} G_{.1}^* + \Gamma \tag{6.45}$$

which is an indefinite spectral-factorization expressed in terms of the standard system model.

6.4 Polynomial Matrix System Description

The polynomial matrix representation of the system model is introduced below. This will enable the optimal filter to be calculated via one spectral-factorization and the solution of two coupled diophantine equations (Grimble, [11]).

The system may now be represented in terms of polynomial matrices. Let the right-coprime polynomial matrices A_o, C_o be defined as:

$$C_oA_o^{-1} = \begin{bmatrix} C_{o1} \\ C_{o2} \end{bmatrix} A_o^{-1} = W_c \text{ where } \overline{W}_c = \overline{C}\,(sI-\overline{A})^{-1} \tag{6.46}$$

and hence let $A_o \triangleq (sI-\overline{A})$, $C_o \triangleq \overline{C}$, $C_{o1} \triangleq \overline{C}_1$, and $C_{o2} \triangleq \overline{C}_2$. The standard system model transfer-function matrices may now be written as:

$$\begin{bmatrix} C_{o1}A_o^{-1}\overline{B}_1 & C_{o1}A_o^{-1} \\ C_{o2}A_o^{-1}\overline{B}_1+\overline{D}_{21} & C_{o2}A_o^{-1} \end{bmatrix} = \begin{bmatrix} G_{11} & G_{12} \\ G_{21} & G_{22} \end{bmatrix}. \tag{6.47}$$

Also write the standard system transfer-function matrices as:

$$G_{.1} = \begin{bmatrix} G_{11} \\ G_{21} \end{bmatrix} \text{ and } G_{.2} = \begin{bmatrix} G_{12} \\ G_{22} \end{bmatrix} = C_oA_o^{-1} \tag{6.48}$$

and note that $\overline{W}_c$ can be written in left-coprime form as:

$$\overline{W}_c = C_oA_o^{-1} = A_2^{-1}C_2. \tag{6.49}$$

6.4.1 *Polynomial J spectral factorization and relationship to Riccati results*

The polynomial form of the J spectral-factor relationship may now be obtained. Substituting from the definition of the polynomial matrix terms in §6.4.1 the spectral-factor relationship (6.42) becomes:

$$\begin{aligned} Y_fJY_f^* &= \overline{W}_cQ\overline{W}_c^{\,*} + R_f + \overline{W}_cS_f + S_f^T\overline{W}_c^{\,*} \\ &= A_2^{-1}(C_2QC_2^* + A_2R_fA_2^* + C_2S_fA_2^* + A_2S_f^TC_2^*)A_2^{*-1} \end{aligned}$$

Clearly the spectral factor Y_f may be written as $Y_f = A_2^{-1}D_f$ where D_f is a polynomial matrix which satisfies:

$$D_fJD_f^* = C_2QC_2^* + A_2R_fA_2^* + C_2S_fA_2^* + A_2S_f^TC_2^* \tag{6.50}$$

The spectral-factor may be related to the state-space results using (6.44):

$$D_f = A_2F_fH_r = (A_2 + C_2K_f)H_r\,. \tag{6.51}$$

It follows that one method of calculating the J spectral-factor is to compute the gain K_f, using (6.38) and to then employ (6.51).

6.4.2 *Diophantine equations*

The diophantine equations which are essential to the derivation of the optimal filter may now be introduced. The solution (G_o, H_o, F_o) with F_o of minimal degree, is obtained from the following equations:

$$G_oD_f^* + A_oF_o = QC_2^* + S_fA_2^* \tag{6.52}$$

$$H_oD_f^* - C_oF_o = R_fA_2^* + S_f^TC_2^* \tag{6.53}$$

where from (6.49) : $C_oA_o^{-1} = A_2^{-1}C_2$. Left multiplying the first of these equations by $A_2^{-1}C_2$ and adding to the second gives:

$$(A_2^{-1}C_2G_o+H_o)D_f^* = A_2^{-1}(C_2QC_2^* + C_2S_fA_2^* + A_2S_f^TC_2^* + A_2R_fA_2^*) \tag{6.54}$$

Substituting from (6.50) and right multiplying by D_f^{*-1} gives:

$$C_2G_o + A_2H_o = D_fJ\,. \tag{6.55}$$

This equation is called the *implied equation* which in some cases is sufficient to determine the H_∞ filter, since only G_o and H_o are needed to compute the filter. However, in general both (6.52) and (6.53) must be solved.

6.4.3 *Structure of the diophantine equation solution for H_o*

The optimization proof which follows employs the fact that the solution of (6.53) gives H_o as a block diagonal matrix. This property is now demonstrated but first note that $\overline{W}_c = \overline{C}\Phi$ is strictly proper and hence from (6.50), as $s \to \infty$: $Y_f = A_2^{-1}D_f \to H_r$. Notice that (6.54) may be written as:

$$A_2^{-1}C_2G_o + H_o = A_2^{-1}C_2QC_2^*D_f^{*-1} + A_2^{-1}C_2S_fY_f^{*-1} + S_f^TC_2^*D_f^{*-1} + R_fY_f^{*-1} \tag{6.56}$$

but from (6.43): $R_f = H_rJH_r^T$ and hence as $s \to \infty$: $H_o \to H_rJ$.

This constant matrix can be confirmed as the solution to (6.53) and (6.55) by noting the degrees of the polynomial matrix terms. That is, from (6.52), (6.53) deg (F_o) = deg (D_f)-1 and from (6.49):

$$C_oA_o^{-1} = \overline{C}\,(sI - \overline{A})^{-1} \;\Rightarrow\; C_o = \overline{C} \text{ and } A_o^{-1} = (sI - \overline{A})^{-1}$$

From equation (6.50), since R_f is full rank, deg (D_f) = deg (A_2), and from (6.53) this implies deg (H_o) = 0. It therefore follows that

$$H_o = H_rJ = \begin{bmatrix} -\gamma I & 0 \\ 0 & R^{1/2} \end{bmatrix}. \tag{6.57}$$

The deg (A_o) = 1 and hence deg (A_oF_o) = deg (D_f) and hence from (6.52) the solution matrix G_o is also constant.

6.5 Solution of the H_∞ Filtering Problem

There are several solution techniques which may be employed to obtain the H_∞ filter but the approach taken below has particular merit since it provides some physical insights into the results. The game problem will be solved in this section. To determine the cost-function to be optimized in the game problem a duality is established with the well known results for the H_∞ state-feedback control problem. The duality relationship and solution are presented in more detail in Grimble [24].

The system equations required for the optimization procedure are developed below based upon the standard model equations (6.16) and (6.17). Letting

$$u = -K_{f1}z - K_{f2}y = -K_f \begin{bmatrix} z \\ y \end{bmatrix} \tag{6.58}$$

$$\begin{bmatrix} z \\ y \end{bmatrix} = \begin{bmatrix} G_{11} \\ G_{21} \end{bmatrix} w + \begin{bmatrix} G_{12} \\ G_{22} \end{bmatrix} u = G_{.1}w + G_{.2}u = (I + G_{.2}K_f)^{-1}G_{.1}\, w \tag{6.59}$$

Let the *filter sensitivity* :

$$M_{.2} \triangleq K_f(I + G_{.2}K_f)^{-1} \tag{6.60}$$

then $u = -M_{.2}G_{.1}w$ and

$$\begin{bmatrix} z \\ y \end{bmatrix} = (G_{.1} - G_{.2}M_{.2}G_{.1})w = \begin{bmatrix} G_{11}-G_{12}M_{.2}G_{.1} \\ G_{21}-G_{22}M_{.2}G_{.1} \end{bmatrix} w. \tag{6.61}$$

The *generalized error* output due to the fictitious measurement noise v, shown in Fig. 6.2, may be found as:

$$z_1 = -G_{12}(I+K_fG_{.2})^{-1}K_{f1}v = -G_{12}S_{.2}K_{f1}v \tag{6.62}$$

where $S_{.2}$ denotes the *sensitivity function*:

$$S_{.2} \triangleq (I + K_fG_{.2})^{-1}. \tag{6.63}$$

The game problem which provides the solution to the control problem (Yaesh and Shaked, [12]) is as follows:

$$J_{c\infty} = (\min_{C_o} \max_{C_1} \frac{1}{2\pi j} \oint_D \mathrm{trace}\{(G_{11}-G_{1.}M_{2.}G_{21})(G_{11}^*-G_{21}^*M_{2.}^*G_{1.}^*)\}$$

$$- \gamma^2 \mathrm{trace}\{C_1S_{2.}G_{21}G_{21}^*S_{2.}^*C_1^*\}\, ds)^{1/2} \tag{6.64}$$

where D denotes the usual D contour taken over the right-half of the s-plane. Duality relationships may be employed, together with this expression, to obtain the cost optimization problem which determines the H_∞ filter.

$$J_{f\infty} = (\min_{K_{f2}} \max_{K_{f1}} \frac{1}{2\pi j} \oint_D \text{trace}\{(G_{11}^* - G_{.1}^* M_{.2}^* G_{12}^*)(G_{11} - G_{12} M_{.2} G_{.1})\}$$

$$- \gamma^2 \text{trace}\{K_{f1}^* S_{.2}^* G_{12}^* G_{12} S_{.2} K_{f1}\}\, ds)^{1/2} \qquad (6.65)$$

6.5.2 *Physical interpretation of the game cost-index*

Consider the physical interpretation of the terms in the integrand of $J_{f\infty}$. The cost integrand may be written as:

$$I_f = \text{trace}\{\Phi_{zz}(s) - \gamma^2 \Phi_{z_1 z_1}(s)\} \qquad (6.66)$$

where

$$\Phi_{zz}(s) \triangleq (G_{11} - G_{12} M_{.2} G_{.1})(G_{11}^* - G_{.1}^* M_{.2}^* G_{12}^*) \qquad (6.67)$$

and

$$\Phi_{z_1 z_1}(s) \triangleq G_{12} S_{.2} K_{f1} K_{f1}^* S_{.2}^* G_{12}^* \qquad (6.68)$$

where Φ_{zz} and $\Phi_{z_1 z_1}$ denote the power spectra of the generalized error z, shown in Fig. 6.2, due to the inputs w and v, respectively. To give the power spectrum (6.66) these signals are assumed to be uncorrelated white noise with identity covariance matrices.

The game problem implied by the duality relationships therefore involves minimization of the error variance:

$$J_{f\infty} = E\{z^T(t) z(t) - \gamma^2 z_1^T(t) z_1(t)\} \qquad (6.69)$$

using K_{f2} and maximization using K_{f1}. The first term is due to the input w which affects both signal and measurement noise. The second term is due to the input v which only adversely effects the generalized error, since it does not contribute to the signal, only to the measurement noise in the fictitious innovations signal loop.

The game problem cost-function can be interpreted as involving the spectrum of the generalized error due to w (which contributes to the process and measurement noise for loop 2) and due to v (which contributes to the measurement noise for the fictitious loop 1). The implementation of the filter does not involve this latter loop since z cannot be measured.

6.5.3 *Cost-function integrand expansion*

The integrand of the cost-function (6.65) may be expanded, by substituting from the equations in the previous section, as follows:

$$I_f = \text{trace}\{\Phi_{zz}(s) - \gamma^2 \Phi_{z_1 z_1}(s)\}$$

$$= \text{trace}\{(G_{11} G_{11}^* - G_{12} M_{.2} G_{.1} G_{11}^* - G_{11} G_{.1}^* M_{.2}^* G_{12}^*$$

$$+ G_{12} S_{.2} K_f (G_{.1} G_{.1}^* + \Gamma) K_f^* S_{.2}^* G_{12}^*\} \qquad (6.70)$$

Substituting from (6.45) obtain:

$$I_f = \text{trace}\{G_{12}M_{.2}Y_fJY_f^*M_{.2}^*G_{12}^* - G_{12}M_{.2}G_{.1}G_{11}^* - G_{11}G_{.1}^*M_{.2}^*G_{12}^* + G_{11}G_{11}^*\}$$

Following a *completing the squares*, philosophy obtain:

$$I_f = \text{trace}\{(G_{12}M_{.2}Y_f - G_{11}G_{.1}^*Y_f^{*-1}J^{-1})J(Y_f^*M_{.2}^*G_{12}^* - J^{-1}Y_f^{-1}G_{.1}G_{11}^*)$$
$$+ G_{11}G_{11}^* - G_{11}G_{.1}^*Y_f^{*-1}J^{-1}Y_f^{-1}G_{.1}G_{11}^*\} \tag{6.71}$$

It is important for the optimization process to observe that the final term:

$$T_3 \triangleq G_{11}(I - G_{.1}^*(Y_fJY_f^*)^{-1}G_{.1})G_{11}^* \tag{6.72}$$

is independent of the filter gain $K_f = [K_{f1} \quad K_{f2}]$. This term does not therefore affect the optimal solution.

6.5.4 *Polynomial form of the cost terms*

The terms in the cost integrand (6.71) may now be evaluated. Using the results from §4.1 obtain:

$$G_{11}G_{.1}^* = G_{11}[G_{11}^* \quad G_{21}^*] = C_{o1}A_o^{-1}[QA_o^{*-1}C_{o1}^* \quad QA_o^{*-1}C_{o2}^* + S]$$
$$= C_{o1}A_o^{-1}(QA_o^{*-1}C_o^* + S_f)$$

hence,

$$G_{11}G_{.1}^*Y_f^{*-1} = C_{o1}A_o^{-1}(QC_2^*D_f^{*-1} + S_fA_2^*D_f^{*-1}). \tag{6.73}$$

This expression involves the transforms of causal (A_o^{-1}) and non-causal (D_f^{*-1}) terms and must be separated into the individual components using a procedure similar to partial fraction expansion (Grimble and Johnson, [13]). This procedure requires the use of the diophantine equations and hence substituting from (6.52):

$$G_{11}G_{.1}^*Y_f^{*-1} = C_{o1}(A_o^{-1}G_o + F_oD_f^{*-1}) \tag{6.74}$$

The sensitivity term in the cost index, after substituting for the polynomial matrices, becomes:

$$G_{12}M_{.2}Y_f = C_{o1}A_o^{-1}M_{.2}A_2^{-1}D_f \tag{6.75}$$

Collecting these results the quadratic term in the cost-function integrand (6.71) may be written as:

$$(G_{12}M_{.2}Y_f - G_{11}G_{.1}^*Y_f^{*-1}J^{-1})$$
$$= C_{o1}(A_o^{-1}M_{.2}A_2^{-1}D_f - A_o^{-1}G_oJ^{-1} - F_oD_f^{*-1}J^{-1}) \tag{6.76}$$

The first term on the right of this expression, involving $A_o^{-1}M_{.2}$, can be simplified, using (6.60) and (6.48):

$$A_o^{-1}M_{.2} = A_o^{-1}K_f(I + G_{.2}K_f)^{-1} = (K_{fd}A_o + K_{fn}C_o)^{-1}K_{fn} \tag{6.77}$$

where $K_f = K_{fd}^{-1}K_{fn}$ and K_{fd}, K_{fn} are polynomial matrices. Recall from the state-space results K_f is known to be a constant matrix but at this point in the polynomial solution K_f has not been shown to be constant, and it must therefore be treated for the present as a transfer-function $K_{fd}^{-1}K_{fn}$.

6.5.5 *Cost-function integrand terms*

The solution proceeds with the definition of stable (T_1) and unstable (T_2) terms within the cost-function. The quadratic term in the cost integrand (6.71) was simplified above using the first of the diophantine equations. The cost term is expanded further below, through the introduction of the second diophantine equation.

The quadratic term in the criterion, using (6.49), (6.76), and (6.77), therefore becomes:

$$G_{12}M_{.2}Y_f - G_{11}G^*_{.1}Y_f^{*-1}J^{-1}$$
$$= C_{01}(K_{fd}A_o + K_{fn}C_o)^{-1}(K_{fn}A_2^{-1}D_f - (K_{fd} + K_{fn}C_oA_o^{-1})G_oJ^{-1})$$
$$-C_{01}F_oD_f^{*-1}J^{-1} \qquad (6.78)$$
$$= C_{01}((K_{fd}A_o+K_{fn}C_o)^{-1}(K_{fn}H_o-K_{fd}G_o) - F_oD_f^{*-1})J^{-1} \qquad (6.79)$$

after substituting for $A_2^{-1}D_f$ from (6.55):

The pole positions of the various terms in the previous expression are important for the optimization argument. Thus, write:

$$G_{12}M_{.2}Y_f - G_{11}G^*_{.1}Y_f^{*-1}J^{-1} = (T_1 + T_2)J^{-1} \qquad (6.80)$$

where T_1 and T_2 are strictly stable and unstable rational functions, respectively, and are defined as:

$$T_1 \triangleq C_{01}(K_{fd}A_o + K_{fn}C_o)^{-1}(K_{fn}H_o-K_{fd}G_o) \qquad (6.81)$$
$$T_2 \triangleq -C_{01}F_oD_f^{*-1}\,. \qquad (6.82)$$

6.5.6 *Contour integral*

The preceding results may now be collected to obtain an expression for the cost-function in terms of the stable (T_1) and unstable (T_2) terms. From (6.78) to (6.82):

$$(G_{12}M_{.2}Y_f - G_{11}G^*_{.1}Y_f^{*-1}J^{-1})J(Y_f^*M^*_{.2}G^*_{12}-J^{-1}Y_f^{-1}G_{.1}G^*_{11})$$
$$= (T_1 + T_2)J^{-1}(T_1^* + T_2^*) \qquad (6.83)$$

The matrices T_1 and T_2 may be partitioned conformally with J in (6.43):

$$T_1 = [T_{11} \;\; T_{12}] \quad \text{and} \quad T_2 = [T_{21} \;\; T_{22}] \qquad (6.84)$$

so that

$$(T_1 + T_2)J^{-1}(T_1^*+T_2^*) = -(T_{11}+T_{21})(T_{11}^*+T_{21}^*)+(T_{12}+T_{22})(T_{12}^*+T_{22}^*) \qquad (6.85)$$

In the evaluation of the contour-integral the cross-product of various terms arise: but T_{11}, T_{12} are strictly stable and T_{21}, T_{22} are strictly unstable. A standard argument, which employs the Residue theorem, may easily be used to show the integral of such terms is zero (Grimble [14]; Grimble and Johnson [13]). The cost-function may therefore be written, using (6.66), (6.71) and (6.85) as:

$$J_{f\infty} = \frac{1}{2\pi j} \oint_D (-\text{trace}\{T_{11}T_{11}^*\} + \text{trace}\{T_{12}T_{12}^*\}$$

$$+ \text{trace}\{T_{22}T_{22}^*\} - \text{trace}\{T_{21}T_{21}^*\} + \text{trace}\{T_3\}) \, ds. \tag{6.86}$$

6.5.7 *Calculation of the optimal gain*

Partitioning the filter gain K_f and the matrix G_o, conformably:

$$K_f = K_{fd}^{-1}K_{fn} = K_{fd}^{-1}[K_{n1} \quad K_{n2}] \quad \text{and } G_o = [G_1 \quad G_2] \tag{6.87}$$

then (6.57), (6.81) and (6.87) give:

$$T_1 = [T_{11} \quad T_{12}] = C_{o1}(K_{fd}A_o + K_{fn}C_o)^{-1}[-\gamma K_{n1} - K_{fd}G_1, \; K_{n2}R^{1/2} - K_{fd}G_2]$$

Recall that $T_2 = [T_{21} \quad T_{22}]$ and from (6.82) T_2 does not depend upon the filter gain K_f. Thus, only the first two terms in (6.86) depend upon K_{f1} and K_{f2} and these terms may be written as:

$$\text{trace}\{-T_{11}T_{11}^* + T_{12}T_{12}^*\}$$

$$= \text{trace}\{C_{o1}(K_{fd}A_o + K_{fn}C_o)^{-1}[-(\gamma K_{n1} + K_{fd}G_1)(\gamma K_{n1} + K_{fd}G_1)^*$$

$$+(K_{n2}R^{1/2} - K_{fd}G_2)(K_{n2}R^{1/2} - K_{fd}G_2)^*](K_{fd}A_o + K_{fn}C_o)^{*-1}C_{o1}^*\} \tag{6.88}$$

Recall that the cost is to be maximized, with respect to $K_{fd}^{-1}K_{n1}$ and is to be minimized, with respect to $K_{fd}^{-1}K_{n2}$. To maximize the cost, noting the presence of the minus sign in (6.88), set

$$\gamma K_{n1} + K_{fd}G_1 = 0 \Rightarrow K_{f1} = K_{fd}^{-1}K_{n1} = -\gamma^{-1}G_1. \tag{6.89}$$

and to minimize the cost let $K_{n2}R^{1/2} - K_{fd}G_2 = 0$

$$\Rightarrow K_{f2} = K_{fd}^{-1}K_{n2} = G_2R^{-1/2}. \tag{6.90}$$

These represent the desired optimal values of the filter gain which may also be written (noting (6.87)) as:

$$K_f = G_oH_o^{-1}. \tag{6.91}$$

The consequence of the previous results is that the condition for optimality (from (6.81)) becomes: $T_1 = 0$ and the solution to the minimax problem may be summarized as:

$$K_f = [K_{f1} \quad K_{f2}] = K_{fd}^{-1}[K_{n1} \quad K_{n2}] = [-\gamma^{-1}G_1, \; G_2R^{-1/2}]. \tag{6.92}$$

Recall that only the gain K_{f2} is actually to be implemented in the model based compensator.

Problem 6.5.1: ***Polynomial Description of the H_∞ Optimal Filtering Problem***

Consider the standard system described in §6.2 which has an output representing the estimation error : $z = \overline{C}_1(x-\hat{x})$ and which includes the model based filter equations (6.6) and (6.7). The H_∞ filter Kalman gain K_{f2}, shown in Fig 6.2, is required to minimize the H_∞ norm of the linear fractional transformation:

$$T_{zw} = G_{11} - G_{12}(I + K_{f2}G_{22})^{-1}K_{f2}G_{21} \tag{6.93}$$

The minimum of the H_∞ cost is denoted by γ_o:

$$\| T_{zw} \|_\infty = \gamma_o$$

The sub-optimal H_∞ filter ensures $\| T_{zw} \|_\infty \le \gamma$ where $\gamma > \gamma_o$.

Let the polynomial matrices A_o, C_o and the left coprime pair A_2, C_2 be introduced.

$$C_o A_o^{-1} = A_2^{-1} C_2 = \begin{bmatrix} G_{12} \\ G_{22} \end{bmatrix} \quad \text{where} \quad C_o = \begin{bmatrix} C_{o1} \\ C_{o2} \end{bmatrix} \tag{6.94}$$

The standard system model is expressed in polynomial matrix form as:

$$\begin{bmatrix} G_{11} & G_{12} \\ G_{21} & G_{22} \end{bmatrix} = \begin{bmatrix} C_{o1} A_o^{-1} \overline{B}_1 & C_{o1} A_o^{-1} \\ C_{o2} A_o^{-1} \overline{B}_1 + \overline{D}_{21} & C_{o2} A_o^{-1} \end{bmatrix} \tag{6.95}$$

The following weights may be defined from this model:

$$Q = \overline{B}_1 \overline{B}_1^T, \; R = \overline{D}_{21} \overline{D}_{21}^T, \; S = \overline{B}_1 \overline{D}_{21}^T, \tag{6.96}$$

$$R_f = \begin{bmatrix} -\gamma^2 I & 0 \\ 0 & R \end{bmatrix} \quad \text{and} \quad S_f = [0 \;\; S]. \tag{6.97}$$

Theorem 6.5.1: ***Polynomial Matrix Solution of the H_∞ Filtering Problem***

Consider the model based filter (6.6), (6.7), represented in polynomial matrix form:

$$\hat{x} = (A_o + K_{f2} C_{02})^{-1} (K_{f2} y_o + \overline{B}_2 u_o) \tag{6.98}$$

$$\hat{z}_o = C_{o1} \hat{x} + \overline{D}_{12} u_o \tag{6.99}$$

The Kalman gain K_{f2} solution to the ***sub-optimal*** H_∞ filtering problem may be found via spectral factorization and diophantine equation calculations, for some $\gamma > \gamma_o$.

J Spectral factorization:

$$D_f J D_f^* = C_2 Q C_2^* + A_2 R_f A_2^* + C_2 S_f A_2^* + A_2 S_f^T C_2^* \tag{6.100}$$

Diophantine equations:

The solution (G_o, H_o, F_o) with F_o of smallest degree, must be obtained of the coupled diophantine equations:

$$G_o D_f^* + A_o F_o = Q C_2^* + S_f A_2^* \tag{6.101}$$

$$H_o D_f^* - C_o F_o = R_f A_2^* + S_f^T C_2^*. \tag{6.102}$$

Partition $G_o = [G_1 \;\; G_2]$ where $G_1 \in P^{n \times p_o}$ and $G_2 \in P^{n \times p}$, then the following results are obtained:

Filter gain:

$$K_{f2} = K_{fd}^{-1}K_{n2} = G_2R^{-1/2} \tag{6.103}$$

Implied equation:

$$C_2G_o + A_2H_o = D_fJ. \tag{6.104}$$

•

6.6 Relationship between Riccati and Diophantine Equation Solutions

The polynomial matrix solution of the H_∞ filtering problem, was completely characterized by a J spectral factorization and by the solution of coupled diophantine equations. The solution of these diophantine equations, in this particular case, was shown to be a constant matrix for G_o and H_o.

Since the solution to the sub-optimal H_∞ filtering problem may be found from either the Riccati or diophantine equations there is clearly a close relationship between the constant matrix solution Y of the former and the polynomial matrix solution (G_o, H_o, F_o) of the latter. The particular relationship is established below following an argument similar to that employed in Grimble [20].

Theorem 6.6.1: ***Relationship between polynomial and state space solution***

Let $Y = Y^T$ denote the positive semi-definite solution of the nxn filter matrix Riccati equation:

$$Y\overline{A}^T + \overline{A}Y + Q - (Y\overline{C}^T + S_f)R_f^{-1}(\overline{C}Y + S_f^T) = 0 \tag{6.105}$$

where the gain:

$$K_f = (Y\overline{C}^T + S_f)R_f^{-1} \quad \text{and} \quad R_f = H_rJH_r^T. \tag{6.106}$$

The minimum-degree solution (G_o, H_o, F_o) of the diophantine equations (6.101) and (6.102) then follows as:

$$G_o = (Y\overline{C}^T + S_f)H_r^{-T} \tag{6.107}$$

$$H_o = H_rJ \tag{6.108}$$

$$F_o = Y\overset{*}{C_2} = YC_2^T(-s). \tag{6.109}$$

•

Proof : Note that the Riccati equation (6.105) follows from (6.30):

$$Y\overline{A}^T + \overline{A}Y + Q - [K_{f1} \;\; K_{f2}]R_f \begin{bmatrix} K_{f1}^T \\ K_{f2}^T \end{bmatrix} = 0$$

and this can be written in the form:

$$Y\overline{A}^T + \overline{A}Y + Q - K_fR_fK_f^T = 0 \tag{6.110}$$

or

$$\Phi Q\Phi^*\overline{C}^T = Y\Phi^*\overline{C}^T + \Phi(Y + K_fR_fK_f^T\Phi^*)\overline{C}^T$$

Now from (6.49) $\overline{C}\Phi = C_0A_0^{-1} = A_2^{-1}C_2$ and from (6.51): $D_f = (A_2 + C_2K_f)H_r$ and thence:

$$\begin{aligned} A_0^{-1}QC_2^*A_2^{*-1} &= YC_2^*A_2^{*-1} + A_0^{-1}K_fR_f - A_0^{-1}S_f + A_0^{-1}K_fR_fK_f^TC_2^*A_2^{*-1} \\ &= YC_2^*A_2^{*-1} - A_0^{-1}S_f + A_0^{-1}K_fH_rJD_f^*A_2^{*-1} \end{aligned} \tag{6.111}$$

hence obtain:

$$\begin{aligned} &A_0YC_2^* + K_fH_rJD_f^* = QC_2^* + S_fA_2^* \\ &(K_fH_rJ)D_f^* + A_0(YC_2^*) = QC_2^* + S_fA_2^* \end{aligned} \tag{6.112}$$

Comparing (6.101) and (6.112) identify

$$G_0 = K_fH_rJ = (Y\overline{C}^T + S_f)H_r^{-T} \quad \text{and} \quad F_0 = YC_2^*. \tag{6.113}$$

Since by assumption $\deg(C_2) < \deg(A_2)$, then $\deg(F_0) < \deg(D_c)$ and the solution satisfies the minimum-degree requirement on F_0.

The second diophantine equation (6.102) is also satisfied by the above definition of F_0. Substituting from (6.113) in (6.102) obtain:

$$\begin{aligned} H_0D_f^* &= C_0F_0 + R_fA_2^* + S_f^TC_2^* = (C_0Y + S_f^T)C_2^* + R_fA_2^* \\ &= R_f(K_f^TC_2^* + A_2^*) = H_rJD_f^* \end{aligned} \tag{6.114}$$

and hence identify :

$$H_0 = H_rJ. \tag{6.115}$$

Note that the computed G_0 and H_0 satisfy the *implied equation*. Substituting for G_0 and H_0 obtain:

$$\begin{aligned} (C_2G_0 + A_2H_0) &= C_2(Y\overline{C}^T + S_f)H_r^{-T} + A_2H_rJ \\ &= (A_2 + C_2K_f)H_rJ = D_fJ. \end{aligned} \tag{6.116}$$

•

6.7 Conclusions

The solution of the H_∞ filtering problem was obtained in polynomial system form. The H_∞ filter was first proposed by Grimble [3] for use in a range of signal processing problems. The filter enables the power spectrum of the estimation error to be reduced to lower values in chosen frequency ranges than is possible with any other filter.

The polynomial solution of the H_∞ filtering problem presented here provides useful insights into the links between the state-space and the frequency domain results. The solution is more general than the previous *equalizing* polynomial solutions of the filtering problem (Grimble, Ho and Elsayed, [18]). For example, the previous work was mainly for scalar filtering problems, whereas the results here apply to the general multi-channel filtering problem.

The recent state-space solution of the standard H_∞ optimal control problem which is based on the solution of a modified Riccati equation, has stimulated the development of several alternative state-space approaches and results. Although many of the previous approaches in H_∞ have been discarded as the subject has evolved, these Riccati results appear to be fundamental and they form the basis of much of the recent published work on the H_∞ approach.

There are now a number of methods of solving the J-spectral factorization problem (Kwakernaak, [15]; Sebek and Kwakernaak, [17]; Sebek [16]) but the link to the Riccati equation results exploited above provides a potentially valuable alternative mechanism. The diophantine equations to be solved are much easier to solve than the related polynomial equations in earlier work, since their solution involves constant rather than polynomial matrices. The previous polynomial solutions were based on the equalizing H_∞-solution which (in a mean square sense) is the worst of all H_∞ filters. The new solution is not in general equalizing and thereby offers an improvement in this regard. The importance of the H_∞ filtering solution will become more evident as uncertain estimation problems are considered (Fu. *et al* [19]). The smoothing and prediction problems (Grimble, [21]) can be solved by a similar approach to that considered above.

References

1. Xie, L., (1991), *H_∞ control and filtering of systems with parameter uncertainty*, PhD Thesis, Dept of Electrical & Computer Eng., Univ. of Newcastle, Australia.

2. Shaked, U., (1990), *H_∞ minimum error state estimation of linear stationary processes*, Technical Notes and Correspondence, IEEE Trans. on Auto. Contr., 35, 5, pp. 554-558.

3. Grimble, M.J., (1987), *H_∞ design of optimal linear filters*, Conference on Mathematical Theory of Networks and Systems, Phoenix, Arizona and published in Linear Circuits, Systems and Signal Processing : Theory and Applications, North Holland, Ed. C.I. Byrnes, C.F. Martin and R.E. Saeks.

4. Elsayed, A., and M.J. Grimble, *A new approach to the H_∞ design of optimal digital linear filters*, IMA Journal of Mathematical Contr., and Information, 6, 233-251.

5. Grimble, M.J. and El Sayed, A., (1990), *Solution of the H_∞ optimal linear filtering problem for discrete-time systems*, IEEE Trans. on Acoustics, Speech and Signal Processing, Vol. 38, No. 7, pp. 1092-1104.

6. Grimble, M.J., (1991), *H_∞ fixed-lag smoothing filter for scalar systems*, IEEE Trans. on Signal Processing, Vol. 39, No. 9, pp. 1955-1963.

7. Doyle, J.C., K. Glover, P.P. Khargonekar, and B.A. Francis, (1989), *State-space solutions to standard H_2 and H_∞ control problems*, IEEE Trans. on Auto. Contr., 34, 8, pp. 831-846.

8. Basar, T., (1990), *Optimum performance levels for H_∞ filters, predictors and smoothers*, Sub. for publication in Systems and Control Letters.

10. Grimble, M.J., (1990), *Polynomial matrix solution of the standard state-feedback H_∞ control problem and relationship to the Riccati equation state-space solution*, Research Report No. ICU/302, University of Strathclyde.

11. Grimble, M.J., (1985), *Polynomial systems approach to optimal linear filtering and prediction*, Int. J. Control, Vol. 41, No. 6, p. 1545-1564.

12. Yaesh, I. and Shaked, U., (1989), *Minimum H_∞ norm regulation of linear discrete-time systems and its relation to linear quadratic discrete games*, Proc. 28th CDC Tampa, Florida, pp. 942-947.

13. Grimble, M.J. and Johnson, M.A., (1988), *Optimal control and stochastic estimation, Theory and Applications*, Volumes 1 and 2, John Wiley, Chichester.

14. Grimble, M.J., (1986), *Multivariable controllers for LQG self-tuning applications with coloured measurement noise and dynamic cost weighting*, Int. J. Systems Science, 17, 4, pp. 543-557.

15. Kwakernaak, H., (1990), *MATLAB macros for polynomial H_∞ control system optimization*, Report No. 881, University of Twente.

16. Sebek, M., (1990), *An algorithm for spectral factorization of polynomial matrices with any signature*, Report No. 912, University of Twente.

17. Sebek M. and Kwakernaak, H., (1991), *J-spectral factorization*, Proc of 30th CDC Conference, Brighton, pp. 1278-1283.

18. Grimble, M.J., D. Ho and Elsayed, A., (1989), *H_∞ robust linear estimator*, IFAC Adaptive Systems in Control and Signal Processing Symposium, Glasgow.

19. Fu M, de Souza C.E. and Xe L., (1990), *H_∞ estimation for uncertain systems*, Technical Report No. EE 9039 University of Newcastle, New South Wales.

20. Grimble, M.J., (1987),. *Relationship between polynomial and state-space solutions of the optimal regulator problem*, Systems and Control Letters, 8, pp. 411-416.

21. Grimble, M.J., (1991), *Polynomial matrix solution to the discrete fixed-lag smoothing problem*, Kybernetika, 27, 3, pp. 190-201.

22. Stoorvogel, A., (1992), *The H_∞ control problem, a state-space approach*, Prentice Hall International Series in Systems and Control Engineering, London.

23. Bryson, A.E., and Ho, Y.C., (1969), *Applied optimal control*, Ginn.

24. Grimble, M.J., (1992), *Polynomial matrix solution of the H_∞ filtering problem and the relationship to Riccati equation state-space results*, IEEE Transactions on Signal Processing.

Chapter 7

n-D Polynomial Equations

M. Šebek

7.1 Introduction

A growing interest has been developed over the past decade into systems that depend on two or more independent variables (denoted usually as 2-D or n-D systems, respectively) due to their potential applications in many engineering areas which include digital image processing [1], seismic data processing [1], control of multipass processes [2], control of time delay systems [3] and robust control to mention just a few.

In addition, the consideration of n-D systems has opened up an entirely new research area whose theoretic interest largely exceeds their immediate application. In fact the standard concepts and techniques of 1-D linear systems are no longer sufficient to analyze the dynamic properties of this class of systems and entirely new notions have been and still have to be introduced.

From the state space theory point of view, the major difference between 1-D and n-D ($n \geq 2$) systems is that in the n-D case a *global state* and a *local state* must be jointly taken into account. The global state which preserves all the past information is of *infinite dimension* while the local state is *finite dimensional* and allow us to express system dynamics by recursive linear difference equations.

A typical example of (the local state space model of) n-D scalar system is the Fornasini-Marchesini's model

$$\begin{aligned} x(k_1+1,\ldots,k_n+1) &= A_1 x(k_1+1,k_2,\ldots,k_n)+\cdots+A_n(k_1,\ldots,k_{n-1},k_n+1)+ \\ &\quad B_1 x(k_1+1,k_2,\ldots,k_n)+\cdots+B_n(k_1,\ldots,k_{n-1},k_n+1), \\ y(k_1,\ldots,k_n) &= Cx(k_1,\ldots,k_n), \end{aligned}$$

where $x(k_1,\ldots,k_n)$ is the local state variable at $(k_1,\ldots,k_n) \in \mathcal{Z}^n$, $u(k_1,\ldots,k_n)$ is the input signal and $y(k_1,\ldots,k_n)$ is the output signal while $A_1,\ldots,A_n, B_1,\ldots,B_n$ and C are real constant matrices of suitable sizes.

The more articulate state structure essentially reflects the more complex algebraic structure exhibited in the ring of polynomials in n indeterminates.

Input-output descriptions of such an n-D systems typically reads

$$y(\mathsf{z}_1,\ldots,\mathsf{z}_n) = w(\mathsf{z}_1,\ldots,\mathsf{z}_n)u(\mathsf{z}_1,\ldots,\mathsf{z}_n)+v(\mathsf{z}_1,\ldots,\mathsf{z}_n)$$

where $u(z_1, \ldots, z_n)$ and $y(z_1, \ldots, z_n)$ are the input and the output sequences, respectively, written as formal power series

$$\begin{aligned} u(z_1, \ldots, z_n) &= \sum_{(k_1,\ldots,k_n)\in\mathcal{Z}^n} u(k_1, \ldots, k_n) z_1^{k_1} \ldots z_n^{k_n} \\ y(z_1, \ldots, z_n) &= \sum_{(k_1,\ldots,k_n)\in\mathcal{Z}^n} y(k_1, \ldots, k_n) z_1^{k_1} \ldots z_n^{k_n} \end{aligned}$$

in the (n) indeterminates $z_1, \ldots, z_n$. The first term on the right hand side expresses the input-output map of the system via the n-D sequence

$$w(z_1, \ldots, z_n) = \sum_{(k_1,\ldots,k_n)\in\mathcal{Z}^n} w(k_1, \ldots, k_n) z_1^{k_1} \ldots z_n^{k_n}$$

called *transfer function* or *impulse response*, while the sequence

$$v(z_1, \ldots, z_n) = \sum_{(k_1,\ldots,k_n)\in\mathcal{Z}^n} v(k_1, \ldots, k_n) z_1^{k_1} \ldots z_n^{k_n}$$

expresses a *free motion* of the system from a non-zero initial (boundary) conditions.

It is often possible to put the composite sequential matrix $[w(z_1, \ldots, z_n), v(z_1, \ldots, z_n)]$ into the form of a left factor coprime n-D polynomial matrix fraction

$$a^{-1}(z_1, \ldots, z_n)\,[b(z_1, \ldots, z_n),\ c(z_1, \ldots, z_n)] = [w(z_1, \ldots, z_n),\ v(z_1, \ldots, z_n)]$$

so that the input-output system description can be rewritten as

$$a(z_1, \ldots, z_n) y(z_1, \ldots, z_n) = b(z_1, \ldots, z_n) u(z_1, \ldots, z_n) + c(z_1, \ldots, z_n).$$

So for the Fornasini-Marchesini's model we simply get

$$w(z_1, z_2) = C(I - A_1 z_1 - \cdots - A_n z_2)^{-1}(B_1 z_1 + \cdots + B_n z_n)$$

and

$$v(z_1, z_2) = C(I - A_1 z_1 - \cdots - A_n z_n)^{-1}\mathcal{X}_0(z_1, \ldots, z_n)$$

where $\mathcal{X}_0(z_1, \ldots, z_n)$ stands for the boundary global state sequence [4].

7.2 n-D Polynomials

Hence, an n-D sequence $s(k_1, \ldots, k_n)$ is expressed as the formal power series

$$s(z_1, \ldots, z_n) = \sum_{(k_1,\ldots,k_n)\in\mathcal{Z}^n} s(k_1, \ldots, k_n) z_1^{k_1} \ldots z_n^{k_n}.$$

The sequence is called *causal* if it is zero outside of the first n-tant of $\mathcal{Z}^n$, that is,

$$s(z_1, \ldots, z_n) = \sum_{(k_1,\ldots,k_n)\in\mathcal{N}^n} s(k_1, \ldots, k_n) z_1^{k_1} \ldots z_n^{k_n}$$

such as the 2-D sequence on Fig 7.1. A finite-extent causal n-D sequence

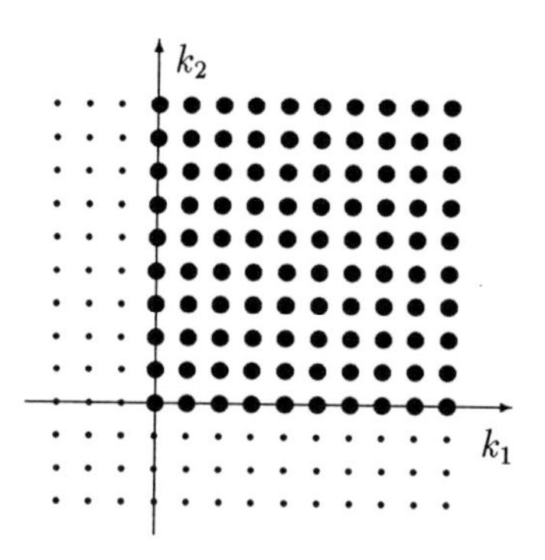

Figure 7.1: Example of a 2-D causal sequence. (The circles represent elements of nonzero values while dots denote elements zero values.)

$$p(z_1, \ldots, z_n) = \sum_{\substack{0 \le k_1 \le m_1 \\ \vdots \\ 0 \le k_n \le m_n}} p(k_1, \ldots, k_n) z_1^{k_1} \ldots z_n^{k_n}$$

is called n-*D polynomial.* As an example, consider the 2-D polynomial

$$p(z_1, z_2) = z_1 + z_2 + z_1 z_2,$$

which is also shown on Fig 7.2. The ring of real n-D polynomials is denoted as usually by $\mathcal{R}[z_1, \ldots, z_n]$. It is often convenient to think of an n-D polynomial

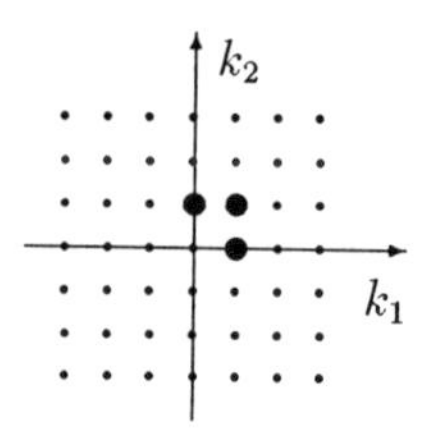

Figure 7.2: 2-D polynomial $p(z_1, z_2)$

$$p(z_1, \ldots, z_n) \in \mathcal{R}[z_1, \ldots, z_n]$$

as of a 1-D polynomial in certain indeterminate, say z_i, over the ring of $(n-1)$-D polynomials in the remaining indeterminates $z_1, \ldots, z_{i-1}, z_{i+1}, \ldots, z_n$, that is

$$\begin{aligned} p(z_1, \ldots, z_n) = \; & p_0(z_1, \ldots, z_{i-1}, z_{i+1}, \ldots, z_n) \\ & + p_1(z_1, \ldots, z_{i-1}, z_{i+1}, \ldots, z_n) z_i \\ & \vdots \\ & + p_m(z_1, \ldots, z_{i-1}, z_{i+1}, \ldots, z_n) z_i^m \end{aligned}$$

The ring of such polynomials is denoted by $\mathcal{R}[z_1, \ldots, z_{i-1}, z_{i+1}, \ldots, z_n][z_i]$.

In particular, the polynomial $p(z_1, \ldots, z_n)$ is called *monic* in z_i (or in $\mathcal{R}[z_1, \ldots, z_{i-1}, z_{i+1}, \ldots, z_n][z_i]$) if

$$p_m(z_1, \ldots, z_{i-1}, z_{i+1}, \ldots, z_n) = 1.$$

However, neither $\mathcal{R}[z_1, \ldots, z_n]$ nor $\mathcal{R}[z_1, \ldots, z_{i-1}, z_{i+1}, \ldots, z_n][z_i]$ is Euclidean for $n \geq 2$. Therefore, to enjoy Euclidean properties, we must embed $\mathcal{R}[z_1, \ldots, z_n]$ into $\mathcal{R}(z_1, \ldots, z_{i-1}, z_{i+1}, \ldots, z_n)[z_i]$, the ring which allows for fractions in all the indeterminates but one (z_i). This "closest larger Euclidean ring" to $\mathcal{R}[z_1, \ldots, z_n]$ ($\mathcal{R}[z_1, \ldots, z_n] \subset \mathcal{R}(z_1, \ldots, z_{i-1}, z_{i+1}, \ldots, z_n)[z_i]$) will be often exercised in what follows. To illustrate this ring, just consider the "polynomial"

$$p(z_1, z_2, z_3) = \frac{1}{z_1 + z_2} + \frac{z_1}{z_2} z_3 \ \in \ \mathcal{R}(z_1, z_2)[z_3].$$

As mentioned earlier, the basic idea of the polynomial approach to n-D systems control and filtering advanced in this book is that the design procedure can be reduced to solving certain linear equations for n-D polynomials or n-D polynomial matrices [5, 6, 7, 8, 9, 3, 10, 11, 12, 13, 14, 15]. The terminology will be taken from the 1-D theory developed during the last two decades but, in fact, it goes far back to the theory of Diophantine equations, for integers and polynomials are similar objects from the abstract-algebraic point of view. For simplicity, we shall survey just scalar linear n-D polynomial equations in this Chapter.

7.3 1-D Polynomial Equations

We begin by reviewing the standard 1-D results which, for convenience, are formulated here for 1-D polynomials over a general field (and not merely the field of real numbers).

Consider the equation

$$a(z_1)x(z_1) + b(z_1)y(z_1) = c(z_1), \tag{7.1}$$

where $a(z_1), b(z_1)$ and $c(z_1)$ are given 1-D polynomials over a field R, i.e., elements of the ring $\mathsf{R}[z_1]$. This ring is clearly Euclidean for R is a field. The equation will be called a *linear 1-D polynomial equation* and by its solution we mean any pair $x(z_1), y(z_1) \in \mathsf{R}[z_1]$ satisfying (7.1).

In the examples of this chapter, we shall mostly consider R being the field of real (1-D till $(n-1)$-D) polynomial fractions $\mathsf{R} = \mathcal{R}(z)$ or, in particular, the field of real numbers $\mathsf{R} = \mathcal{R}$. Most of the $\mathsf{R}[z_1]$ results presented in this Section are almost identical with the $\mathcal{R}[z_1]$ results published in the popular book [16].

So the following fundamental theorem is well known.

Theorem 7.1 (Solvability) *Equation (7.1) has a solution if and only if* $\gcd(a, b) | c$.

PROOF OF 7.1: Only if: Let x', y' be a solution of (7.1). Then with the notation (7.2)

$$g(\bar{a}x' + \bar{b}y') = c$$

so that $g|c$.

If: Let $\gcd(a, b)|c$ and denote

$$g = \gcd(a, b), \quad c = g\bar{c}.$$

As the ring is Euclidean, polynomials $p, q \in \mathsf{R}[\mathsf{z}_1]$ always exist [16] such that

$$ap + bq = g.$$

Multiplying by $\bar{c}$, we obtain

$$a(p\bar{c}) + b(q\bar{c}) = c$$

and hence we have constructed a solution $p\bar{c}, q\bar{c}$ of (7.1). □

If $g = \gcd(a, b)$ then we denote by $\bar{a}$ and $\bar{b}$ the *coprime parts* of a and b, respectively, that is coprime polynomials such that

$$a = g\bar{a} \quad \text{and} \quad b = g\bar{b}. \tag{7.2}$$

Equation (7.1) is linear and hence its general solution can be obtained from a particular solution as follows.

Theorem 7.2 (General Solution: 1-D case) *Let x', y' be a particular solution of (7.1). Then with notation (7.2) the general solution of (7.1) reads*

$$\begin{aligned} \mathrm{x} &= \mathrm{x}' - \bar{\mathrm{b}}\mathrm{t} \\ \mathrm{y} &= \mathrm{y}' + \bar{\mathrm{a}}\mathrm{t} \end{aligned} \tag{7.3}$$

for an arbitrary polynomial $\mathrm{t} \in \mathsf{R}[\mathsf{z}_1]$.

◇

PROOF OF 7.2: By assumption $ax' + by' = c$ so that

$$a(x - x') = -b(y - y'). \tag{7.4}$$

Now the polynomials $\bar{a}$ and $\bar{b}$ defined in (7.2) are coprime and satisfy $a\bar{b} = b\bar{a}$. As a result $\bar{b}|(x - x')$ and $\bar{a}|(y - y')$, that is,

$$\begin{aligned} x - x' &= -\bar{b}t \\ y - y' &= \bar{a}t \end{aligned}$$

for some polynomial t. To obtain any solution of (7.4) we let t range over $\mathsf{R}[\mathsf{z}_1]$ and the claim follows. □

There are many methods how to obtain a solution of (7.1) since R is a field. For example, we can apply the transformation

$$\begin{bmatrix} a & b \\ 1 & 0 \\ 0 & 1 \end{bmatrix} \longrightarrow \begin{bmatrix} g & 0 \\ p & r \\ q & s \end{bmatrix}, \tag{7.5}$$

and calculate a greatest common divisor g of a and b along with two pairs of coprime polynomials p, q and r, s satisfying the relations

$$\begin{aligned} ap + bq &= g \\ ar + bs &= 0 \end{aligned} \tag{7.6}$$

We also calculate $\bar{c} = c/g$. Then, by comparing (7.6) with Theorem 7.2, we can identify $x' = p\bar{c}$, $y' = q\bar{c}$ and $-\bar{b} = r$, $\bar{a} = s$. Hence the general solution of the equation (7.1) is obtained from

$$\begin{aligned} x &= p\bar{c} + rt \\ y &= q\bar{c} + st \end{aligned} \tag{7.7}$$

where $t \in \mathsf{R}[\mathsf{z}_1]$ is an arbitrary polynomial. The $\bar{c}$ must be a polynomial; this forms a useful check on solvability of the equation.

Example 7.3 Solve the equation

$$(1 + \mathsf{z}_1\mathsf{z}_2 + \mathsf{z}_2^2 + \mathsf{z}_1\mathsf{z}_2^2)\, x + (\mathsf{z}_2 + \mathsf{z}_1\mathsf{z}_2)\, y = 2 + \mathsf{z}_1 + \mathsf{z}_2$$

in the ring $\mathcal{R}(\mathsf{z}_2)[\mathsf{z}_1]$. After rewriting it properly to the form

$$\left((1 + \mathsf{z}_2^2) + (\mathsf{z}_2 + \mathsf{z}_2^2)\mathsf{z}_1\right) x + (\mathsf{z}_2 + \mathsf{z}_2\mathsf{z}_1)y = (2 + \mathsf{z}_2) + \mathsf{z}_1,$$

we apply the transformation (7.5)

$$\begin{bmatrix} (1 + \mathsf{z}_2^2) + (\mathsf{z}_2 + \mathsf{z}_2^2)\mathsf{z}_1 & \mathsf{z}_2 + \mathsf{z}_2\mathsf{z}_1 \\ 1 & 0 \\ 0 & 1 \end{bmatrix} \longrightarrow \begin{bmatrix} 1 & 0 \\ 1/1 - \mathsf{z}_2^2 & -\mathsf{z}_2 - \mathsf{z}_2\mathsf{z}_1 \\ -1/1 - \mathsf{z}_2 & (1 + \mathsf{z}_2) + (\mathsf{z}_2 + \mathsf{z}_2^2)\mathsf{z}_1 \end{bmatrix}$$

to calculate the polynomials

$$\begin{aligned} g &= 1 \\ p &= \frac{1}{1 - \mathsf{z}_2^2}, & q &= \frac{-1}{1 - \mathsf{z}_2} \\ r &= -\mathsf{z}_2 - \mathsf{z}_2\mathsf{z}_1, & s &= (1 + \mathsf{z}_2^2) + (\mathsf{z}_2 + \mathsf{z}_2^2)\mathsf{z}_1. \end{aligned}$$

Since

$$\bar{c} = 2 + \mathsf{z}_2 + \mathsf{z}_1,$$

the equation is solvable and its general solution (in $\mathcal{R}(\mathsf{z}_2)[\mathsf{z}_1]$) reads

$$\begin{aligned} x &= \frac{2 + \mathsf{z}_1 + \mathsf{z}_2}{1 - \mathsf{z}_2^2} - (\mathsf{z}_2 + \mathsf{z}_2\mathsf{z}_1)t \\ y &= -\frac{2 + \mathsf{z}_1 + \mathsf{z}_2}{1 - \mathsf{z}_2} + (1 + \mathsf{z}_2^2 + (\mathsf{z}_2 + \mathsf{z}_2^2)\mathsf{z}_1)t \end{aligned}$$

with an arbitrary polynomial $t \in \mathcal{R}(\mathsf{z}_2)[\mathsf{z}_1]$.

◇

Example 7.4 As another example, consider the same equation in the ring $\mathcal{R}(z_1)[z_2]$, that is

$$\left(1 + z_1 z_2 + (1 + z_1) z_2^2\right) x + (1 + z_1) z_2 \, y = (2 + z_1) + z_2.$$

Now transformation (7.5) reads

$$\begin{bmatrix} 1 + z_1 z_2 + (1 + z_1) z_2^2 & (1 + z_1) z_2 \\ 1 & 0 \\ 0 & 1 \end{bmatrix} \longrightarrow \begin{bmatrix} 1 & 0 \\ 1 & -(1 + z_1) z_2 \\ -z_1/(1 + z_1) - z_2 & 1 + z_1 z_2 + (1 + z_1) z_2^2 \end{bmatrix}$$

so that we have calculated the polynomials

$$\begin{aligned} & g = 1 \\ & p = 1, \qquad q = \frac{-z}{(1 + z_1)} - z_2 \\ & r = -(1 + z_1) z_2, \quad s = 1 + z_1 z_2 + (1 + z_1) z_2^2. \end{aligned}$$

As

$$\bar{c} = 2 + z_1 + z_2,$$

the equation is solvable and has the general solution (in $\mathcal{R}(z_1)[z_2]$)

$$\begin{aligned} x = &\ 2 + z_1 + z_2 & & - \ (1 + z_1) z_2 t \\ y = &\ -\frac{2 z_1 + z_1^2}{1 + z_1} - \frac{2 + 3 z_1 + z_1^2}{1 + z_1} z_2 + z_2^2 & & + \ (1 + z_1 z_2 + (1 + z_1) z_2^2) t. \end{aligned}$$

where $t \in \mathcal{R}(z_1)[z_2]$.

◇

In applications, among many solutions of the equation we are usually to find a particular solution satisfying certain additional properties due to the nature of the problem at hand. Perhaps the most important particular solution is one which minimizes the degree of one polynomial, say x. It is called the *minimum-degree solution* with respect to x and can be easily found.

Let x', y' be a particular solution of (7.1). By Theorem 7.2 the general solution takes the form (7.3) for any $t \in \mathsf{R}[z_1]$. By polynomial division in $\mathsf{R}[z_1]$, reduce x' modulo $\bar{b}$

$$x' = \bar{b} u + v$$

with $\deg v < \deg \bar{b}$. Then

$$x = v - \bar{b}(t - u)$$

and the minimum-degree solution x, y with respect to x becomes

$$\begin{aligned} x &= v \\ y &= y' + \bar{a} u \end{aligned}$$

on putting $t = u$. Hence, the minimum-degree solution with respect to x is distinguished by the
property

$$\deg x < \deg \bar{b}.$$

The fellow polynomial y then satisfies

$$\deg y \leq \deg a + p - 1,$$

where

$$p = \max\{0, \deg c - \deg a - \deg b + 1\}.$$

Dually, the minimum-degree solution with respect to y can be derived featuring

$$\deg y < \deg \bar{a}.$$

and

$$\deg x \leq \deg b + p - 1.$$

Note that minimum-degree solutions with respect to x and with respect to y may not in general be identical; however, either solution is unique. Nonetheless, whenever

$$\deg a + \deg b - 1 \geq \deg c,$$

then the two minimum-degree solutions (with respect to x and with respect to y) coincide and are called simply the *minimum-degree* solution.

Example 7.5 Consider again the equation from Example 7.4 and determine its minimum-degree solution with respect to x. The general solution in the ring $\mathcal{R}(\mathsf{z}_1)[\mathsf{z}_2]$ was found to be

$$\begin{aligned} x = {} & 2 + \mathsf{z}_1 + \mathsf{z}_2 && - \quad (1 + \mathsf{z}_1)\mathsf{z}_2 t \\ y = {} & -\frac{2\mathsf{z}_1 + \mathsf{z}_1^2}{1 + \mathsf{z}_1} - \frac{2 + 3\mathsf{z}_1 + \mathsf{z}_1^2}{1 + \mathsf{z}_1}\mathsf{z}_2 + \mathsf{z}_2^2 && + \quad (1 + \mathsf{z}_1\mathsf{z}_2 + (1 + \mathsf{z}_1)\mathsf{z}_2^2)t. \end{aligned}$$

To calculate the minimum-degree solution, we divide the polynomial $\bar{b} = (1 + \mathsf{z}_1)\mathsf{z}_2$ into $x' = 2 + \mathsf{z}_1 + \mathsf{z}_2$. We obtain the quotient

$$u = \frac{-1}{1 + \mathsf{z}_1}$$

and the reminder

$$v = 2 + \mathsf{z}_1.$$

Thus, the minimum-degree solution[1] in $\mathcal{R}(\mathsf{z}_1)[\mathsf{z}_2]$ results from the substitution $t = u = -1/(1 + \mathsf{z}_1)$ into (7.7) and equals

$$\begin{aligned} x &= 2 + \mathsf{z}_1 \\ y &= -1 - \mathsf{z}_1 - 2\mathsf{z}_2 - \mathsf{z}_1\mathsf{z}_2. \end{aligned}$$

At the same time, this is the minimum-degree solution in $\mathcal{R}(\mathsf{z}_1)[\mathsf{z}_2]$ with respect to y for $\deg_{\mathsf{z}_2} a + \deg_{\mathsf{z}_2} b - 1 = 2 > \deg_{\mathsf{z}_2} c = 1$.

◇

[1]Notice that, accidentally, $x, y \in \mathcal{R}[\mathsf{z}_1, \mathsf{z}_2]$ ($\subset \mathcal{R}(\mathsf{z}_1)[\mathsf{z}_2]$). In fact, computations in $\mathcal{R}(\mathsf{z}_1)[\mathsf{z}_2]$ are often used to get the solution in $\mathcal{R}[\mathsf{z}_1, \mathsf{z}_2]$ as we shall see in Section 7.4.

Thus, as the analysis above reveals, there exists a unique solution $x, y \in \mathsf{R}[z_1]$ such that $\deg x < \deg \bar{b}$ and, consequently, at least one solution with $\deg x < \deg b$ whenever the equation is solvable. When forming the matrices

$$A_0 = \underbrace{\left.\begin{bmatrix} a_0 & \dots & a_k & & & & \\ & a_0 & \dots & a_k & & & \\ & & \ddots & & \ddots & & \\ & & & a_0 & \dots & a_k, 0, \dots, 0 \end{bmatrix}\right\} l,}_{p+k+l} \tag{7.8}$$

$$B_p = \left.\begin{bmatrix} b_0 & \dots & b_l & & & \\ & b_0 & \dots & b_l & & \\ & & \ddots & & \ddots & \\ & & & b_0 & \dots & b_l \end{bmatrix}\right\} k+p, \tag{7.9}$$

$$C_p = \underbrace{\begin{bmatrix} c_0, \dots, c_m, 0, \dots, 0 \end{bmatrix}}_{p+k+l}, \tag{7.10}$$

where we have denoted

$$\begin{aligned} k &= \deg a, \\ l &= \deg b, \\ m &= \deg c, \\ p &= \max\{0, m - (k + l - 1)\}, \end{aligned} \tag{7.11}$$

along with the matrices

$$\begin{aligned} X_p &= \begin{bmatrix} x_0, \dots, x_{l-1} \end{bmatrix}, \\ Y_p &= \begin{bmatrix} y_0, \dots, y_{k+p-1} \end{bmatrix}, \end{aligned} \tag{7.12}$$

the original scalar polynomial equation (in $\mathsf{R}[z_1]$) (7.1) can be transformed into a matrix equation

$$\begin{bmatrix} X & Y \end{bmatrix} \begin{bmatrix} A_0 \\ B_p \end{bmatrix} = C_p. \tag{7.13}$$

in the field R. Clearly (7.1) is solvable whenever (7.13) is and vice versa. This maneuver suggests an alternative to the solvability condition of Theorem 7.1:

Theorem 7.6 (Solvability) *Equation (7.1) is solvable if and only if (7.13) is.* ◇

If $b = \bar{b}$, then the solution of (7.13) is unique and generates the minimum-degree solution of (7.1) with respect to x.

Similar matrix equation can be derived for the minimum-degree solution with respect to y. Such a way, yet another solvability condition to (7.1) can be obtained.

To complete the discussion, notice that the matrix

$$\begin{bmatrix} A_0 \\ B_p \end{bmatrix} \tag{7.14}$$

becomes square if $p = 0$ and, in addition, nonsingular if a and b are coprime (in $\mathcal{R}(z_1)[z_2]$). In such a case, (7.13) is simply solvable by

$$\begin{bmatrix} X & Y \end{bmatrix} = C \begin{bmatrix} A_0 \\ B_0 \end{bmatrix}^{-1}. \tag{7.15}$$

This relation gives directly rise to the (unique) minimum-degree solution of (7.1).

Example 7.7 Consider once again the equation (7.3) from Example 7.4. Because $k = 2$, $l = m = 1$ and $p = 0$, it gives rise to the matrices

$$\begin{aligned} A_0 &= \begin{bmatrix} 1 & z_1 & 1+z_1 \end{bmatrix} \\ B_0 &= \begin{bmatrix} 0 & 1+z_1 & 0 \\ 0 & 0 & 1+z_1 \end{bmatrix} \\ C_0 &= \begin{bmatrix} 2+z_1 & -1 & 0 \end{bmatrix} \end{aligned}$$

and the desired solution is expected to come from the matrices

$$\begin{aligned} X_0 &= [x_0] \\ Y_0 &= \begin{bmatrix} y_0 & y_1 \end{bmatrix}. \end{aligned}$$

As $p = 0$, the matrix $\begin{bmatrix} A_0 \\ B_0 \end{bmatrix}$ is square. Further it is nonsingular (for a and b are coprime in the ring $\mathcal{R}(z_1)[z_2]$) so that

$$\begin{bmatrix} X_0 & Y_0 \end{bmatrix} = C_0 \begin{bmatrix} A_0 \\ B_0 \end{bmatrix}^{-1} = \begin{bmatrix} 2+z_1 & -1-z_1 & -2-z_1 \end{bmatrix}.$$

This yields the same solution as in Example 7.5

$$\begin{aligned} x &= 2 + z_1, \\ y &= -1 - z_1 - 2z_2 - z_1 z_2. \end{aligned}$$

◇

7.4 2-D Polynomial Equations

Let us now appeal the 2-D version of the scalar linear equation [6]. The ring $\mathcal{R}[z_1, z_2]$ is not endowed by Euclidean division and, hence, one can hardly expect that most of the results presented in Section 7.3 keep holding. Surprisingly, some of them stand up all the same, occasionally under more restrictive conditions. In fact, we often benefit from thinking of 2-D polynomials as of elements of $\mathcal{R}[z_1][z_2]$ or $\mathcal{R}[z_2][z_1]$, i.e., as of 1-D polynomials over a Euclidean ring.[2] This enables one to perform Euclidean division among coefficients, at least.

Afresh we study the equation

$$a(z_1, z_2)x(z_1, z_2) + b(z_1, z_2)y(z_1, z_2) = c(z_1, z_2) \tag{7.16}$$

where now $a(z_1, z_2), b(z_1, z_2)$ and $c(z_1, z_2)$ are given polynomials from $\mathcal{R}[z_1, z_2]$ and, naturally, the solution x, y is to be found within the same ring. The first difference between 1-D and 2-D equations materializes when verifying Theorem 7.1.

Theorem 7.8 (Solvability) $\gcd(a, b)|c$ *whenever (7.16) is solvable.*

◇

PROOF OF 7.8: For a greatest common divisor $d = \gcd(a, b)$ (and, in fact, for any common divisor at all), the solvability of (7.16) implies $ax + by = d(\bar{a}x + \bar{b}y) = c$ so that $d|c$.

□

Hence, the divisibility condition remains necessary. On the other hand, it is *not sufficient* any longer.

Example 7.9 To see that the condition of Theorem 7.6 is not sufficient, just check that the simple 2-D equation

$$z_1\, x + z_2\, y = 1, \tag{7.17}$$

is without solution despite that the polynomials $a = z_1$ and $b = z_2$ are factor coprime (clearly $\gcd(z_1, z_2) = 1$). Indeed, on substituting $z_1 = z_2 = 0$, the left hand side of (7.17) vanishes while the right hand side remains nonzero. Hence, a and b are *not* zero coprime having the zero $(z_1, z_2) = (0, 0)$ in common.

◇

Roughly speaking, the equation (7.1) is solvable provided that every common zero of a and b is a zero of c with the right multiplicity (the famous Fundamental Theorem of Noether). In the case of a constant right hand side (such as in (7.17)), this property is relatively easy to test.

Theorem 7.10 *The 2-D equation*[3]

$$a(z_1, z_2)x(z_1, z_2) + b(z_1, z_2)y(z_1, z_2) = 1 \tag{7.18}$$

is solvable if and only if the polynomials a *and* b *have* no *zero in common.*

◇

PROOF OF 7.10: This is a direct consequence of the famous Hilbert Nullstelensatz.

□

---

[2]The former ring, $\mathcal{R}[z_1][z_2]$, will be preferred throughout this section. It goes without saying, however, that the roles of z_1 and z_2 can be always completely interchanged. This would yield a dual collection of theorems and examples.

[3]The equation (7.18) is often called *Bezout equation*.

However, the more general version of Theorem 7.8, the Fundamental Theorem of Noether, is very difficult to inspect practically. Computationally workable solvability conditions are more elaborate and will be derived later on.

For simplicity, we shall assume from now on, that any possible common factor of a and b has already been canceled from (7.16) so that

Assumption 7.11 *the polynomials* a *and* b *are* factor coprime,

which calls for

$$a = \bar{a} \text{ and } b = \bar{b}.$$

This is quite fair for the impact of possible common factors, which was discussed in the last section, remains unchanged regardless on n. Thus, we can concentrate on new phenomena arising from the 2-D nature.

When supposing Assumption 7.11, the general solution of (7.16) is as follows.

Theorem 7.12 (General Solution: 2-D case) *Let* a *and* b *be factor coprime polynomials and* x′, y′ *be a particular solution of (7.16). Then the general solution of (7.16) reads*

$$\begin{aligned} \mathrm{x} &= \mathrm{x}' - \mathrm{bt} \\ \mathrm{y} &= \mathrm{y}' + \mathrm{at} \end{aligned} \tag{7.19}$$

for an arbitrary polynomial $\mathrm{t} \in \mathsf{R}[\mathsf{z}_1, \mathsf{z}_2]$.

◇

PROOF OF 7.12: The proof is identical with that of Theorem 7.2. □

In contrast to the 1-D case, we postpone the general question of solvability until at the existence the minimum-degree solution will be clarified. In fact, all the 2-D solvability test published thus far consist of checking the existence of the minimum-degree solution. Lacking of the Euclidean algorithm type tool, one should not expect the existence of the minimum-degree solution in 2-D at all. Notwithstanding, it appears to exist under certain quite general conditions. To see this, let us write the given polynomials as elements of the ring $\mathcal{R}[\mathsf{z}_1][\mathsf{z}_2]$

$$\begin{aligned} a(\mathsf{z}_1, \mathsf{z}_2) &= a_0(\mathsf{z}_1) + a_1(\mathsf{z}_1)\mathsf{z}_2 + \cdots + a_k(\mathsf{z}_1)\mathsf{z}_2^k \\ b(\mathsf{z}_1, \mathsf{z}_2) &= b_0(\mathsf{z}_1) + b_1(\mathsf{z}_1)\mathsf{z}_2 + \cdots + b_l(\mathsf{z}_1)\mathsf{z}_2^l \\ c(\mathsf{z}_1, \mathsf{z}_2) &= c_0(\mathsf{z}_1) + c_1(\mathsf{z}_1)\mathsf{z}_2 + \cdots + c_m(\mathsf{z}_1)\mathsf{z}_2^m \end{aligned} \tag{7.20}$$

where $k = \deg_{\mathsf{z}_2} a$, $l = \deg_{\mathsf{z}_2} b$ and $m = \deg_{\mathsf{z}_2} c$. Further denote

$$d_{\mathrm{LC}}(\mathsf{z}_1) = \gcd(a_k(\mathsf{z}_1), b_l(\mathsf{z}_1)) \in \mathcal{R}[\mathsf{z}_1] \tag{7.21}$$

and

$$\begin{aligned} a_k(\mathsf{z}_1) &= \bar{a}_k\,(\mathsf{z}_1)d_{\mathrm{LC}}(\mathsf{z}_1) \\ b_l(\mathsf{z}_1) &= \bar{b}_l(\mathsf{z}_1)\,d_{\mathrm{LC}}(\mathsf{z}_1) \end{aligned} \tag{7.22}$$

where $\bar{a}_k(\mathsf{z}_1), \bar{b}_l(\mathsf{z}_1) \in \mathcal{R}[\mathsf{z}_1]$, too. Then there are sufficient conditions for the minimum-degree solution in z_2 similar to those derived in Section 7.3.

Theorem 7.13 (Minimum-degree solution) *Let the equation (7.16) be solvable and let*

$$\mathrm{m} \leq \mathrm{k} + \mathrm{l} - 1. \tag{7.23}$$

If either

$$\mathrm{d}_{\mathrm{LC}}(\mathrm{z}_1) = 1 \tag{7.24}$$

or, at least[4],

$$\gcd\left(\mathrm{d}_{\mathrm{LC}}(\mathrm{z}_1),\ \mathrm{a}_{\mathrm{k}-1}(\mathrm{z}_1)\bar{\mathrm{b}}_{\mathrm{l}}(\mathrm{z}_1) - \bar{\mathrm{a}}_{\mathrm{k}}(\mathrm{z}_1)\mathrm{b}_{\mathrm{l}-1}(\mathrm{z}_1)\right) = 1 \ \textit{and}\ \ \mathrm{d}_{\mathrm{LC}}(\mathrm{z}_1)|\mathrm{c}_{\mathrm{k}+\mathrm{l}-1}(\mathrm{z}_1), \tag{7.25}$$

then there exists a unique solution $\mathrm{x}(\mathrm{z}_1, \mathrm{z}_2), \mathrm{y}(\mathrm{z}_1, \mathrm{z}_2)$ *of (7.16) such that*

$$\begin{aligned} \deg_{\mathrm{z}_2}\mathrm{x} &\leq \mathrm{l} - 1 \\ \deg_{\mathrm{z}_2}\mathrm{y} &\leq \mathrm{k} - 1 \end{aligned} \tag{7.26}$$

◇

PROOF OF 7.13: Write

$$\begin{aligned} x' &= x'_0 + x'_1\mathrm{z}_2 + \cdots + x_p\mathrm{z}_2^p \\ y' &= y'_0 + y'_1\mathrm{z}_2 + \cdots + y_q\mathrm{z}_2^q, \end{aligned}$$

where $x'_i, y'_j \in \mathcal{R}[\mathrm{z}_1]$, for a solution x', y' of (7.16) and suppose that $p > l - 1$. Quoting from (7.19), any x of the form

$$x = x'_0 + x'_1\mathrm{z}_2 + \cdots + x_p\mathrm{z}_2^p + (b_0 + b_1\mathrm{z}_2 + \cdots + b_l\mathrm{z}_2^l)t$$

with an arbitrary $t \in \mathcal{R}[\mathrm{z}_1, \mathrm{z}_2]$ is also a solution of (7.16). An x of lower degree ($\deg_{\mathrm{z}_2} x < p$) can be evidently found if and only if

$$b_l | x'_p. \tag{7.27}$$

We are now to show that this is always the case when (7.23) and (7.24) or (7.25) are satisfied. To do this we equate the coefficients of the highest (i.e, $p + k$) power of z_2 in (7.16) to get

$$a_k x'_p + b_l y'_q = 0 \tag{7.28}$$

which implies

$$b_l | a_k x'_p. \tag{7.29}$$

Now (7.24) causes (7.29) to yield (7.28) immediately. If, however, only (7.25) is satisfied, then (7.29) proves just $\bar{b}_l | x'_p$ so that

$$x'_p = \bar{b}_l \tag{7.30}$$

for a polynomial $u \in \mathcal{R}[\mathrm{z}_1]$ whereby

$$y'_q = -\bar{a}_k u \tag{7.31}$$

as well. Nevertheless, equating further the coefficients of z_2^{l+p-1} in (7.16) gives

$$a_{k-1}x'_p + a_k x'_{p-1} + b_{q-1}y'_q + b_l y'_{q-1} = \begin{cases} 0 & \text{for } p > l \\ c_{k+l-1} & \text{for } p = l \end{cases} \tag{7.32}$$

[4] Here, if necessary, we take $c_i = 0$ for $i > m$.

and, on inserting (7.30) and (7.31),

$$(a_{p-1}\bar{b}_q - \bar{a}_p b_{q-1})u + d_{\mathrm{LC}}(\bar{a}_k x_{q-1} + \bar{b}_l)y_{p-1} = \begin{cases} 0 \\ c_{k+l-1} \end{cases} \qquad (7.33)$$

from which, by (7.25),

$$d_{\mathrm{LC}}|u,$$

so that, in view of (7.31), the desired (7.30) results again.

We summarize what was done so far: For any solution x', y' with $\deg_{\mathbf{z}_2} x' > l-1$ one can employ (7.27) to compute another solution x, y with $\deg_{\mathbf{z}_2} x < \deg_{\mathbf{z}_2} x'$. After repeating this procedure until $\deg_{\mathbf{z}_2} x \leq l-1$, the minimum $\mathbf{z}_2$-degree solution results. It is easy to see, that under (7.23) always $q = k + p - l$ so that for the desired solution we have $\deg_{\mathbf{z}_2} y \leq k-1$. □

Example 7.14 As an example, examine the equation

$$(1 + \mathbf{z}_1 + \mathbf{z}_2)x - (\mathbf{z}_1 + \mathbf{z}_2)y = 1.$$

Here simply $k = 1$ and $a_k = 1$ while $l = 1$ with $b_l = 1$ and $m = 0$ with $c_0 = 1$. Hence, (7.23) and (7.24) are satisfied. The equation has the unique minimum degree solution (7.26)

$$x = 1, \quad y = 1.$$

Its general solution reads, e.g.,

$$\begin{aligned} x &= 1 + (\mathbf{z}_1 + \mathbf{z}_2)t \\ y &= 1 + (1 + \mathbf{z}_1 + \mathbf{z}_2)t. \end{aligned}$$

◇

Since the assumptions (7.23)- (7.25) make it possible to estimate the minimal $\mathbf{z}_2$-degree of the solution, they simplify both solvability conditions and numerical procedure as we shall see in a while. The assumptions are satisfied in many practical problems. Even if not, they may hold in the complementary ring $\mathcal{R}[\mathbf{z}_2][\mathbf{z}_1]$ and the role of $\mathbf{z}_1$ and $\mathbf{z}_2$ can be completely interchanged.

Example 7.15 In the case of the equation from Example 7.14, Theorem 7.13 applies also in the ring $\mathcal{R}[\mathbf{z}_2][\mathbf{z}_1]$. Indeed, the dual conditions are satisfied so that the minimum $\mathbf{z}_1$-degree solution exists. By a coincidence, it equals the minimum $\mathbf{z}_2$-degree solution found in Example 7.14, that is,

$$x = 1, \quad y = 1.$$

In particular, the condition (7.24) of Theorem 7.13 is met if either a or b is monic in $\mathcal{R}[z_1][z_2]$, that is, if either $a_k(z_1) = 1$ or $b_l(z_1) = 1$. Recall that any polynomial

$$e(z_1, z_2) = e(\bar{z}_1, \bar{z}_2)$$

can be made monic on substituting

$$\begin{aligned} z_1 &= c_1\bar{z}_2 + \bar{z}_1 \\ z_2 &= c_2\bar{z}_2 \end{aligned} \tag{7.34}$$

with appropriately chosen nonzero real constants c_1 and c_2. Such a substitution makes it always possible to satisfy the condition (7.24) of Theorem 7.13.

Let us now investigate what happens, when the conditions of Theorem 7.13 are infringed. So if the right hand side polynomial c is of higher degree which goes beyond the bound (7.23) by

$$p = m - (k + l - 1) \geq 0, \tag{7.35}$$

there always exist a solution with

$$\begin{aligned} \deg_{z_2} x &= l - 1 + p &= m - k \\ \deg_{z_2} y &= k - 1 + p &= m - l. \end{aligned} \tag{7.36}$$

Hence, a possible violation of (7.23) causes no problems and we are still able to estimate the minimum z_2-degree solution. No such bound is known, however, when (7.24) is violated by

$$\gcd(d_{\mathrm{LC}}, a_{k-1}\bar{b}_l - \bar{a}_k b_{l-1}) \neq 1.$$

Example 7.16 Just consider the equation

$$(1 + z_2 + (-1 + z_1)z_2)\,x + (z_1z_2 + (-1 + z_1)z_2)\,y = (1 + z_1)z_2^3$$

having the solution

$$\begin{aligned} x &= z_2^3 \\ y &= z_2^2 - z_2^3, \end{aligned}$$

the z_2-degree of which cannot be further reduced. This is clear from a glance at the general solution

$$\begin{aligned} x &= z_2^3 &&- (z_1z_2 + (-1 + z_1)z_2)t \\ y &= z_2^2 - z_2^3 &&+ (1 + z_2 + (-1 + z_1)z_2)t. \end{aligned}$$

◇

Although the study of minimum degree solution presented above is interesting by itself for its relation to minimum order controllers, its importance stems from the fact that it implies constructive necessary and sufficient solvability conditions to the polynomial equation (7.16).

The underlaying philosophy is as follows. Once an *a priori* degree bound of the solution is known, then only existence of this (minimum-degree) solution can be tested instead of existence of any solution. If the minimum-degree solution exists, then the equation is solvable, of course. If it fails to exist, however, then the equation has *no solution at all* due to Theorem 7.13!

We shall discuss two ways of solving of the equation (7.16), or, as the case may be, of verifying its unsolvability. The first approach is based on the 1-D matrix polynomial equation with matrices created from 1-D polynomial coefficients of the given polynomials.

More concretely, let Assumptions 7.11 be satisfied and let us define 1-D polynomial matrices

$$A(z_1) = \left.\begin{bmatrix} a_0 & \dots & a_k & & & \\ & a_0 & \dots & a_k & & \\ & & \ddots & & \ddots & \\ & & & a_0 & \dots & a_k \end{bmatrix}\right\} l, \tag{7.37}$$

$$B(z_1) = \left.\begin{bmatrix} b_0 & \dots & b_l & & & \\ & b_0 & \dots & b_l & & \\ & & \ddots & & \ddots & \\ & & & b_0 & \dots & b_l \end{bmatrix}\right\} k, \tag{7.38}$$

$$C(z_1) = \underbrace{\begin{bmatrix} c_0, \dots, c_m, 0, \dots, 0 \end{bmatrix}}_{k+l}, \tag{7.39}$$

where the same notation has been used as in (7.11) and (7.20). Then the necessary and sufficient solvability condition for (7.16) is as follows.

Theorem 7.17 (Solvability) *Let Assumption 7.11 be satisfied. Then the equation (7.16) is solvable is and only if the square matrix*

$$\begin{bmatrix} A(z_1) \\ B(z_1) \end{bmatrix} \tag{7.40}$$

is a right divisor of the matrix $C(z_1)$, *that is, there exist a 1-D polynomial matrix* $\bar{C}(z_1)$ *such that*

$$C(z_1) = \bar{C}(z_1) \begin{bmatrix} A(z_1) \\ B(z_1) \end{bmatrix}. \tag{7.41}$$

The minimum z_2*-degree solution of (7.16)*

$$\begin{aligned} x(z_1, z_2) &= x(z_1)_0 + x_1(z_1)z_2 + \dots + x_{l-1}(z_1)z_2^{l-1} \\ y(z_1, z_2) &= y_0(z_1) + y_1(z_1)z_2 + \dots + y_{k-1}(z_1)z_2^{k-1} \end{aligned} \tag{7.42}$$

is given by the matrices

$$\begin{aligned} X(z_1) &= \begin{bmatrix} x_0, \dots, x_{l-1} \end{bmatrix}, \\ Y(z_1) &= \begin{bmatrix} y_0, \dots, y_{k-1} \end{bmatrix}, \end{aligned} \tag{7.43}$$

computed via

$$\begin{bmatrix} X(z_1) & Y(z_1) \end{bmatrix} = \bar{C}(z_1) \tag{7.44}$$

PROOF OF 7.17: In view of Theorem 7.13, (7.16) is solvable if and only if it has a minimum z_2-degree solution (7.42). When substituting this solution, (7.16) transforms into the matrix equation

$$\begin{bmatrix} X & Y \end{bmatrix} \begin{bmatrix} A \\ B \end{bmatrix} = C. \tag{7.45}$$

in the ring $\mathcal{R}[z_1]$. The theorem then follows from the solvability condition of (7.45). □

Example 7.18 As an example, consider again the equation from Example 7.3 with

$$\begin{aligned} a(z_1, z_2) &= 1 + z_1 z_2 + (1 + z_1) z_2 \\ b(z_1, z_2) &= (1 + z_1) z_2 \\ c(z_1, z_2) &= 2 + z_1 + z_2 \end{aligned}$$

which gives rise to the following matrices:

$$\begin{aligned} A(z_1) &= \begin{bmatrix} 1 & z_1 & 1 + z_1 \end{bmatrix} \\ B(z_1) &= \begin{bmatrix} 0 & 1 + v & 0 \\ 0 & 0 & 1 + v \end{bmatrix} \\ C(z_1) &= \begin{bmatrix} 2 + z_1 & -1 & 0 \end{bmatrix}. \end{aligned}$$

Since $\begin{bmatrix} A \\ B \end{bmatrix}$ is upper triangular, one easy calculates

$$\begin{bmatrix} X & Y \end{bmatrix} = \bar{C} = \begin{bmatrix} 2 + z_1 & -1 - z_1 & -2 - z_1 \end{bmatrix}.$$

Hence, the equation is solvable and its minimum z_2-degree solution was just found to be

$$\begin{aligned} x &= 2 + z_1 \\ y &= -1 - z_1 - 2z_2 - z_1 z_2. \end{aligned}$$

◇

When Theorem 7.13 does not apply but another degree estimate is available, we can proceed similarly. Knowing that certain (not necessarily unique) solution solution up to a given degree must exist whenever one exist at all, one can check existence of this particular solution simply by transforming the scalar 2-D equation (7.16) into a 1-D matrix polynomial equation.

So it is for the solution (7.36) where we take p as in (7.35) and define the matrices

$$A_p(z_1) = \left.\begin{bmatrix} a_0 & \dots & a_k & & & \\ & a_0 & \dots & a_k & & \\ & & \ddots & & \ddots & \\ & & & a_0 & \dots & a_k \end{bmatrix}\right\} l + p, \tag{7.46}$$

$$B_p(\mathsf{z}_1) = \left[\begin{array}{cccccc} b_0 & \dots & b_l & & & \\ & b_0 & \dots & b_l & & \\ & & \ddots & & \ddots & \\ & & & b_0 & \dots & b_l \end{array}\right]\Bigg\} k+p, \tag{7.47}$$

$$C_p(\mathsf{z}_1) = \underbrace{\left[\, c_0, \dots, c_m, 0, \dots, 0 \,\right]}_{p+k+l}, \tag{7.48}$$

as well as

$$\begin{aligned} X_p(\mathsf{z}_1) &= \left[\, x_0, \dots, x_{l+p-1} \,\right], \\ Y_p(\mathsf{z}_1) &= \left[\, y_0, \dots, y_{k+p-1} \,\right]. \end{aligned} \tag{7.49}$$

The original scalar 2-D equation is now solvable if and only if the 1-D matrix polynomial equation (in z_1)

$$\left[\, X_p \;\; Y_p \,\right] \left[\begin{array}{c} A_p \\ B_p \end{array}\right] = C_p. \tag{7.50}$$

is solvable, that is, if and only if a greatest common right divisor of A_p and B_p is a right divisor of C_p. Moreover, any solution of (7.50) yields a solution of (7.1) via (7.49) and all the resulting solutions evidently satisfy (7.36).

When denoting A, B in (7.37)-(7.39) by A_0, B_0, respectively, then Theorem 7.17 cover the limit case of the general situation discussed above. Namely, the relation (7.44) enumerates the solution of (7.50) for $p = 0$. In this particular case, the solution of (7.50) becomes unique for the matrix

$$\left[\begin{array}{c} A_0 \\ B_0 \end{array}\right] = \left[\begin{array}{c} A \\ B \end{array}\right]$$

is square. In contrast, any $p > 0$ gives rise to the tall matrix

$$\left[\begin{array}{c} A_p \\ B_p \end{array}\right]. \tag{7.51}$$

Example 7.19 As an example, consider the equation

$$\mathsf{z}_1\, x + \mathsf{z}_2\, y = \mathsf{z}_1 \mathsf{z}_2^2. \tag{7.52}$$

By inspection, $k = 0, l = 1$ and $m = 2$ so that $p = 2$ and the matrix equation (7.50) reads

$$X_2 \left[\begin{array}{ccc} \mathsf{z}_1 & 0 & 0 \\ 0 & \mathsf{z}_1 & 0 \\ 0 & 0 & \mathsf{z}_1 \end{array}\right] + Y_2 \left[\begin{array}{ccc} 0 & 1 & 0 \\ 0 & 0 & 1 \end{array}\right] = \left[\begin{array}{ccc} 0 & 0 & \mathsf{z}_1 \end{array}\right].$$

Its solution

$$\begin{aligned} X_2 &= \left[\begin{array}{ccc} x_0 & x_1 & x_2 \end{array}\right] &= \left[\begin{array}{ccc} 0 & -1 & 1 \end{array}\right] \\ Y_2 &= \left[\begin{array}{cc} y_0 & y_1 \end{array}\right] &= \left[\begin{array}{cc} \mathsf{z}_1 & 0 \end{array}\right] \end{aligned}$$

gives rise to the desired solution of (7.49)

$$\begin{aligned} x &= -\mathsf{z}_2 + \mathsf{z}_2^2 \\ y &= \mathsf{z}_1. \end{aligned}$$

◇

Example 7.20 As another example, let us examine the equation

$$(1 + z_2 + (z_1 - 1)z_2^2)x + (z_1 z_2 + (z_1 - 1)z_2^2)y = (1 + z_1)z_2^3 \tag{7.53}$$

Here $k = 2$, $l = 2$ and $m = 3$, so that $p = 0$. Observing that

$$\begin{bmatrix} A \\ B \end{bmatrix} = \begin{bmatrix} 1 & 1 & z_1 - 1 & 0 \\ 0 & 1 & 1 & z_1 \\ 0 & z_1 & z_1 - 1 & 0 \\ 0 & 0 & z_1 & z_1 - 1 \end{bmatrix}$$

is not a right divisor of

$$C = \begin{bmatrix} 0 & 0 & 0 & 1 + z_1 \end{bmatrix},$$

one could conclude that (7.53) is unsolvable. However, this is a mistake. Theorem 7.13 fails due to the violation of (7.25) so that the unsolvability of (7.45) does not imply unsolvability of (7.53) at all. Indeed, a higher degree solution exists, such as

$$\begin{aligned} x &= z_2^3 \\ y &= z_2^2 - z_2^3 \end{aligned}$$

which was found in Example 7.16. ◇

Now we describe yet another approach to solving (7.16). Let us think of (7.16) as of an equation in the ring $\mathcal{R}(z_1)[z_2]$. First of all, a and b are factor coprime in $\mathcal{R}[z_1, z_2]$ (by Assumption (7.11)) and consequently, they are coprime in $\mathcal{R}(z_1)[z_2]$ as well. Second, the ring $\mathcal{R}(z_1)[z_2]$ is Euclidean so that all the results described in section 7.3 apply. Namely, there exist unique solution of (7.16) in $\mathcal{R}(z_1)[z_2]$ for which

$$\deg_{z_2} x < \deg_{z_2} b. \tag{7.54}$$

On the other hand, (7.16) possesses a unique solution in $\mathcal{R}[z_1, z_2]$ with the same property whenever Theorem 7.13 holds. As $\mathcal{R}[z_1, z_2]$ is a subset of $\mathcal{R}(z_1)[z_2]$, we conclude that these two minimum degree solutions coincide. This means that the minimum degree solution in $\mathcal{R}(z_1)[z_2]$ is, in fact, from $\mathcal{R}[z_1, z_2]$ whenever (7.16) is in $\mathcal{R}[z_1, z_2]$ solvable. If, on the contrary, the minimum degree solution in $\mathcal{R}(z_1)[z_2]$ contains a fraction in z_1, (7.16) cannot be in $\mathcal{R}[z_1, z_2]$ solvable. This observation is summarized as follows.

Theorem 7.21 () *Let the conditions of Theorem 7.13 be satisfied and let $\tilde{x}, \tilde{y}$ be the minimum degree solution of (7.16) in the ring $\mathcal{R}(z_1)[z_2]$. Then the equation (7.16) is solvable in $\mathcal{R}[z_1, z_2]$ if and only if*

$$\tilde{x}, \tilde{y} \in \mathcal{R}[z_1, z_2].$$

◇

Example 7.22 To illustrate Theorem 7.21, consider again the equation (7.17)

$$z_1\, x + z_2\, y = 1 \tag{7.55}$$

which is unsolvable in $\mathcal{R}[z_1, z_2]$ (Example 7.9). One easy finds its general solution in $\mathcal{R}(z_1)[z_2]$ to be

$$\begin{aligned} \tilde{x} &= \frac{1}{z_1} - z_2\tilde{t} \\ \tilde{y} &= 0 + z_1\tilde{t} \end{aligned}$$

so that its minimum degree solution in this ring reads

$$\tilde{x} = \frac{1}{z_1}, \quad \tilde{y} = 0.$$

Indeed, $\tilde{x} \notin \mathcal{R}[z_1, z_2]$ which implies that *no solution exists in* $\mathcal{R}[z_1, z_2]$ at all.

◇

Example 7.23 Now consider again the equation

$$(1 + z_1z_2 + z_2^2 + z_1z_2^2)\, x + (z_2 + z_1z_2)\, y = 2 + z_1 + z_2.$$

and proceed, for change, in the ring $\mathcal{R}(z_2)[z_1]$. The general solution in this ring was found in Example 7.3 to be

$$\begin{aligned} \tilde{x}' &= \frac{2 + z_1 + z_2}{1 - z_2^2} - (z_2 + z_2z_1)\tilde{t} \\ \tilde{y}' &= -\frac{2 + z_1 + z_2}{1 - z_2} + (1 + z_2^2 + (z_2 + z_2^2)z_1)\tilde{t} \end{aligned} \tag{7.56}$$

with arbitrary $\tilde{t} \in \mathcal{R}(z_2)[z_1]$. By division, we observe (see Example 7.5) that the minimum z_2-degree solution in the ring $\mathcal{R}(z_2)[z_1]$ results from the substitution of

$$\tilde{t} = -\frac{1}{1 + z_1}$$

and reads

$$\begin{aligned} \tilde{x} &= 2 + z_1 \\ \tilde{y} &= -1 - z_1 - 2z_2 - z_1z_2. \end{aligned}$$

As this $\mathcal{R}(z_1)[z_2]$ solution has happened to be from $\mathcal{R}[z_1, z_2]$, it clearly equal the desired $\mathcal{R}[z_1, z_2]$ solution, that is,

$$\begin{aligned} x &= \tilde{x} = 2 + z_1 \\ y &= \tilde{y} = -1 - z_1 - 2z_2 - z_1z_2. \end{aligned}$$

◇

7.5 *n*-D Polynomial Equations

Although the advance from 2-D polynomials to 3-D and higher dimensions brings some additional *principal* differences in general, none of them seems to influence directly our analysis of n-D scalar polynomial equations [17, 18, 19]. On the other hand, the use of various rings of the type $\mathcal{R}[z_1, \ldots, z_{i-1}, z_{i+1}, \ldots, z_n][z_i]$ does not help so much as the "coefficient ring" $\mathcal{R}[z_1, \ldots, z_{i-1}, z_{i+1}, \ldots, z_n]$ is not Euclidean either for $n > 2$. Needless to say, the computational complexity increases rapidly with every dimension added.

Once again, we study the equation

$$a(z_1,\ldots,z_n)x(z_1,\ldots,z_n)+b(z_1,\ldots,z_n)y(z_1,\ldots,z_n)=c(z_1,\ldots,z_n) \tag{7.57}$$

where now $a(z_1,\ldots,z_n), b(z_1,\ldots,z_n)$ and $c(z_1,\ldots,z_n)$ are given polynomials from $\mathcal{R}[z_1,\ldots,z_n]$ and, naturally, the solution $x(z_1,\ldots,z_n), y(z_1,\ldots,z_n)$ is to be found within the same ring.

Some of the 2-D results keep holding and will be just repeated here without proofs. Namely, the divisibility condition is not sufficient but remains necessary.

Theorem 7.24 (Solvability) $\gcd(\mathrm{a},\mathrm{b})|\mathrm{c}$ *whenever (7.57) is solvable.*

◇

Again, equation (7.57) is solvable provided that, roughly, every common zero of a and b is a zero of c with the right multiplicity. In the particular case of a constant right hand side (n-D *Bezout equation*), this property can be precisely expressed as follows.

Theorem 7.25 *The n-D equation*

$$\mathrm{a}(z_1,\ldots,z_n)\mathrm{x}(z_1,\ldots,z_n)+\mathrm{b}(z_1,\ldots,z_n)\mathrm{y}(z_1,\ldots,z_n)=1$$

is solvable if and only if the polynomials a *and* b *have* no *zero in common.*

◇

For simplicity, we shall again assume that

Assumption 7.26 *The polynomials* a *and* b *are* factor coprime,

which implies $a=\bar{a}$ and $b=\bar{b}$.

Theorem 7.27 (General Solution: n-D case) *Let* a *and* b *be factor coprime polynomials and* x′, y′ *be a particular solution of (7.57). Then the general solution of (7.57) reads*

$$\begin{aligned} \mathrm{x} &= \mathrm{x}' - \mathrm{bt} \\ \mathrm{y} &= \mathrm{y}' + \mathrm{at} \end{aligned} \tag{7.58}$$

for an arbitrary polynomial $\mathrm{t} \in \mathrm{R}[z_1,\ldots,z_n]$.

◇

To see this, let us write the given polynomials as elements of the ring $\mathcal{R}[z_1,\ldots,z_{i-1},z_{i+1},\ldots,z_n][z_i]$

$$\begin{aligned} a(z_1,\ldots,z_n) &= a_0+a_1z_i+\cdots+a_kz_i^k \\ b(z_1,\ldots,z_n) &= b_0+b_1z_i+\cdots+b_lz_i^l \\ c(z_1,\ldots,z_n) &= c_0+c_1z_i+\cdots+c_mz_i^m \end{aligned} \tag{7.59}$$

where $k=\deg_{z_i}a$, $l=\deg_{z_i}b$ and $m=\deg_{z_i}c$. The "coefficients" a_j, b_j, c_j range the ring $\mathcal{R}[z_1,\ldots,z_{i-1},z_{i+1},\ldots,z_n]$ of course. Further denote

$$d_{\mathrm{LC}}=\gcd(a_k,b_l)\in\mathcal{R}[z_1,\ldots,z_{i-1},z_{i+1},\ldots,z_n] \tag{7.60}$$

and

$$
\begin{aligned}
a_k &= \bar{a}_k \, d_{\mathrm{LC}} \\
b_l &= \bar{b}_l \, d_{\mathrm{LC}}
\end{aligned}
\tag{7.61}
$$

where $\bar{a}_k, \bar{b}_l \in \mathcal{R}[\mathrm{z}_1, \ldots, \mathrm{z}_{i-1}, \mathrm{z}_{i+1}, \ldots, \mathrm{z}_n]$, too. Then one can generalize Theorem 7.13 as follows.

Theorem 7.28 (Minimum-degree solution) *Let the equation (7.57) be solvable and let*

$$
\mathrm{m} \leq \mathrm{k} + \mathrm{l} - 1. \tag{7.62}
$$

If either

$$
\mathrm{d}_{\mathrm{LC}} = 1 \tag{7.63}
$$

or, at least,

$$
\gcd\left(\mathrm{d}_{\mathrm{LC}},\ \mathrm{a}_{\mathrm{k}-1}\bar{\mathrm{b}}_{\mathrm{l}} - \bar{\mathrm{a}}_{\mathrm{k}}\mathrm{b}_{\mathrm{l}-1}\right) = 1 \text{ and } \mathrm{d}_{\mathrm{LC}} | \mathrm{c}_{\mathrm{k}+\mathrm{l}-1}, \tag{7.64}
$$

then there exists a unique solution x, y *of (7.57) such that*

$$
\begin{aligned}
\deg_{\mathrm{z}_i} \mathrm{x} &\leq \mathrm{l} - 1 \\
\deg_{\mathrm{z}_i} \mathrm{y} &\leq \mathrm{k} - 1
\end{aligned}
\tag{7.65}
$$

◇

Again, this theorem is most useful if a or b is monic in $\mathcal{R}[\mathrm{z}_1][\mathrm{z}_2]$, that is, if either $a_k(\mathrm{z}_1, \ldots, \mathrm{z}_{i-1}, \mathrm{z}_{i+1}, \ldots, \mathrm{z}_n) = 1$ or $b_l(\mathrm{z}_1, \ldots, \mathrm{z}_{i-1}, \mathrm{z}_{i+1}, \ldots, \mathrm{z}_n) = 1$.

However, this can be always arranged:[5] If just the choice of i does not help, then the substitution

$$
\begin{array}{lclcl}
\mathrm{z}_1 & = & c_1 \bar{\mathrm{z}}_i & + & \bar{\mathrm{z}}_1 \\
 & \vdots & & & \\
\mathrm{z}_{i-1} & = & c_{i-1} \bar{\mathrm{z}}_i & + & \bar{\mathrm{z}}_{i-1} \\
\mathrm{z}_i & = & c_i \bar{\mathrm{z}}_i & & \\
\mathrm{z}_{i+1} & = & c_{i+1} \bar{\mathrm{z}}_i & + & \bar{\mathrm{z}}_i + 1 \\
 & \vdots & & & \\
\mathrm{z}_n & = & c_n \bar{\mathrm{z}}_n & + & \bar{\mathrm{z}}_n
\end{array}
\tag{7.66}
$$

will do the job. A proper choice of nonzero real constants $c_1, \ldots, c_n$ makes, say,

$$
a_k(\bar{z}_1, \ldots, \bar{\mathrm{z}}_{i-1}, \bar{\mathrm{z}}_{i+1}, \ldots, \bar{\mathrm{z}}_n) = 1
$$

and, hence, the polynomial

$$
a(\bar{z}_1, \ldots, \bar{\mathrm{z}}_n)
$$

monic in $\mathcal{R}[\bar{z}_1, \ldots, \bar{\mathrm{z}}_{i-1}, \bar{\mathrm{z}}_{i+1}, \ldots, \bar{\mathrm{z}}_n][\bar{\mathrm{z}}_i]$.

More generally, for

$$
p = \max\{0, m - (k + l - 1)\}, \tag{7.67}
$$

and (7.63)-(7.64) satisfied, there always exist a solution with

$$
\begin{aligned}
\deg_{\mathrm{z}_i} x &= l - 1 + p &= m - k \\
\deg_{\mathrm{z}_i} y &= k - 1 + p &= m - l.
\end{aligned}
\tag{7.68}
$$

[5] Confer the substitution (7.34).

By use of matrices (in $\mathcal{R}[\mathsf{z}_1, \ldots, \mathsf{z}_{i-1}, \mathsf{z}_{i+1}, \ldots, \mathsf{z}_n]$)

$$A_p = \left[\begin{array}{cccccc} a_0 & \ldots & a_k & & & \\ & a_0 & \ldots & a_k & & \\ & & \ddots & & \ddots & \\ & & & a_0 & \ldots & a_k \end{array}\right] \left.\vphantom{\begin{array}{c}a\\a\\a\\a\end{array}}\right\} l+p, \tag{7.69}$$

$$B_p = \left[\begin{array}{cccccc} b_0 & \ldots & b_l & & & \\ & b_0 & \ldots & b_l & & \\ & & \ddots & & \ddots & \\ & & & b_0 & \ldots & b_l \end{array}\right] \left.\vphantom{\begin{array}{c}a\\a\\a\\a\end{array}}\right\} k+p, \tag{7.70}$$

$$C_p = \underbrace{\left[\begin{array}{c} c_0, \ldots, c_m, 0, \ldots, 0 \end{array}\right]}_{p+k+l}, \tag{7.71}$$

as well as

$$\begin{aligned} X_p &= \left[\begin{array}{c} x_0, \ldots, x_{l+p-1} \end{array}\right], \\ Y_p &= \left[\begin{array}{c} y_0, \ldots, y_{k+p-1} \end{array}\right], \end{aligned} \tag{7.72}$$

the original scalar n-D equation transforms into $(n-1)$-D matrix polynomial equation (in the indeterminates $\mathsf{z}_1, \ldots, \mathsf{z}_{i-1}, \mathsf{z}_{i+1}, \ldots, \mathsf{z}_n$)

$$\left[\begin{array}{cc} X_p & Y_p \end{array}\right] \left[\begin{array}{c} A_p \\ B_p \end{array}\right] = C_p. \tag{7.73}$$

In fact, (7.57) is solvable if and only if (7.73) is. In contrast to the last section, this test on solvability becomes useful only if one can solve the $(n-1)$-D but matrix equation (7.73). In any case, the original scalar n-D problem has been reduced into a $n-1$-D problem. also the second method of the last section can be generalized in higher dimensions.

Theorem 7.29 *Let (7.62)-(7.64) be satisfied and let* $\tilde{\mathsf{x}}, \tilde{\mathsf{y}}$ *be the minimum degree solution of (7.57) in the ring* $\mathcal{R}(\mathsf{z}_1, \ldots, \mathsf{z}_{i-1}, \mathsf{z}_{i+1}, \ldots, \mathsf{z}_n)[\mathsf{z}_i]$. *Then (7.57) is solvable in* $\mathcal{R}[\mathsf{z}_1, \ldots, \mathsf{z}_n]$ *if and only if*

$$\tilde{\mathsf{x}}, \tilde{\mathsf{y}} \in \mathcal{R}[\mathsf{z}_1, \ldots, \mathsf{z}_n].$$

◇

Example 7.30 As an example, check out the 3-D polynomial equation

$$(1 + \mathsf{z}_1\mathsf{z}_2 + \mathsf{z}_3)\, x + \mathsf{z}_1\mathsf{z}_2\mathsf{z}_3\, y = \mathsf{z}_1\mathsf{z}_2 + \mathsf{z}_1^2\mathsf{z}_2^2 \tag{7.74}$$

Via 2-D polynomial matrices

$$\begin{aligned} A_0 &= \left[\begin{array}{cc} 1 + \mathsf{z}_1\mathsf{z}_2 & 1 \end{array}\right] \\ B_0 &= \left[\begin{array}{cc} 0 & \mathsf{z}_1\mathsf{z}_2 \end{array}\right] \\ C_0 &= \left[\begin{array}{cc} \mathsf{z}_1\mathsf{z}_2 + \mathsf{z}_1^2\mathsf{z}_2^2 & 0 \end{array}\right] \end{aligned}$$

it is converted into the following 2-D matrix polynomial equation

$$[x_0]\begin{bmatrix} 1+z_1z_2 & 1 \end{bmatrix} + [y_0]\begin{bmatrix} 0 & z_1z_2 \end{bmatrix} = \begin{bmatrix} z_1z_2+z_1^2z_2^2 & 0 \end{bmatrix}.$$

Various methods to solve such a 2-D matrix equations are discussed elsewhere. In this particular example, however, the solution is almost trivial

$$[x_0] = [z_1z_2], \quad [y_0] = [-1].$$

These matrices yield the desired solution of (7.74)

$$\begin{aligned} x &= z_1z_2 \\ y &= -1. \end{aligned}$$

To exercise the second method, perform the transformation

$$\begin{bmatrix} (1+z_1z_2+z_3) & z_1z_2z_3 \\ 1 & 0 \\ 0 & 1 \end{bmatrix} \longrightarrow \begin{bmatrix} 1 & 0 \\ 1/(1+z_1z_2) & -z_1z_2z_3 \\ -1/z_1z_2(1+z_1z_2) & 1+z_1z_2+z_3 \end{bmatrix}$$

to obtain the minimum degree solution of (7.74) in the ring $\mathcal{R}(z_1,z_2)[z_3]$

$$\begin{aligned} x &= z_1z_2 \\ y &= -1. \end{aligned}$$

This, of course, is also the minimum z_3-degree solution in the ring $\mathcal{R}[z_1,z_2,z_3]$. Needless to say, the both methods result in the same minimum solution. In addition, the general solution in $\mathcal{R}[z_1,z_2,z_3]$ reads

$$\begin{aligned} x &= z_1z_2 &+& \; z_1z_2z_3\,t \\ y &= -1 &+& \; (1+z_1z_2+z_3)t. \end{aligned}$$

7.6 Conclusions

Linear equations in scalar n-D polynomials has been surveyed from both theoretical and computational point of view. As a preparatory step, standard 1-D results have been first rephrased for 1-D polynomial over a general field. Then 2-D polynomial equations have been treated in detail. The main differences between the 1-D and 2-D cases have been pointed out. Finally, all results ave been generalized for n-D polynomials. This Chapter provides useful tools for control and filter design in 2-D and n-D systems.

Bibliography

[1] D. E. Dudgeon and R. M. Mersereau, *Multidimensional Digital Signal Processing*. Englewood Cliffs: Prentice-Hall, 1984.

[2] E. Rogers and D. Owens, "2D transfer-function based controller design for differential repetitive processes," in *Preprints of the 2nd IFAC Workshop on System Structure and Control*, (Prague, Czechoslovakia), pp. 420–423, Sept. 1992.

[3] M. Šebek, "Asymptotic tracking in 2-D and delay–differential systems," *Automatica (IFAC)*, vol. 24, pp. 711–713, Sept. 1988.

[4] M. Šebek, E. Fornasini, and M. Bisiacco, "Controllability and reconstructibility conditions for 2-D systems," *IEEE Transactions on Automatic Control*, vol. AC–33, pp. 496–499, May 1988.

[5] M. Šebek, "2-D exact model matching," *IEEE Transactions on Automatic Control*, vol. AC-28, pp. 215–217, Feb. 1992.

[6] M. Šebek, "2-D polynomial equations," *Kybernetika (Prague)*, vol. 19, no. 3, pp. 212–224, 1983.

[7] M. Šebek, "Author's reply to 'Comments on 2-D exact model matching'," *IEEE Transactions on Automatic Control*, vol. AC–29, pp. 374–379, Apr. 1984.

[8] M. Šebek, "On 2-D pole placement," *IEEE Transactions on Automatic Control*, vol. AC–30, pp. 819–722, Aug. 1985.

[9] M. Šebek, "Model matching of 2-D multi–input multi–output systems," *Kybernetika (Prague)*, vol. 23, no. 4, pp. 319–328, 1987.

[10] M. Šebek, "Polynomial approach to pole placement in MIMO n-D systems," *Kybernetika (Prague)*, vol. 25, no. 4, pp. 271–277, 1989.

[11] M. Šebek, "Polynomial solution of 2-D Kalman-Bucy filtering problem," *IEEE Transactions on Automatic Control*, 1992. To appear.

[12] M. Šebek, "Stochastic LQ-optimal control for 2-D systems," *Automatica (IFAC)*, 1993. Submitted for publication.

[13] M. Šebek, "Invariant polynomials assignment for multi-input multi-output n-D systems," in *Computing and Computers for Control Systems* (P. Borne *et al.*, eds.), pp. 361–363, J.C. Baltzer AG, Scientific Publishing Co., 1989.

[14] M. Šebek, "2-D Kalman-Bucy filtering problem: 2-D polynomial solution," in *New Trends in Systems Theory* (A. P. G. Conte and B. Wyman, eds.), Vol. 7 in the Series Progress in Systems and Control Theory, pp. 660–667, Boston: Birkhäuser, 1991.

[15] M. Šebek, "Multi-variable 2D Kalman-Bucy filtering problem via polynomial matrix techniques," in *Recent Advances in Mathematical Theory of Systems, Control Networks and Signal Processing II.* (H. Kimura and S. Kodama, eds.), pp. 147–153, Osaka, Japan: Mita Press Co., Ltd., 1992.

[16] V. Kučera, *Discrete Linear Control: The Polynomial Approach.* Chichester: Wiley, 1980.

[17] M. Šebek, "n-D matrix polynomial equations," *IEEE Transactions on Automatic Control*, vol. AC–33, pp. 499–502, May 1988.

[18] M. Šebek, "One more counterexample in n-D systems," *IEEE Transactions on Automatic Control*, vol. AC–33, pp. 502–503, May 1988.

[19] M. Šebek, "Two–sided equations and skew primeness for n-D polynomial matrices," *System and Control Letters*, vol. 12, pp. 331–337, 1989.

Chapter 8

Eigenstructure Assignment in Linear Systems by State Feedback

P. Zagalak, V. Kučera and J. J. Loiseau

8.1 Introduction

Assignment of invariant factors in linear systems has been intensively studied in control theory for more than two decades since it is of great importance in many areas of this theory. For instance, such classical tasks as linear quadratic control and deadbeat control lead to specific requirements for poles placement of closed-loop systems.

The problem of pole placement is also closely related to basic concepts of linear systems theory such as system poles and zeros, controllability, etc. and in fact forms the basis of an approach that is nowadays frequently referred to as a structural approach. From this point of view, Rosenbrock's theorem [24] is a very nice specimen of this approach and its further generalization to the realm of implicit systems and uncontrollable systems brings a lot of insight into the structure of linear systems.

The main purpose of this contribution is to present just the above mentioned generalizations that were developed during the last thirteen years. They represent very strong results that are of fundamental theoretical importance and, at the same time, represent a so-called polynomial approach (see [10], [14] and the references therein).

8.2 Notation and Basic Concepts

We shall consider a linear, time-invariant system described by

$$E\dot{x} = Ax + Bu \tag{8.1}$$

where E and A are $n \times n$ matrices and B is an $n \times m$ matrix over $\mathbb{R}$, the field of real numbers. Without any loss of generality, we suppose that rank $B = m$. We divide the systems described by (8.1) into two classes:

- *Explicit systems*, which are characterized by the property $detE \neq 0$ (synonyms: state-space systems, classical systems,...) and

- *Implicit systems* (singular systems, generalized state-space systems, descriptor systems,...) with the property $detE = 0$.

For these two classes of systems, we shall investigate the effect of state feedback

$$u = -K\dot{x} + Fx + v, \tag{8.2}$$

where K and F are $m \times n$ matrices over $\mathbf{R}$, upon the system (8.1). Applying the feedback (8.2) around the system (8.1), we obtain the closed-loop system

$$(E + BK)\dot{x} = (A + BF)x + Bv. \tag{8.3}$$

The form of the closed-loop system hints that the best way to describe differences between the systems (8.1) and (8.3) is to consider changes in the eigenstructure of the pencil $sE - A$ brought about by the transformation (8.2). To that end, we take the Laplace transform of (8.1) with the zero initial condition $x(0^-) = 0$ to get

$$(sE - A)X(s) = BV(s)$$

or

$$\begin{bmatrix} sE - A & -B \end{bmatrix} \begin{bmatrix} X(s) \\ V(s) \end{bmatrix} = 0. \tag{8.4}$$

This equation is of fundamental importance in the study of state feedback, as we shall see later. But first we introduce some basic concepts.

8.2.1 Regularity and regularizability

The system (8.1) is said to be *regular* if $det(sE - A)$ is not identically equal to zero. The most important property of regular systems is that they have a transfer function,

$$T(s) := (sE - A)^{-1}B$$

It can be readily seen that explicit systems are always regular.

As we are interested in studying the properties of state feedback (8.2), our first and natural question is whether regularity is preserved under the action of state feedback. Unfortunately, the answer is a negative. To remedy this situation, we shall define the concept of *regularizability* [23]. This property is of course invariant under state feedback (8.2).

Definition 1 *The system (8.1) is said to be regularizable if there exists a state feedback (8.2) such that the closed-loop system (8.3) is regular.*

8.2.2 Poles and zeros

There are many definitions of zeros and poles in the literature; we favour the following ones.

Definition 2 *The finite pole structure of (8.1) is given by the zeros of the invariant polynomials of $sE - A$.*

As far as the infinite pole structure of (8.1) is concerned, the situation is a little bit more complicated. To that end, we first make a mention of the Smith-McMillan form at infinity, see [14],[29] and the references therein.

Proposition 1 *Let $H(s)$ be an $m \times n$ matrix over $\mathbb{R}(s)$. Then there exist biproper matrices $B_1(s)$ and $B_2(s)$ such that*

$$H(s) = B_1(s)diag[s^{n_1}, ..., s^{n_k}, 0, \dots 0]B_2(s)$$

where $k = rankH(s)$ and the integers n_i, $i = 1, 2, ..., k$ are nonincreasigly ordered.

The positive integers n_i are called *infinite pole orders* of $H(s)$ and the negative integers n_i are said to be *infinite zero orders* of $H(s)$.

Definition 3 *The infinite pole structure of (8.1) is given as the infinite zero structure of the matrix $sE - A$.*

Now we define the system matrix $P(s)$ of (8.1), see [24], as

$$P(s) := \begin{bmatrix} sE - A & -B \\ I_n & 0 \end{bmatrix}.$$

Then *the finite zero structure of (8.1)* is given by the zeros of the invariant polynomials of $P(s)$ and the *an infinite zero structure of (8.1)* is defined as the infinite zero structure of $P(s)$.

When considering the effect of state feedback (8.2) upon the system (8.1), we can readily see, using the above definitions, that these two structures, i.e. the pole and the zero structures, can be changed. This problem is closely related to the concept of controllability. But it is also known [23] that the notion of controllability depends on the type of state feedback (8.2). Since we are mainly interested in changing the pole structure of (8.1), we shall first investigate the effect of state feedback (8.2) where $K = 0$. Such a feedback is usually said to be *a proportional (P) state feedback* while, when $K \neq 0$ and $F \neq 0$, we speak about *a proportional-and-derivative (PD) state feedback*. When $F = 0$, the feedback (8.2) is said to be a (pure) *derivative (D) state feedback*.

8.2.3 Reachability and controllability

We recall first the definitions of reachability and controllability given in [21].

Definition 4 *Given points x_0 and x_T in $\mathbb{R}^n$, then x_T is said to be reachable from x_0 if there exists a finite time $T \geq 0$ and an n-times continuously differentiable input u(t) defined on $< 0, T >$ such that there exists a continuously differentiable trajectory $x(t)$ of (8.1) defined on $< 0, T >$ with no impulsive behaviour at $t = 0$ and $x(0^-) = x_0$, $x(T) = x_T$.*

The set $\Re$ of all states reachable from the point $x_0 = 0$ forms a subspace of $\mathbb{R}^n$ and is called *a reachable subspace of (8.1)*. If $\mathbb{R}^n$ is the reachable subspace of (8.1), then the system (8.1) is said to be *reachable.*
We define *a controllable subspace of (8.1)* as a subset of all x_0 from which x_T can be reached. Hence controllability means the reachability of the origin. The system (8.1) is said to be *controllable* if $\Re + Ker E = \mathbb{R}^n$.

It easily follows that reachability implies controllability, but not vice versa. The following two propositions sum up some known criteria of reachability and controllability[16].

Proposition 2 *The following statements are equivalent:*

(a) the system (8.1) is reachable;

(b) rank $\begin{bmatrix} sE - A & -B \end{bmatrix} = n$ *for all* s *(finite and infinite) and rank* $[\ E \quad B\ ] = n$.

Proposition 3 *The following statements are equivalent:*

(a) the system (8.1) is controllable;

(b) $E\Re = ImE$*;*

(c) rank $[sE - A \quad B] = n$ *for all* s *(finite and infinite).*

8.2.4 External description

The description of system (8.1) we have used up to now is very often called an internal description. But the relationship (8.4) shows that the vector $\begin{bmatrix} X(s) \\ V(s) \end{bmatrix}$, which reflects the input-output or external behaviour of (8.1), lies in $Ker[sE-A \ -B]$. Hence, under the assumption of the controllability of (8.1), a basis of $Ker[sE - A \quad - B]$ serves as a generator of all possible input-output vectors $\begin{bmatrix} X(s) \\ V(s) \end{bmatrix}$. Therefore, such a basis is an alternative description of system (8.1). We shall refer it to as *an external description* of (8.1).

The existence of external description can be proven on the basis of the canonical form of (8.1) under the action of transformations described by

$$(E, A, B) \longrightarrow (WEV, W(A + BF)V, WBU) \tag{8.5}$$

where F is arbitrary and U, V, W are invertible matrices; see [19] for details.

Theorem 1 *Given* E, A, B *in (8.1), there exist polynomial matrices* $N(s)$ *and* $D(s)$ *of respective sizes* $n \times m$ *and* $m \times m$ *such that*

(i) $\begin{bmatrix} N(s) \\ D(s) \end{bmatrix}$ *is a minimal polynomial basis of* $Ker[sE - A \quad - B]$*;*

(ii) $N(s)$ *is a minimal polynomial basis of* $KerP(sE-A)$ *where* P *is a maximal anihilator of* B*;*

(iii) $\begin{bmatrix} N(s) \\ D(s) \end{bmatrix}$ *is decreasingly column-degree ordered.*

Proof.
Let (E_c, A_c, B_c) be the canonical form of the triple (E, A, B) under the transformations (8.5).
We have

$$W\begin{bmatrix} sE-A & -B \end{bmatrix}\begin{bmatrix} V & 0 \\ FV & U \end{bmatrix} = \begin{bmatrix} sE_c - A_c & -B_c \end{bmatrix},$$

which implies

$$Ker\begin{bmatrix} sE-A & -B \end{bmatrix} = \begin{bmatrix} V & 0 \\ FV & U \end{bmatrix} Ker\begin{bmatrix} sE_c - A_c & -B_c \end{bmatrix}. \tag{8.6}$$

From $B_c = WBU$, we get that $P_cW = TP$ where P and P_c are canonical projections $\mathbb{R}^n \longrightarrow \mathbb{R}^n/ImB$ and $\mathbb{R}^n \longrightarrow \mathbb{R}^n/ImB_c$, respectively, and T is invertible. Hence

$$P_c(sE_c - A_c) = TP(sE-A)V,$$

which implies

$$KerP_c(sE_c - A_c) = VKerP(sE-A). \tag{8.7}$$

Consider now the triple $(E_c\ A_c, B_c)$. By [18, Theorem 3.1] we know that only two types of blocks in $[sE_c - A_c\ \ -B_c]$ affect $Ker[sE_c - A_c\ \ -B_c]$, particularly the blocks

$$\left[\begin{array}{cccc|c} s & -1 & & & 0 \\ & \ddots & \ddots & & \vdots \\ & & \ddots & -1 & \vdots \\ & & & s & 0 \end{array}\right] \tag{8.8}$$

of the sizes $\sigma_i \times (\sigma_i + 1)$ and the blocks

$$\left[\begin{array}{cccc|c} s & -1 & & & 0 \\ & \ddots & \ddots & & \vdots \\ & & s & -1 & 0 \end{array}\right] \tag{8.9}$$

of the sizes $(\gamma_i - 1) \times (\gamma_i + 1)$, associated with two types of column minimal indices σ_i and γ_i.

Now we construct the matrices

$$N_c(s) = diag[S_{\sigma_1}, S_{\sigma_2}, \ldots, S_{\gamma_1}, S_{\gamma_2}, \ldots]$$

and

$$D_c(s) = diag[s^{\sigma_1}, s^{\sigma_2}, \ldots, 0, 0, \ldots]$$

where $S_k = [1, s, \ldots, s^{k-1}]^T$. Obviously, the matrix $\begin{bmatrix} N_c(s) \\ D_c(s) \end{bmatrix}$ is a minimal basis of $Ker[sE_c - A_c \quad -B_c]$ and $N_c(s)$ is a minimal basis of $KerP_c(sE_c - A_c)$. The claim follows from (8.6) and (8.7) if we put

$$\begin{bmatrix} N(s) \\ D(s) \end{bmatrix} := \begin{bmatrix} V & 0 \\ FV & U \end{bmatrix} \begin{bmatrix} N_c(s) \\ D_c(s) \end{bmatrix}. \tag{8.10}$$

□

Definition 5 *The matrices $N(s)$ and $D(s)$ satisfying (i), (ii) and (iii) of Theorem 1 are said to form a normal external description of (8.1).*

Two points should be stressed at this moment:

Remark 1 A normal external description of (8.1) is not unique. If $N(s), D(s)$ and $N'(s), D'(s)$ form normal external descriptions of (8.1), then they are related by

$$\begin{bmatrix} N(s) \\ D(s) \end{bmatrix} = \begin{bmatrix} N'(s) \\ D'(s) \end{bmatrix} U(s)$$

where $U(s)$ is a unimodular matrix.

Remark 2 The relationship (8.5) shows that a normal external description of (8.1) is derived under the action of P state feedback. Hence, it cannot be used when investigating the effect of PD state feedback. For this and other reasons, we shall refer this description to as a *P-normal external description.*

A P-normal external description can now be used to define the controllability and reachability indices. We define

$$k_i = deg_{ci}N(s)$$

and

$$c_i = deg_{ci} \begin{bmatrix} N(s) \\ D(s) \end{bmatrix}$$

for $i = 1, 2, \ldots, m$ where $deg_{ci}(.)$ denotes the degree of the i-th column.

Definition 6 *The integers $r_i = 1 + k_i, i = 1, 2, \ldots, m$ are said to be reachability indices of (8.1) while the integers $c_i, i = 1, 2, \ldots, m$ are said to be controllability indices of (8.1). When $c_i > k_i$, then c_i is said to be a proper controllability index; otherwise it is said to be a nonproper one.*

There are many other ways to define the reachability and controllability indices and we refer the reader to [4], [16], [21] for details.

Theorem 2 *The system (8.1) is reachable if and only if*

$$\sum_{i=1}^{m} r_i = n \tag{8.11}$$

and controllable if and only if

$$\sum_{i=1}^{m} c_i = rankE. \tag{8.12}$$

Proof.
Consider again the canonical form (E_c, A_c, B_c) of (8.1) and the matrix $\begin{bmatrix} N_c(s) \\ D_c(s) \end{bmatrix}$ as in the proof of Theorem 1.

We define

$$r_i = 1 + deg_{ci} N_c(s)$$

and

$$c_i = deg_{ci} \begin{bmatrix} N_c(s) \\ D_c(s) \end{bmatrix}$$

for $i = 1, 2, \ldots, m$.

It can be readily seen that there is a one-to-one correspondence between the blocks (8.8) and (8.9) and the diagonal terms of $N_c(s)$ and $D_c(s)$. As the matrix $N_c(s)$ is column reduced with column indices $\sigma_1 - 1, \sigma_2 - 1, \ldots, \gamma_1 - 1, \gamma_2 - 1, \ldots$ it follows that $\sigma_1 + \sigma_2 + \ldots + \gamma_1 + \gamma_2 + \ldots = n$. We put $k_1 = \sigma_1 - 1, k_2 = \sigma_2 - 1, \ldots$. By the same arguments we can show that $\begin{bmatrix} N_c(s) \\ D_c(s) \end{bmatrix}$ is also column reduced with column indices $\sigma_1, \sigma_2, \ldots, \gamma_1 - 1, \gamma_2 - 1, \ldots$ and their sum is equal to $rank E$. We put $c_1 = \sigma_1, c_2 = \sigma_2, \ldots$. If we premultiply $\begin{bmatrix} N_c(s) \\ D_c(s) \end{bmatrix}$ by $\begin{bmatrix} V & 0 \\ FV & U \end{bmatrix}$, the integers r_i's and c_i's are left unchanged, which proves (8.11) and (8.12).

□

The effect of P state feedback upon the controllable system (8.1) can also be described using a P- normal external description. We have

$$\begin{bmatrix} sE - A & -B \end{bmatrix} \begin{bmatrix} I_n & 0 \\ F & I_m \end{bmatrix} = \begin{bmatrix} sE - A - BF & -B \end{bmatrix} \tag{8.13}$$

and

$$\begin{bmatrix} I_n & 0 \\ -F & I_m \end{bmatrix} \begin{bmatrix} N(s) \\ D(s) \end{bmatrix} = \begin{bmatrix} N(s) \\ D(s) - FN(s) \end{bmatrix} \tag{8.14}$$

$$[sE - A - BF \quad -B] \begin{bmatrix} N(s) \\ D(s) - FN(s) \end{bmatrix} = 0. \tag{8.15}$$

Lemma 1 *Let $N(s), D(s)$ form a P-normal external description of the controllable system (8.1). Then the matrices $N(s), D(s) - FN(s)$ form a P-normal external description of the closed-loop system under the action of P state feedback and*

$$deg_{ci} \begin{bmatrix} N(s) \\ D(s) - FN(s) \end{bmatrix} = c_i, \quad i = 1, 2, \ldots, m. \tag{8.16}$$

We omit the very simple proof. Some remarks are now in order:

Remark 3 Lemma 1 states the invariance of controllability and reachability under the action of P state feedback.

Remark 4 If the system (8.1) is neither reachable nor controllable, then the matrices $N(s)$ and $D(s)$ reflect only the reachable part of (8.1), i.e.,

$$\sum_{i=1}^{m} r_i = dim\Re$$

and

$$\sum_{i=1}^{m} c_i = dimE\Re$$

Remark 5 It follows from the proof of Theorem 2, see (8.10), that the matrix $N(s)$ is irreducible for all finite s. This property of $N(s)$ is characteristic for systems of the form (8.1) and means that such systems have no finite zeros. Moreover, the matrix $N(s)$ form a polynomial basis of the reachability subspace of (8.1).

Remark 6 As far as the infinite zeros of (8.1) are concerned, it can be readily seen that the system can have zeros at infinity. A very simple analysis shows that their structure is given by positive integers $z_1 \leq z_2 \leq \ldots \leq z_q$ where q is the number of the proper controllability indices of (8.1) and $z_i = 1$ for $i = 1, 2, \ldots, q$.

Remark 7 In many cases we want to make the system regular by P state feedback. Then regularizability should also be assumed for (8.1). But we can easily see that the controllability of (8.1) is sufficient to assume since this property is stronger than the regularizability of (8.1).

8.3 Eigenstructure Assignment by P State Feedback

8.3.1 Fundamental theorem of P state feedback for explicit systems

This classical result was first proved by Rosenbrock [24]. The proof we shall present here is based on the properties of P-normal external description, which is a special case of matrix fraction description, the tool used in [11] for the proof of this theorem.

To begin with, let (8.1) represent an explicit controllable system, which is, for the sake of simplicity, written in the form

$$\dot{x} = Ax + Bu, \;\; A \in \mathbb{R}^{n \times n}, \;\; B \in \mathbb{R}^{n \times m}, \;\; rankB = m. \tag{8.17}$$

Let further

$$u = Fx + v, \;\; F \in \mathbb{R}^{m \times n} \tag{8.18}$$

be a P state feedback applied to (8.17) yielding the closed-loop system

$$\dot{x} = (A + BF)x + Bv. \tag{8.19}$$

Problem 1 *Given a controllable system (8.17) with controllability indices* $c_1 \geq c_2 \geq \ldots \geq c_m$ *and a list of monic polynomials* $\psi_1(s), \psi_2(s), \ldots, \psi_n(s)$ *such that* $\psi_{i+1}(s)$ *divides* $\psi_i(s), i = 1, 2, \ldots, n-1$ *(we adopt the notation* $\psi_1(s) \geq \psi_2(s) \geq \ldots \geq \psi_n(s)$ *where* $\psi_i(s) \geq \psi_{i+1}(s)$ *means that* $\psi_{i+1}(s)$ *divides* $\psi_i(s)$*).*

Find necessary and sufficient conditions for the existence of a feedback (8.18) such that $\psi_i(s), i = 1, 2, \ldots, n$ *are the invariant polynomials of (8.19), i.e., those of* $sI_n - A - BF$*. Give a procedure to calculate one such gain.*

Before stating the main theorem of this section, we adopt the following notation [17]. Let $\{\alpha_i\}_m, \{\beta_i\}_m$ denote nonincreasingly ordered lists of m nonnegative integers that have the same sums. The set of all such sequences can be ordered with respect to the relation

$$\{\alpha_i\}_m \succ \{\beta_i\}_m \Longleftrightarrow \sum_{i=1}^{k} \alpha_i \geq \sum_{i=1}^{k} \beta_i \Longleftrightarrow \sum_{i=k}^{m} \alpha_i \leq \sum_{i=k}^{m} \beta_i, \quad k = 1, 2, \ldots, m.$$

Obviously, this set forms a lattice and has both maximal and minimal elements. The maximal element is the sequence $\{S, 0, 0, \ldots, 0\}$ where $S = \sum_{i=1}^{m} \alpha_i$ and the minimal element is the sequence $\{\underbrace{t+1, t+1, \ldots, t+1}_{r}, \underbrace{t, t, \ldots, t}_{m-r}\}$ where $S = tm + r, r < m$.

If the sequence $\{\beta_i\}_n, n \geq m$ is compared with $\{\alpha_i\}_m$, then we put $\alpha_{m+1} = \ldots = \alpha_n = 0$, i.e., the shorter sequence is completed by zeros to the number n.

Given two sequences $\{\alpha_i\}_n$, $\{\beta_i\}_m$ whose sums need not be equal, we denote by $\{\alpha_i\}_n \geq \{\beta_i\}_m$ the inequalities

$$\sum_{i=j}^{k} \alpha_i \geq \sum_{i=j}^{k} \beta_i, \quad j = 1, 2, \ldots, k = max(m, n)$$

where again the shorter sequence is completed by zeros.

Theorem 3 *Problem 1 has a solution if and only if*

$$\{c_i\}_m \prec \{deg\psi_i(s)\}_n. \tag{8.20}$$

P r o o f. (Necessity).

Let $F \in \mathbf{R}^{m \times n}$ be such that $\psi_1(s), \psi_2(s), \ldots, \psi_n(s)$ are the invariant polynomials of $sI_n - A - BF$ and let $N(s), D(s)$ form a P-normal external description of (8.17). Then, by Lemma 1, $N(s)$ and $D(s) - FN(s)$ form a P normal external description of (8.19) and the nonunit invariant polynomials of $sI_n - A - BF$ are exactly those of $D(s) - FN(s)$, by Lemma 4 of the Appendix. Since $sI_n - A - BF$ is of full rank and $D(s) - FN(s)$ is $m \times m$, it follows that $\psi_{m+1}(s) = \ldots = \psi_n(s) = 1$.

By Lemma 1, the matrix $D(s) - FN(s)$ is column reduced with column degrees $c_i, i = 1, 2, \ldots, m$. Then the product $\psi_{k+1}(s) \ldots \psi_m(s)$ is the greatest common divisor of all minors of order $m - k$ of $D(s) - FN(s)$. The degree of this divisor cannot exceed the sum $c_{k+1} + c_{k+2} + \ldots + c_m$. Hence

$$\sum_{i=k+1}^{m} deg\psi_i(s) \leq \sum_{i=k+1}^{m} c_i, \quad k = 0, 1, \dots, m-1$$

with equality holding for $k = 0$.

These inequalities together with equalities $\psi_m(s) = \psi_{m+1}(s) = \dots = \psi_n(s) = 1$ imply the inequalities (8.20).

(Sufficiency).
Given monic polynomials $\psi_1(s) \geq \psi_2(s) \geq \dots, \geq \psi_n(s)$ that satisfy (8.20) , we are going to construct $F \in \mathbb{R}^{m \times n}$ such that $\psi_i(s)$'s are the invariant polynomials of $sI_n - A - BF$.

To that end, we first construct the matrix

$$\bar{M}(s) := diag[\psi_1(s), \psi_2(s), \dots, \psi_m(s)]$$

It is to be noted that from (8.20) we have $deg\psi_i(s) = 0$ for $i = m+1, \dots, n$ and hence

$$\psi_{m+1}(s) = \psi_{m+2}(s) = \dots = \psi_n(s) = 1.$$

Then by the inequalities (8.20) and by Lemma 5 of the Appendix, there exist unimodular matrices $U(s)$ and $V(s)$ such that the matrix

$$M(s) = U(s)\bar{M}(s)V(s)$$

is column reduced with column indices equal to the controllability indices $c_1, c_2, \dots, c_n$ of (8.17).

As the matrix $M(s)$ characterizes the desired dynamics of the closed-loop system (8.19), it is possible to get the matrix F from a solution of the equation

$$XD(s) + YN(s) = M(s). \tag{8.21}$$

Indeed, since the matrices $D(s), N(s)$ and $M(s)$ satisfy the assumptions of Lemma 7 of the Appendix, there exists a constant solution pair X, Y of (8.21) with X invertible.

We put $F = -X^{-1}Y$. The claim follows from Lemma 4 of the Appendix as a result of the fact that both $D(s) - FN(s)$ and $sI_n - A - BF$ have the same nonunit invariant polynomials. Hence, $\psi_i(s), i = 1, 2, \dots, n$ are the invariant polynomials of $sI_n - A - BF$.

□

The sufficiency part of Theorem 2 provides us a very efficient way of the construction of feedback gain F and we summarize here the major steps of this construction.

Given $N(s)$ and $D(s)$, a P-normal external description of (8.17), and monic polynomials $\psi_1(s) \geq \psi_2(s) \geq \ldots \psi_n(s)$ satisfying (8.20), find F.

- Construct a column reduced matrix $M(s)$ having the same column degrees as $D(s)$ and the invariant polynomials $\psi_1(s), \psi_2(s), \ldots, \psi_m(s)$.
- Find a constant solution pair X, Y with X invertible of the equation (8.21).
- Put $F = -X^{-1}Y$.

It is to be noted that the feedback gain F given by the above procedure is by no means the only one that solves the problem. The freedom in choosing the matrix F is mainly embodied in the variety of ways by that we can obtain a column reduced matrix $M(s)$.

Theorem 2 has some interesting consequences, that we shall present in the form of corollaries.

Corollary 1 *Given a controllable system (8.17) with controllability indices* $c_1 = c_2 = \ldots = c_k = t+1$ *and* $c_{k+1} = c_{k+2} = \ldots = c_m = t$ *where* $n = tm + k, k < m$. *Then for every set of monic polynomials* $\psi_1(s) \geq \psi_2(s) \geq \ldots \geq \psi_m(s) \geq \psi_{m+1}(s) = \ldots = \psi_n(s) = 1$ *such that* $\sum_{i=1}^{m} deg\psi_i(s) = n$, *there exists a state feedback (8.18) which assigns the invariant polynomials* $\psi_i(s)$*'s to the closed-loop system (8.19).*

Corollary 2 *Let (8.17) be a controllable system. Then there always exists a feedback (8.18) such that the closed-loop system (8.19) is cyclic.*

The very simple proofs of the above corollaries are left to the reader.

The formulation of Theorem 2 is made from the view-point of control theory. But Theorem 2 can be stated in other ways emphasizing other aspects. For instance, the following theorem will stress the property of the equation (8.21) to have a constant solution pair X, Y with X nonsingular.

Theorem 4 *Let* $N(s)$ *and* $D(s)$ *form a P-normal external description of (8.17) with* $c_i = deg_{ci}D(s)$ *and let* $M(s)$ *be an* $m \times m$ *polynomial matrix. Then the equation*

$$XD(s) + YN(s) = M(s)$$

possesses a constant solution X, Y *with* X *nonsingular if and only if* $M(s)$ *is column reduced with* $deg_{ci}M(s) = c_i, i = 1, 2, \ldots, m$.

The proof is again very simple and is omitted. The following formulation of Rosenbrock's theorem is a bit unusual.

Theorem 5 *Let $D(s) \in \mathbb{R}^{m \times m}[s]$ be a column reduced matrix with column degrees $c_1 \geq c_2 \geq \ldots \geq c_m$ and let $\psi_1(s) \geq \psi_2(s) \geq \ldots \geq \psi_m(s)$ be monic polynomials. Then there exists a biproper matrix $B(s)$ such that $\psi_i(s)'s$ are the invariant polynomials of the matrix*

$$M(s) := B(s)D(s) \tag{8.22}$$

if and only if

$$\{deg\psi_i(s)\}_m \succ \{c_i\}_m \tag{8.23}$$

P r o o f.(Necessity).

Clearly, the matrix $M(s)$ is polynomial, since its invariant factors are just the polynomials $\psi_i(s)'s$, column reduced, and has the same column degrees as $D(s)$ because $B(s)$ is biproper. Hence, by Lemma 7 of the Appendix, the degrees of $\psi_i(s)'s$ and $c_i's$ verify the inequalities (8.23).

(Sufficiency).

We define first the matrix

$$\bar{M}(s) := diag[\psi_1(s), \ldots, \psi_m(s)]$$

By Lemma 2 of the Appendix, there exist unimodular matrices $U(s)$ and $V(s)$ such that the matrix

$$M(s) = U(s)\bar{M}(s)V(s)$$

is column reduced with column degrees $c_1 \geq c_2 \geq \ldots \geq c_m$. Put

$$B(s) = M(s)D^{-1}(s)$$

This matrix is clearly biproper and satisfies (8.22).

□

8.3.2 Fundamental theorem of P state feedback for explicit and uncontrollable systems

The case of eigenstructure assignment in explicit uncontrollable systems is a little bit more involved since we shall need some new mathematical tools, especially the concept of polynomial path [26].

To begin with, consider the system (8.17) which is not completely controllable. Then there exists a similarity transformation [8], [23] such that the matrices A and B in (8.17) can be brought into the form

$$A \sim \bar{A} := \begin{bmatrix} A_p & A_{pq} \\ 0 & A_q \end{bmatrix}, B \sim \tilde{B} := \begin{bmatrix} B_p \\ 0 \end{bmatrix}$$

where A_p, A_q and A_{pq} are matrices of respective sizes $p \times p, q \times q$ and $p \times q, p+q = n$, and B is a $p \times m$ matrix. We shall assume, that B_p is of full column rank. If not, an additional transformation acting on the input space can be used to achieve this property.

If we now apply a state feedback of the form (8.18) to $(\bar{A}, \tilde{B})$, then the controllable part (A_p, B_p) of $(\bar{A}, \tilde{B})$ is modified while the noncontrollable part, described by A_q, is left unchanged. Of course, the invariant factors of the overall system are modified, too. And that is the matter in question we are interested.

In order that to make the problem more transparent, we shall amend the matrix $\tilde{B}$ such that we obtain a new matrix $\bar{B}$,

$$\bar{B} := \begin{bmatrix} B_p & 0 \\ 0 & I_q \end{bmatrix},$$

and next we define a new input $\bar{u} := \begin{bmatrix} u \\ 0 \end{bmatrix}$.

This matrix makes our system completely controllable, which enable us to use the same approach as in the case of controllable systems. Moreover, the subsystem (A_q, I_q) is not effected by the input $\bar{u}$. Thus, adopting the relationships (8.13), (8.14) and (8.15) to our case, we get

$$\left[\begin{array}{cc|cc} sI_p - A_p & -A_{pq} & -B_p & 0 \\ 0 & sI_q - A_q & 0 & -I_q \end{array}\right] \left[\begin{array}{cc|cc} I_p & 0 & & \\ 0 & I_q & & 0 \\ \hline F_p & F_q & I_m & 0 \\ 0 & 0 & 0 & I_q \end{array}\right] \times$$

$$\left[\begin{array}{cc|cc} I_p & 0 & & \\ 0 & I_q & & 0 \\ \hline -F_p & -F_q & I_m & 0 \\ 0 & 0 & 0 & I_q \end{array}\right] \left[\begin{array}{c|c} N_m(s) & N_q(s) \\ \hline D_m(s) & D_{mq}(s) \\ 0 & sI_q - A_q \end{array}\right] = 0 = \tag{8.24}$$

$$= \left[\begin{array}{cc|cc} sI_p - A_p - B_pF_p & -A_{pq}F_q & -B_p & 0 \\ 0 & sI_q - A_q & 0 & -I_q \end{array}\right] \times$$

$$\times \left[\begin{array}{c|c} N_m(s) & N_q(s) \\ \hline D_m(s) - F\,N_m(s) & D_{mq}(s) - F\,N_q(s) \\ 0 & sI_q - A_q \end{array}\right]$$

where $F := [F_p \ F_q]$. Notice that the matrices

$$N(s) := [N_m(s) \ N_q(s)]$$

and

$$D(s) := \begin{bmatrix} D_m(s) & D_{mq}(s) \\ 0 & sI_q - A_q \end{bmatrix}$$

do not form a P-normal external description of $(\bar{A}, \bar{B})$, the tool that plays a crucial role in our approach. To remedy that, we shall divide the matrix $D_{mq}(s)$ by $sI_q - A_q$, which yields some constant matrix D_0 as the remainder. But, as we have used some elementary row operations for this purpose, we must apply the inverse operations to $\begin{bmatrix} sI_n - \bar{A} & -\bar{B} \end{bmatrix}$ to keep the equality

$$\begin{bmatrix} sI_n - \bar{A} & -\bar{B} \end{bmatrix} \begin{bmatrix} N(s) \\ D(s) \end{bmatrix} = 0.$$

This will change the matrix $\bar{B}$ to the form $\bar{B}(s) := \begin{bmatrix} B_p & B_pX(s) \\ 0 & I_q \end{bmatrix}$ where $X(s)$ is a polynomial matrix coming from the division algorithm we have performed. Fortunately, we can do that because of the special form of the input $\bar{u}$, which means that the dynamical behaviour of $(\bar{A}, \bar{B}(s))$ is the same as that of $(\bar{A}, \tilde{B})$. Hence, we can consider the matrix $D(s)$ in the form in which $D_{mq}(s)$ is a constant matrix D_0 and $D_m(s)$ is column reduced with column degrees $c_1 \geq c_2 \geq \ldots \geq c_m$.

The matrices $N(s)$ and $D(s)$ do not form a P-normal external description in the sense of Definition 5, but they are quite sufficient for our purpose since the matrix $D(s)$ reflects all necessary information regarding the eigenstructure assignbility.

Now, if we consider the matrices $N(s)$ and $D_F(s) := D(s) - \bar{F}N(s)$ which describe the closed-loop system $(\bar{A} + \bar{B}\bar{F}, \bar{B})$ and where $\bar{F} := \begin{bmatrix} F \\ 0 \end{bmatrix}$, everything what has been said about $N(s)$ and $D(s)$ can be applied to $N(s)$ and $D_F(s)$. To sum up the above observations, we can state the following:

- $c_1 \geq c_2 \geq \ldots \geq c_m \geq \underbrace{1 = \ldots = 1}_{q}$ are the column degrees of both $D(s)$ and $D_F(s)$ where $c_1, c_2, \ldots, c_m$ are the controllability indices of $(\bar{A}, \tilde{B})$;
- the matrices $D_{mq}(s)$ and $D_{mq}(s) - FN_q(s)$ can be considered to be constant matrices;
- the invariant polynomials, say $\psi_1(s) \leq \psi_2(s) \leq \ldots \leq \psi_{m+q}(s)$, of $D_F(s)$ depend on the invariant polynomials $\alpha_1(s) \leq \alpha_2(s) \leq \ldots \leq \alpha_q(s)$ of $sI_q - A_q$ which describes the dynamics of the noncontrollable part of $(\bar{A}, \tilde{B})$;
- the invariant polynomials of $sI_q - A_q$ cannot be modified by state feedback;
- the matrices $N_m(s), D_m(s)$ and $N_m(s), D_m(s) - FN_m(s)$ form P-normal external descriptions of $(\bar{A}, \tilde{B})$ and $(\bar{A} + \tilde{B}\bar{F}, \tilde{B})$, respectively.

The reader could notice that the invariant factors are now increasingly ordered. The purpose of that is to simplify the subsequent notation.

The above preliminary considerations enable us to state the problem of eigenstructure assignment in linear, explicit and uncontrollable systems.

Problem 2 *Let (8.17) be an uncontrollable system having the controllability indices* $c_1 \geq c_2 \geq \ldots \geq c_m$ *and let* $\alpha_1(s) \leq \alpha_2(s) \leq \ldots \leq \alpha_q(s)$ *be the invariant polynomials of* $sI_q - A_q$*, the noncontrollable part of* $sI_n - A$*. Let* $\psi_1^*(s) \leq \psi_2^*(s) \leq \ldots \leq \psi_n^*(s), n = p + q$*, be monic polynomials.*

Does there exist a state feedback (8.18) such that the polynomials $\psi_i^*(s)$*'s will be the invariant factors of the closed-loop system (8.19)?*

Problem 2 is equivalent, on the basis of Lemma 1 of the Appendix, to the problem of finding the matrices $D_m(s) - FN_m(s)$ and $D_{mq}(s) - FN_q(s)$ such that the matrix $D_F(s)$ will have prescribed invariant polynomials $\psi_i(s)$'s. These polynomials, completed by 1's to the number n, define the polynomials $\psi_i^*(s)$'s. Moreover, the matrix $D_F(s)$, when column reduced, should be of the column degrees

$c_1 \geq c_2 \geq \ldots \geq c_m \geq 1 = \ldots = 1$. We shall find a solution to this problem,which will be subsequently referred to as Problem 2a.

It is to be noted that Problem 2 was solved in [30] and was treated as a special case of matrix completion problems (see [27], [31] and references therein).

As the first step to characterize all possible sets of $\psi_i(s)$'s, we need the following Lemma 2 that introduces the concept of a polynomial path.

Lemma 2 *Let $A(s)$ be a $q \times q$ polynomial matrix having the invariant polynomials $\alpha_1(s) \leq \alpha_2(s) \leq \ldots \leq \alpha_q(s)$ and let $\psi_1(s) \leq \psi_2(s) \leq \ldots \leq \psi_{m+q}(s)$ be monic polynomials. Then there exist polynomial matrices $P(s)$ and $Q(s)$ such that the $(m+q) \times (m+q)$ matrix $\begin{bmatrix} P(s) & Q(s) \\ 0 & A(s) \end{bmatrix}$ has $\psi_i(s)$'s as its invariant polynomials if and only if*

$$\phi_i(s) \leq \alpha_i(s) \leq \phi_{i+m}(s), \quad i = 1, 2, \ldots, q. \tag{8.25}$$

Proof.
Without any loss of generality we can suppose that $A(s) = diag[\alpha_q(s), \ldots, \alpha_1(s)]$. Let $m = 1$. Then, by Lemma 5 of the Appendix, there exists a $1 \times (q+1)$ polynomial vector $\omega^1(s) = (\omega^1_{q+1}(s), \ldots, \omega^1_1(s))$ such that the matrix

$$D_1(s) := \left[\begin{array}{cc} \multicolumn{2}{c}{\omega^1(s)} \\ \hline 0 & A(s) \end{array} \right]$$

has the invariant factors $\phi^1_1(s) \leq \phi^1_2(s) \leq \ldots \leq \phi^1_{q+1}(s)$ if and only if

$$\phi^1_i(s) \leq \alpha_i(s) \leq \phi^1_{i+1}(s), \quad i = 1, 2, \ldots, q. \tag{8.26}$$

Now we define the $(q+j) \times (q+j)$ matrix

$$D_j(s) := \left[\begin{array}{cc} \multicolumn{2}{c}{\omega^j(s)} \\ \hline 0 & D_{j-1}(s) \end{array} \right]$$

where $\omega^j(s)$ is a $1 \times (q+j)$ polynomial vector and the matrix $D_{j-1}(s)$ has the invariant factors $\phi^{j-1}_1(s) \leq \ldots \leq \phi^{j-1}_{q+j-1}(s)$. Then, as above, the invariant polynomials $\phi^j_1(s) \leq \ldots \leq \phi^j_{q+j}$ of $D_j(s)$ satisfy

$$\phi^j_i(s) \leq \phi^{j-1}_i(s) \leq \phi^j_{i+1}(s), i = 1, 2, \ldots, q + j - 1. \tag{8.27}$$

For $j = m$, it follows from (8.27) and (8.26) that

$$\phi^m_i(s) \leq \ldots \leq \phi^1_i(s) \leq \alpha_i(s) \leq \phi^1_{i+1}(s) \leq \ldots \leq \phi^m_{i+m}(s)$$

for $i = 1, 2, \ldots, q$, which verifies (8.25).

□

Definition 7 *The system of the above polynomials* $\phi_i^j(s), i = 1,2,\ldots,q+j, j = 0,1,\ldots,m$, *where* $\phi_i^0(s) = \alpha_i(s), i = 1,2,\ldots,q$ *and* $\phi_i^m(s) = \psi_i(s), i = 1,2,\ldots,q+m$ *and (8.25) holds, is said to be a polynomial path from the polynomials* $\alpha_i(s)$*'s to the polynomials* $\psi_i(s)$*'s.*

In fact, Lemma 2 states that there exists a polynomial path [26] from $\alpha_i(s)$'s to $\psi_i(s)$'s if and only if the conditions (8.25) are satisfied. The problem now is how to construct a polynomial path from $\alpha_i(s)$'s to $\psi_i(s)$'s. Let $\psi_i(s) \leq \alpha_i(s) \leq \psi_{i+m}(s)$ for $i = 1,2,\ldots,q$. We define the polynomials

$$\beta_i^j(s) := lcm(\alpha_{i-j}(s), \psi_i(s)) \tag{8.28}$$

for $i = 1,2,\ldots,q+j$ and $j = 0,1,\ldots,m$ where $\alpha_i(s) := 1, i < 1$. Clearly, $\beta_i^0(s) = \alpha_i(s), i = 1,2,\ldots,q$ and $\beta_i^m(s) = \psi_i(s)$ for $i = 1,2,\ldots,q+m$.

Proposition 4 *The above polynomials* $\beta_i^j(s)$*'s satisfy the following conditions:*

(p1) $\beta_i^j(s) \leq \phi_i^j(s), i = 1,2,\ldots,q+j, j = 0,1,\ldots,m$ *where* $\phi_i^j(s)$*'s form any polynomial path from* $\alpha_i(s)$*'s to* $\psi_i(s)$*'s;*

(p2) $\beta_i^j(s) \leq \alpha_i(s) \leq \beta_{i+j}^j(s), i = 1,2,\ldots,q, j = 0,1,\ldots,m;$

(p3) $\psi_i(s) \leq \beta_i^j(s) \leq \psi_{i+m-j}(s), i = 1,2,\ldots,q+j, j = 0,1,\ldots,m;$

(p4) Let $\omega_i^j(s) := \frac{\beta_1^j(s),\ldots\beta_i^j(s)}{\beta_1^{j-1}(s)\ldots\beta_{i-1}^{j-1}(s)} \in \mathbb{R}[s], i = 1,2,\ldots,q+j$ *and* $j = 1,2,\ldots,m$, *then* $\omega_{q+j}^j(s) \leq \omega_{q+j+1}^{j+1}(s), j = 1,2,\ldots,m-1.$

The proof of Proposition 4 follows from the properties of the least common multiple of two polynomials and is left to the reader; see also [26] and [31].

Definition 8 *The polynomials* $\beta_i^j(s)$*'s defined by (8.28) are said to form a minimal path from* $\alpha_i(s)$*'s to* $\psi_i(s)$*'s.*

The property of a path to be the minimal path is based on (p1) of Proposition 4 and the relationship (8.28) shows how to construct it. Now we are ready to solve Problem 2a.

Theorem 6 *Problem 2a has a solution if and only if*

$$\psi_i(s) \leq \alpha_i(s) \leq \psi_{i+m}(s), \qquad i = 1,2,\ldots,q \tag{8.29}$$

and

$$\sum_{i=1}^{q} deg\alpha_i(s) + \sum_{i=1}^{j} c_{m-i+1} \geq \sum_{i=1}^{q+j} deg\beta_i^j(s), j = 1,2,\ldots,m \tag{8.30}$$

with equality for $j = m$.

Proof.
To prove necessity, let F be a feedback gain such that $D_F(s)$ has the invariant polynomials $\psi_i(s)$'s. Then the conditions (8.29) are clearly necessary, by Lemma 2.

As it has been explained above, the matrix $D_F(s)$ can be made column reduced with column indices $c_1 \geq c_2 \geq \ldots \geq c_m \geq 1 = \ldots = 1$ and we have, by Lemma 7 of the Appendix, that

$$\sum_{i=1}^{j} deg\psi_i(s) \geq \sum_{i=1}^{j} c'_{m+q-i+1}, j = 1, 2, \ldots, m+q$$

where $c'_1 = c_1, c'_2 = c_2, \ldots, c'_m = c_m, c'_{m+1} = 1, \ldots, c'_{m+q} = 1$ and equality holds for $j = m + q$.

These inequalities can be rewritten into the form

$$\sum_{i=1}^{q+j} deg\psi_i(s) \leq \sum_{i=1}^{j} c_{m-i+1} + \sum_{i=1}^{q} deg\alpha_i(s), \tag{8.31}$$

$j = 1, 2, \ldots, m$ with equality for $j = m$.

Now, since there exists the minimal path $\beta_i^j(s)$'s from $\alpha_i(s)$'s to $\psi_i(s)$'s, we have by (p3) of Proposition 4

$$\sum_{i=q+j}^{q+m} deg\beta_i^j(s) \geq \sum_{i=q+j}^{q+m} deg\psi_i(s) \tag{8.32}$$

As the equality holds in (8.31) for $j = m$, these inequalities can be reordered such that

$$\sum_{i=j+q}^{q+m} deg\psi_i(s) \geq \sum_{i=j}^{m} c_{m-i+j} + \sum_{i=1}^{q} deg\alpha_i(s) \tag{8.33}$$

$j = 1, 2, \ldots, m$ with equality for $j = 1$. Hence, combining (8.32) and (8.33) and going back to the original ordering, the inequalities (8.30) follow.

To prove sufficiency, let the polynomials $\alpha_i(s)$'s and $\psi_i(s)$'s satisfying (8.29) and (8.30) be given. We are going to find a feedback gain F such that the matrix $D_F(s)$ will have the invariant polynomials $\psi_i(s)$'s.

Let $\beta_i^j(s), i = 1, 2, \ldots, q + j, j = 0, 1, \ldots, m$ be again the minimal path from $\alpha_i(s)$'s to $\psi_i(s)$'s. We define the $(q + m) \times (q + m)$ matrix

$$\bar{D}(s) := \begin{bmatrix} \bar{Q}(s) & P(s) \\ 0 & R(s) \end{bmatrix} := \left[\begin{array}{cccc|ccc} \omega_{q+m}^m(s), & \cdots & & \omega_{q+1}^m(s) & \cdots & \cdots & \omega_1^m(s) \\ & & \ddots & \vdots & & & \\ & & & \omega_{q+1}^1(s) & \cdots & \cdots & \omega_1^1(s) \\ \hline & & & & \alpha_q(s) & & \\ & & 0 & & & \ddots & \\ & & & & & & \alpha_1(s) \end{array}\right]$$

where $\omega_i^j(s)$'s are defined by Proposition 4. It is easy to see that $\psi_i(s)$'s are the invariant factors of $\bar{D}(s)$ and the only thing to do is to bring $\bar{D}(s)$ to a column reduced form with column degrees $c_1 \geq c_2 \geq \ldots \geq c_m \geq 1 = \ldots = 1$. We begin with the matrix $\bar{Q}(s)$.

Using elementary row operations, we can make $\bar{Q}(s)$ column reduced with the column degrees $deg\omega_{q+j}^j(s), j = 1, 2, \ldots, m$. Since $\omega_{q+j}^j(s) \leq \omega_{q+j+1}^{j+1}(s), j = 1, 2, \ldots, m-1$ and, by (p4) of Proposition 4,

$$\sum_{j=1}^{k} deg\omega_{q+j}^j(s) = \sum_{j=1}^{q+k} deg\beta_j^k(s) - \sum_{j=1}^{q} deg\alpha_j(s), k = 1, 2, \ldots, m$$

with equality for $k = m$, it follows from (8.32) that

$$\sum_{i=1}^{k} deg\omega_{q+i}^i(s) \leq \sum_{i=1}^{k} c_{m-i+1}, k = 1, 2, \ldots, m$$

with equality for $k = m$. Hence, we can apply Lemma 2 of the Appendix to bring this matrix to a column reduced form with column indices equal to c_i's. Let $Q(s)$ denote such a matrix.

Applying again the same lemma to the matrix $R(s)$, we can bring it into the form $sI_q - A_q, A_q \in \mathbf{R}^{q \times q}$. Finally, dividing $P(s)$ (which has been modified by the previous operations) by $sI_q - A_q$, we obtain some constant matrix, say P_0.

All the above operations do not change the invariant factors of $\bar{D}(s)$ and we can put

$$D_F(s) := \begin{bmatrix} Q(s) & P_0 \\ 0 & sI_q - A_q \end{bmatrix}.$$

This matrix has the desired properties, i.e., it is column reduced with column indices $c_1, c_2, \ldots, c_m, 1, \ldots, 1$, and has $\psi_i(s)$'s as its invariant factors.

Now the rest of the proof is the same as that of Theorem 3. A state feedback gain F is calculated from a constant solution X and Y to the equation

$$XD(s) + YN(s) = D_F(s) \tag{8.34}$$

where X is invertible. We put $F = -X^{-1}Y$.

□

Remark 9. Theorem 3 is a partial case of the more general Theorem 6. Indeed, if the system $(\bar{A}, \tilde{B})$ is controllable, there are no $\alpha_i(s)$'s since $q = 0$, $\beta_i^j(s) = \psi_i(s)$ for $i = j = 1, 2, \ldots, m$ and the inequalities (8.30) reduce to (8.20). The inequalities (8.29) disappear since there are no $\alpha_i(s)$'s.

Remark 10. A more detailed analysis of the equation (8.34) shows that the matrix X is of the form $X = \begin{bmatrix} X_m & 0 \\ 0 & I_q \end{bmatrix}$ and $Y = \begin{bmatrix} Y_m \\ 0 \end{bmatrix}$, which can mean some simplification in calculating a state feedback gain F.

8.3.3 Fundamental theorem of P state feedback for linear implicit systems

The realm of implicit systems is much richer than that of explicit systems. No wonder, it has been tempting to generalize the Rosenbrock's theorem also in the case of implicit systems.

In this chapter, we shall consider a controllable system (8.1) with the matrix E of rank $r < n$ and we also assume that rank $B = m$. It was already mentioned that such systems have finite as well as infinite poles and infinite zeros. Therefore, when moving finite poles to infinity, a special care must be taken since poles might be shifted to the position in which the system has zeros.

This situation cannot happen in the case of explicit systems because there is no chance to move the poles to infinity due to keeping the controllability of (8.17). Recall that the explicit system (8.17) has always zeros at infinity. In the case of implicit systems, the situation is the same; no cancelation can occur. This fact influences the form of inequalities (8.20).

Another peculiarity occuring in the implicit systems lies in what the regularity is not invariant under the action of P state feedback. Thus, there is no reason to assume the regularity of (8.1) and, of course, the regularity of the closed loop system

$$E\dot{x} = (A + BF)x + Bv \tag{8.35}$$

is also not required.

Problem 3 *Let (8.1) be a controllable system with controllability indices $c_1 \geq c_2 \geq \ldots \geq c_m$ and let $\psi_1(s) \geq \psi_2(s) \geq \ldots \geq \psi_t(s)$ be monic polynomials. Further, let $d_1 \geq d_2 \geq \ldots \geq d_p$ be positive integers and let q denote the number of the proper controllability indices.*

Then the problem of eigenstructure assignment is the following. Find necessary and sufficient conditions under which there exists a feedback gain F such that the finite pole structure of (8.23) will be given by $\psi_i(s)'$ and the infinite pole structure will be given by $d_i's$.

The generalization of Rosenbrock's theorem has been studied in [12], [22], [33], [34], [35] and many others (see references therein). The following theorem represents the most general version of this kind of fundamental theorem and can be found in [35].

Theorem 7 *Problem 3 has a solution if and only if*

$$n - m + p + q \leq t \leq n \tag{8.36}$$

and

$$\{deg\psi_i(s) + d_i\}_t \leq \{c_i^*\}_{t+m-n} \quad for \quad t < n \tag{8.37}$$

or

$$\{deg\psi_i(s) + d_i\}_n \succ \{c_i\}_m \quad for \quad t = n. \tag{8.38}$$

where $c_1^* \geq c_2^* \geq \ldots \geq c_{t+m-n}^*$ is a subset of $\{c_i\}_m$ that includes all the proper controllability indices and $d_i := 0$ for $i > p$.

P r o o f. (Necessity).

Suppose there exists a feedback gain F such that the eigenstructure of (8.35) is given by the polynomials $\psi_i(s)'s$ and integers $d_i's$. We shall verify (8.36), (8.37) and (8.38).

To this end, we apply first the conformal mapping

$$s = \frac{1 + aw}{w} \tag{8.39}$$

where a is any nonzero complex number which is neither a pole nor a zero of (8.35). This mapping sends the point $s = a$ to $w = \infty$ and $s = \infty$ to $w = 0$. This means that both the finite and infinite poles of (8.35) are now at finite positions, which enables us to handle these two structures in a uniform way.

Let $N(s), D(s)$ be a P- normal external description of (8.1). Then, $N(s), D(s) - FN(s)$ form a P- normal external description of (8.35), by Lemma 1. Applying (8.39) to $N(s)$ and $D(s) - FN(s)$, we define

$$\begin{bmatrix} Q(w) \\ P(w) \end{bmatrix} := \begin{bmatrix} N\left(\frac{1+aw}{w}\right) \\ D\left(\frac{1+aw}{w}\right) \quad +FN\left(\frac{1+aw}{w}\right) \end{bmatrix} \times diag\left[w^{c_1}, \ldots, w^{c_m}\right] \tag{8.40}$$

This matrix is polynomial over $\mathbb{R}[w]$, irreducible and column reduced with column degrees $c_1, c_2, \ldots, c_m$.

Similarly we define

$$[K(w)\ L(w)] := diag\left[\underbrace{w, \ldots, w\}}_{n+m-q}, 1, \ldots, 1\right] \times \left[\frac{1 + aw}{w}E - A - BF \quad -B\right] \tag{8.41}$$

where the matrix E is, without any loss of generality, assumed to be of the form

$$E = \begin{bmatrix} E_1 \\ 0 \end{bmatrix} \tag{8.42}$$

with E_1 of full row rank $r = n + m - q$.

Clearly, the matrix (8.41) is also polynomial over $\mathbb{R}[w]$, irreducible, and row reduced. Moreover,

$$[K(w)\ L(w)] \begin{bmatrix} Q(w) \\ P(w) \end{bmatrix} = 0$$

By Lemma 1 of the Appendix, there exist unimodular matrices $V_1(w), V_2(w)$ and $U_2(w)$ such that

$$\begin{bmatrix} U_2^{-1}(w) & 0 \\ 0 & V_1(w) \end{bmatrix} \begin{bmatrix} Q(w) \\ P(w) \end{bmatrix} V_2(w) = \begin{bmatrix} S_Q \\ S_P \end{bmatrix}$$

where the matrices S_Q and S_P assume the diagonal form of S_N and S_D shown in Lemma 1. Hence, the invariant polynomials of S_Q and S_P reveal the zero structure and pole structure of (8.35), respectively. More precisely, we have

$$S_Q = \begin{bmatrix} diag[\overbrace{1, \ldots, 1}^{t+m-n-q}, \overbrace{w, \ldots, w}^{q} 1, 1, \ldots, 1] \\ 0 \end{bmatrix},$$

$$S_P = diag\,[\pi_1(w), \ldots, \pi_{t+m-n}(w), 0, \ldots, 0]$$

where

$$\pi_i(w) = \left(\begin{array}{ll} \bar{\psi}_i(w)w^{d_i} & , i = 1, 2, \ldots, p \\ \bar{\psi}_i(w) & , i = p+1, \ldots, t+m-n \end{array} \right.$$

and $n-t$ is the rank definiciency of both $P(w)$ and $K(w)$. The polynomial $\bar{\psi}_i(w)$ is related to $\psi_i(s)$ by

$$\psi_i(s) = k_i(s-a)^{deg\bar{\psi}_i(w)}\bar{\psi}_i\left(\frac{1}{s-a}\right)$$

where k_i is a scaling constant, $i = 1, 2, \ldots, t+m-n$.

The matrices S_Q and S_P are right comprime, which implies that $\pi_i(w)$ is not divisible by w for $i = t+m-n-q+1, \ldots, t+m-n$. Thus, $p+q \leq t+m-n$ and the inequalities (8.36) are verified.

In order that to prove the inequalities (8.37) and (8.38), we divide first, by some column permutations, the matrix $P(w)$ into three parts. The first part, say $P_1(w)$, will consist of all the columns of $P(w)$ corresponding to the proper controllability indicies. There are q such columns in $P(w)$ and they are linearly independent over $\mathbb{R}[w]$, i.e., they form a column reduced submatrix of $P(w)$. We define $P_2(w)$ by

$$rank\,[P_1(w)\ P_2(w)] = t+m-n,$$

which means that $[P_1(w)\ P_2(w)]$ comprises all the linearly independent columns of $P(w)$. The third part $P_3(w)$ simply makes up of the rest of the columns of $P(w)$, i.e.,

$$rank\,[P_1(w)\ P_2(w)\ P_3(w)] = rank\,[P_1(w)P_2(w)]\,.$$

The inequalities (8.37) can now be easily verified. Apply Lemma 7 of the Appendix to $P_1(w)$ to get

$$\left\{deg\tilde{\psi}_i(w)\right\}_q \leq \{\tilde{c}_i\}_q$$

where $\tilde{\psi}_1(w) \geq \tilde{\psi}_2 \geq \ldots \geq \tilde{\psi}_q(w)$ are the invariant polynomials of $P_1(w)$ and $\tilde{c}_1 \geq \tilde{c}_2 \geq \ldots \geq \tilde{c}_q$ are the proper controllability indices of (8.1).

Next we apply, column by column, Corollary 1 of the Appendix to $[P_1(w)\ P_2(w)]$ so that we obtain

$$\left\{deg\hat{\psi}_i(w)\right\}_{t+m-n} \leq \{\hat{c}_i\}_{t+m-n}$$

where $\hat{\psi}_1(w) \geq \hat{\psi}_2(w) \ldots \geq \hat{\psi}_{t+m-n}(w)$ are the invariant polynomials of $[P_1(w)P_2(w)]$. The column degrees of $P_2(w)$ are equal to the corresponding non-proper controllability indices, say $c_i^{'}, i = 1, 2, \ldots, t+m-n-q$, and the list $\hat{c}_1 \geq \hat{c}_2 \geq \ldots \geq \hat{c}_{t+m-n}$ is given by reordering the union of both $\tilde{c}_i$ and $c_i^{'}$ in the nonincreasing order.

The last step consists in applying Corollary 2 of the Appendix to $[P_1(w)\ P_2(w)\ P_3(w)]$. We get

$$\left\{deg\bar{\psi}_i(w)\right\}_{t+m-n} \leq \{\bar{c}_i\}_{t+m-n}$$

where $\bar{\psi}_1(w) \geq \bar{\psi}_2(w) \geq \ldots \geq \bar{\psi}_{t+m-n}(w)$ are the invariant polynomials of $[P_1(w)\ P_2(w)\ P_3(w)]$ and $\bar{c}_i = \hat{c}_i, i = 1, 2, \ldots, t+m-n$.

If we now complete the list $\bar{\psi}_i(w)$ by 1's to the number t and put

$$c_i^* = \begin{cases} \bar{c}_i, & i = 1, 2, \ldots, t+m-n \\ 0, & i = t+m-n+1, \ldots, t, \end{cases}$$

the inequalities (8.37) follow. If $t = n$, we get the inequalities (8.38).

(Sufficiency).

Given a list of monic polynomials $\psi_1(s) \geq \ldots \geq \psi_t(s)$ and a list of positive integers $d_1 \geq d_2 \geq \ldots \geq d_p$. Let $c_1^* \geq c_2^* \geq \ldots \geq c_{t+m-n}^*$ be a subset of the controllability indices of (8.1) that includes all the proper controllability indices of (8.1) such that the conditions (8.36), (8.37) or (8.38) are satisfied. We are going to construct a matrix F such that the eigenstructure of (8.35) will be given by $\psi_i(s)'$s and d_i's.

Apply first the conformal mapping (8.39) to the polynomials $\psi_i(s), i = 1, 2, \ldots, t$ to get

$$\bar{\psi}_i(w) = k_i w^{deg\psi_i(s)} \psi_i\left(\frac{1+aw}{w}\right)$$

where k_i makes $\bar{\psi}_i(w)$ monic, $i = 1, 2, \ldots, t$.

Let $N(s), D(s)$ be a P-normal external description of (8.1). Applying (8.39) to $\begin{bmatrix} N(s) \\ D(s) \end{bmatrix}$ and $[sE - A \; -B]$ where E assumes the form (8.42), we obtain, by the same way as in (8.40) and (8.41), the matrices $\begin{bmatrix} Q'(w) \\ P'(w) \end{bmatrix}$ and $[K'(w) \; L'(w)]$, respectively, satisfying

$$\left[K'(w) \; L'(w)\right] \begin{bmatrix} Q'(w) \\ P'(w) \end{bmatrix} = 0. \tag{8.43}$$

Of course, the above matrices have exactly the same properties as those in (8.40) and (8.41).

We define the polynomials

$$\phi_i(w) = \begin{cases} \bar{\psi}_i(w) w^{d_i} & , i = 1, 2, \ldots, p \\ \bar{\psi}_i(w) & , i = p+1, \ldots, t+m-n \end{cases} \tag{8.44}$$

and next we shall construct a special matrix, say $M'(w)$, having $\phi_i(w)'$s as its invariant polynomials. To this end, suppose first that $t < n$ (i.e., $m > t+m-n$) and define the $m \times (t+m-n)$ matrix

$$\bar{M}(w) := \left[\begin{array}{ccc} \omega_1(w) & & \\ \phi_1(w) & \ddots & \\ & \ddots & \omega_{t+m-n}(w) \\ & & \phi_{t+m-n}(w) \\ \hline & 0 & \end{array}\right]$$

where $\omega_i(w)'$s is any set of monic polynomials such that

(i) $\omega_1(w) \geq \omega_2(w) \geq \ldots \geq \omega_{t+m-n}(w)$,

(ii) $\omega_i(w) \geq \phi_i(w), i = 1, 2, \ldots, t+m-n$,

(iii) $\{deg\omega_i(w)\}_{t+m-n} \leq \{c_i^*\}_{t+m-n}$.

Obviously, such polynomials $\omega_i(w)$'s exist. If $t = n$, put

$$\bar{M}(w) := diag[\phi_1(w), \ldots, \phi_m(w)]$$

since $\sum_{i=1}^{m} deg\phi_i(w) = \sum_{i=1}^{m} c_i$.

It should be clear that $\phi_i(w)$'s are the invariant polynomials of $\bar{M}(w)$. Now we apply Lemma 2 of the Appendix to $\bar{M}(w)$ to obtain a column reduced matrix, say $\hat{M}(w)$, with column degrees $c_1^*, c_2^*, \ldots, c_{t+m-n}^*$.

Further we define an $m \times (t+m-n)$ matrix $\hat{Q}(w)$ such that $\hat{Q}(w)$ consists of those columns of $Q'(w)$ that correspond to the controllability indices $c_1^*, c_2^*, \ldots, c_{t+m-n}^*$ and consider the matrix

$$\begin{bmatrix} \hat{Q}(w) \\ \hat{M}(w) \end{bmatrix}$$

If the above matrix is irreducible, we define the $m \times m$ matrix

$$\hat{M}(w) := \begin{bmatrix} \hat{M}(w) & 0 \end{bmatrix}.$$

If not, there is a zero of $\begin{bmatrix} \hat{Q}(w) \\ \hat{M}(w) \end{bmatrix}$ at $w = 0$. This implies a zero column among the first p columns of the matrix $\begin{bmatrix} \hat{Q}(0) \\ \hat{M}(0) \end{bmatrix}$. As there are q nonunit invariant polynomials of $\hat{Q}(w)$ equal to w and the operations implied by Lemma 2 of the Appendix cannot change the linear independency of the last $t+m-n-p$ columns of $\hat{M}(0)$, due to (8.36), there exists (see [34] for more details) a nonsingular matrix T over $\mathbb{R}$ such that the matrix $\begin{bmatrix} \hat{Q}(w) \\ \hat{M}(w)T \end{bmatrix}$ is irreducible. Put

$$\tilde{M}(w) := \begin{bmatrix} \hat{M}(w)T & 0 \end{bmatrix}$$

and

$$\tilde{Q}(w) := \begin{bmatrix} \hat{Q}(w) & Q^*(w) \end{bmatrix}$$

where $Q^*(w)$ denotes the rest of $Q'(w)$ after extracting $\hat{Q}(w)$.

Let S be a permutation matrix restoring the original order of the columns of $Q'(w)$. Then we define

$$M'(w) := \tilde{M}(w)S.$$

In either case the matrix $\begin{bmatrix} Q'(w) \\ M'(w) \end{bmatrix}$ is irreducible and column reduced with column degrees $c_1, c_2, \ldots, c_m$.

Now we apply the inverse transformation of (8.39) to $\begin{bmatrix} Q'(w) \\ M'(w) \end{bmatrix}$ in order that to get

$$\begin{bmatrix} N(s) \\ M(s) \end{bmatrix} := \begin{bmatrix} Q'\left(\frac{1}{s-a}\right) \\ M'\left(\frac{1}{s-a}\right) \end{bmatrix} diag\left[(s-a)^{c_1}, \ldots, (s-a)^{c_m}\right] \tag{8.45}$$

This matrix is polynomial over $\mathbb{R}[s]$, irreducible, and column reduced with column degrees $c_1, c_2, \ldots, c_m$. Moreover, the submatrix of $M(s)$ formed by those columns of $M(s)$ whose column degrees match the proper controllability indices is also column reduced. Hence, by Lemma 4 of the Appendix, there exist matrices X, Y, Z over $\mathbb{R}$, X and Z invertible, such that

$$XD(s) + YN(s) = M(s)Z. \tag{8.46}$$

Put $F = -X^{-1}Y$. The matrix F qualifies as a feedback gain assigning the desired eigenstructure. This immediately follows from Lemma 1 of the Appendix.

□

The sufficiency part of the proof gives us a method for the construction of F, which is summarized below.

a) Given $E, A, B, \psi_1(s) \geq \psi_2(s) \geq \ldots \geq \psi_t(s)$ and $d_1 \geq d_2 \geq \ldots \geq d_p$, calculate $N(s)$ and $D(s)$, a P-normal external description of (8.1).

b) Read out $c_1, c_2, \ldots, c_m$, the column degrees of $\begin{bmatrix} N(s) \\ D(s) \end{bmatrix}$, and identify the subset of q proper controllability indices.

c) Choose $c_1^* \geq c_2^* \geq \ldots \geq c_{t+m-n}^*$ and check the existence of F using (8.36), (8.37) or (8.38).

d) Using (8.40) and (8.41), calculate matrices $Q'(w)$ and $P'(w)$.

e) Construct an $m \times m$ matrix $M'(w)$ having the invariant polynomials (8.44) and make the composite matrix $\begin{bmatrix} Q'(w) \\ M'(w) \end{bmatrix}$ irreducible and column reduced with column degrees $c_1, c_2, \ldots, c_m$.

f) Find matrices X, Y, Z over $\mathbb{R}$, X and Z invertible, such that (8.46) holds.

g) Put $F = -X^{-1}\,Y$.

□

The matrix F computed by this procedure is of course not unique. The same remark, as that regarding the procedure for the computation of feedback gain in the case of explicit systems, can be applied to this procedure.

Remark 8 It can be readily seen that Theorem 4 is a special case of Theorem 7. Indeed, in the case of explicit systems, the pencil $sI_n - A - BF$ is regular for all F

and hence, $t = n, p = 0$, and $q = m$. This implies that (8.36) is trivially satisfied and (8.38) becomes (8.20) with $c_i^* = c_i$ for $i = 1, 2, \ldots, m$. But there are another special cases that we shall discuss just now.

Remark 9 (Regularization and finite eigenstructure assignment [12])

It could be desirable to assign only a finite eigenstructure to the closed loop system (8.35) since infinite poles produce an impulsive behaviour, the thing we would like to avoid. In this case we have $t = n$ and $p = 0$. This implies that (8.36) is trivially satisfied ($q \leq m$) and there are no d_i's. Hence, the inequalities (8.38) are in the form (8.20).

To sum up, the inequalities (8.20) characterize the dynamics of all explicit systems obtainable from a controllable system (8.1) via P state feedback.

Remark 10 (Regularization and infinite eigenstructure assignment [34]).

If, on the contrary, just the impulsive behaviour of (8.1) is required, we have to assign no finite poles to (8.35), i.e., $\psi_i(s) = 1$ for $i = 1, 2, \ldots, n$. Then (8.36) and (8.38) are of the form

$$p + q \leq m$$

and

$$\{c_i\}_m \prec \{d_i\}_p ,$$

respectively.

These inequalities characterize the dynamics of all pure implicit systems obtainable from a controllable system (8.1) via P state feedback.

□

To illustrate the construction of F, consider the following

Example 1. Let

a)

$$E = \begin{bmatrix} 0 & 1 & 0 & 0 & 0 & 0 \\ 0 & 0 & 1 & 0 & 0 & 0 \\ 0 & 0 & 0 & 1 & 0 & 0 \\ 0 & 0 & 0 & 0 & 0 & 0 \\ 0 & 0 & 0 & 0 & 0 & 1 \\ 0 & 0 & 0 & 0 & 0 & 0 \end{bmatrix}, \quad A = \begin{bmatrix} 1 & 0 & 0 & 0 & 0 & 0 \\ 0 & 1 & 0 & 0 & 0 & 0 \\ 0 & 0 & 1 & 0 & 0 & 0 \\ 0 & 0 & 0 & 0 & 0 & 0 \\ 0 & 0 & 0 & 0 & 1 & 0 \\ 0 & 0 & 0 & 0 & 0 & 0 \end{bmatrix},$$

$$B = \begin{bmatrix} 0 & 0 & 0 \\ 0 & 0 & 1 \\ 0 & 0 & 0 \\ 1 & 0 & 0 \\ 0 & 0 & 0 \\ 0 & 1 & 0 \end{bmatrix},$$

$\psi_1(s) = s$, $\psi_2(s) = 1$, and $d_1 = 1$. A P–normal external description of (E, A, B) can be found in the form

$$N(s) = \begin{bmatrix} 0 & 0 & s \\ 0 & 0 & 1 \\ s & 0 & 0 \\ 1 & 0 & 0 \\ 0 & s & 0 \\ 0 & 1 & 0 \end{bmatrix}, D(s) = \begin{bmatrix} 0 & 0 & 0 \\ 0 & 0 & 0 \\ s^2 & 0 & -1 \end{bmatrix}$$

b) The controllability indices of (E, A, B) are $c_1 = 2, c_2 = 1, c_3 = 1$ and there is one proper controllability index $c_1 = 2$.

c) With respect to the polynomials $\psi_1(s), \psi_2(s)$ and d_1, we choose $c_1^* = 2$ and $c_2^* = 1$. Then the inequalities (8.36) and (8.37) are satisfied.

d) Using the conformal mapping $s = \frac{1+w}{w}$, we obtain the matrices

$$Q'(w) = \begin{bmatrix} 0 & 0 & 1+w \\ 0 & 0 & w \\ w+w^2 & 0 & 0 \\ w & 0 & 0 \\ 0 & 1+w & 0 \\ 0 & w & 0 \end{bmatrix}, P'(w) = \begin{bmatrix} 0 & 0 & 0 \\ 0 & 0 & 0 \\ (1+w)^2 & 0 & -w \end{bmatrix}$$

e) We get first $\psi_1(w) = 1 + w, \psi_2(w) = 1$ and then, by (8.44) , $\phi_1(w) = w + w^2, \phi_2(w) = 1$. Put

$$\bar{M}(w) = \begin{bmatrix} w+w^2 & 0 \\ 0 & w \\ 0 & 1 \end{bmatrix},$$

which gives

$$M'(w) = \begin{bmatrix} w+w^2 & 0 & 0 \\ w & w & 0 \\ 1 & 1 & 0 \end{bmatrix}$$

Then the composite matrix $\begin{bmatrix} Q'(w) \\ M'(w) \end{bmatrix}$ is irreducible and column reduced with column degrees equal to c_1, c_2, c_3.

f) Using (8.45), we get

$$M(s) = \begin{bmatrix} s & 0 & 0 \\ s-1 & 1 & 0 \\ (s-1)^2 & s-1 & 0 \end{bmatrix}$$

and then, calculating matrices X, Y and Z, the step (g) gives

$$F = \begin{bmatrix} 0 & 0 & -1 & 0 & 0 & 0 \\ 0 & 0 & -1 & 1 & 0 & -1 \\ 0 & -1 & 2 & -1 & -1 & 1 \end{bmatrix}$$

The matrix F can be calculated directly from the equation

$$\begin{bmatrix} I_6 & 0 \\ -F & I_3 \end{bmatrix} \begin{bmatrix} N(s) \\ D(s) \end{bmatrix} = \begin{bmatrix} N(s) \\ M(s) \end{bmatrix}.$$

□

8.4 Eigenstructure Assignment by PD State Feedback

The effect of PD state feedback upon the system (8.1) is much complicated than that of P state feedback. We now deal with a special kind of dynamic feedback, i.e., the number of poles of closed-loop system (8.3) can be, in certain limits, increased. When writting the effect of PD state feedback upon (8.1) in the form

$$[sE - A \quad -B] \begin{bmatrix} I_n & 0 \\ -sK + F & I_m \end{bmatrix} = [s(E + BK) - (A + BF) \quad -B] \quad (8.47)$$

it easily follows that the maximal number of poles, the system (8.3) can be assigned, is n.

It has been already mentioned that the notion of controllability under PD state feedback is different from that regarding the controllability under P state feedback. To investigate this question more deeply, we shall introduce the concept of extended system. To this end, we adjoin to the state x its derivative $y = \dot{x}$ and obtain a new system described by

$$\begin{bmatrix} E & 0 \\ I_n & 0 \end{bmatrix} \begin{bmatrix} \dot{x} \\ \dot{y} \end{bmatrix} = \begin{bmatrix} A & 0 \\ 0 & I_n \end{bmatrix} \begin{bmatrix} x \\ y \end{bmatrix} + \begin{bmatrix} B \\ 0 \end{bmatrix} u \quad (8.48)$$

Then, the action of PD state feedback can be represented by the action of P state feedback

$$u = [F \quad -K] \begin{bmatrix} x \\ y \end{bmatrix} + v \quad (8.49)$$

upon the system (8.48). The following relationship clearly shows this action

$$\begin{bmatrix} sE - A & 0 & -B \\ sI_n & -I_n & 0 \end{bmatrix} \begin{bmatrix} I_n & 0 & 0 \\ 0 & I_n & 0 \\ F & -K & I_m \end{bmatrix} =$$

$$= \begin{bmatrix} sE - A - BF & BK & -B \\ sI_n & -I_n & 0 \end{bmatrix} \quad (8.50)$$

Lemma 3 *The nonunit invariant polynomials of the matrix*

$$\begin{bmatrix} sE - A & 0 \\ sI_n & -I_n \end{bmatrix} \quad (8.51)$$

correspond to those of $sE - A$.

Lemma 4 *Let $\alpha_1 \geq \alpha_2 \geq \ldots \geq \alpha_e$ be infinite zero orders of (8.51). Then the infinite zero orders $\beta_1 \geq \beta_2 \geq \ldots \geq \beta_k$ of $sE - A$ are given by*

$$\beta_i = \alpha_i - 1, i = 1, 2, \ldots, k$$

wher k is the largest integer i such that $\alpha_k < 1$.

The proofs of the above lemmas easily follow from the definition of the infinite zero structure of $sE - A$ (see Proposition 1) and is left to the reader.

Considering the closed-loop pencil (8.47) again, Lemma 4 give us a correspondence between the zero structure of $s(E + BK) + (A + BF)$ and that of the matrix

$$\begin{bmatrix} sE - A - BF & BK \\ sI_n & -I_n \end{bmatrix}$$

Proposition 5 *Let (8.1) be a reachable system and let $N(s), D(s)$ be its P-normal external description. Then the extended system (8.48) is controllable with controllability indices that are equal to the reachability indices of (8.1) and the matrices $\begin{bmatrix} N(s) \\ sN(s) \end{bmatrix}, D(s)$ form a P-normal external description of (8.48).*

Proof.
If $N(s)$ and $D(s)$ form a P normal external description of (8.1), then, follows from

$$\begin{bmatrix} sE - A & 0 & -B \\ sI_n & -I_n & 0 \end{bmatrix} \begin{bmatrix} N(s) \\ sN(s) \\ D(s) \end{bmatrix} = 0,$$

the matrices $\begin{bmatrix} N(s) \\ sN(s) \end{bmatrix}$ and $D(s)$ form a P normal external description of (8.48). The controllability of (8.48) follows immediately from (8.12), Theorem 2.

□

Notice that the system (8.48) has no proper controllability indices.

If the system (8.1) is only controllable, then the system (8.48) is not controllable; more precisely, the pencil

$$\begin{bmatrix} sE - A & 0 & -B \\ sI_n & -I_n & 0 \end{bmatrix}$$

can have some infinite zeros. But still the controllability indices of (8.48) are equal to the reachability indices of (8.1).

Definition 9 *Given a controllable system (8.1), the matrices $\begin{bmatrix} N(s) \\ sN(s) \end{bmatrix}$ and $D(s)$ are said to form a PD-normal external description of (8.1) and the integers*

$$k_i = deg_{ci} [sN(s)]$$

are said to be controllability indices of (8.1) under the action of PD state feedback.

Now we can state

Problem 4 *Let (8.1) be a reachable system with reachability indices* $r_1 \geq r_2 \geq \ldots \geq r_m$ *and let* $\psi_1(s) \geq \psi_2(s) \geq \ldots \geq \psi_t$ *be monic polynomials. Let furher* $d_1 \geq d_2 \geq \ldots \geq d_p$ *be positive integers and let* $r_1^* \geq r_2^* \geq \ldots \geq r_{t+m-n}^*$ *form a subset of the reachability indices of (8.1).*

Find necessary and sufficient conditions for there to exist a state feedback (8.2) such that the finite and infinite pole structure of (8.3) will be given by $\psi_i(s)'s$ *and* $d_i's$*, respectively.*

Applying Theorem 7 to the extended system (8.48), we immediately obtain [36]

Theorem 8 *The Problem 4 has a solution if and only if*

$$n - m + p \leq t \leq n \tag{8.52}$$

and

$$\{deg\psi_i(s) + d_i\}_t \leq \{r_i^*\}_{t+m-n} \cdot \tag{8.53}$$

Remark 11 Theorem 8 shows that PD state feedback is really very powerful tool to modify the pole structure of (8.1). Neither reachability nor controllability are preserved under the action of PD state feedback, which is natural since PD state feedback is a special kind of dynamic feedback.

Remark 12 It should be emphasized again that the assumption of reachability is fundamental in our considerations. To establish a similar theorem like Theorem 8 for the unreachable system (8.1) is still an open problem. To calculate PD state-feedback gains K and F, we can use the same procedure as in the case of P state feedback.

Bibliography

[1] Armentano, V.A. Eigenvalue Placement for Generalized Linear Systems. Systems & Control Letters, 4, 1984, 199-202.

[2] Brunovský P. A Classification of Linear Controllable Systems. Kybernetika, 3, 1970, 173-187.

[3] Cobb, D. Feedback and Pole Placement in Descriptor Variable Systems. Int. J. Control, Vol. 6, 1981, 1135-1146.

[4] Cobb, D. Controllability, Observability, and Duality in Singular Systems. IEEE Trans. Aut. Control, AC-29, 1984, 1076-1082.

[5] Dickinson, B.W. On the Fundamental Theorem of Linear State Variable Feedback. IEEE Trans. Aut. Control, AC-19, 1974, 577-579.

[6] Flamm, D.S. A New Proof of Rosenbrock's Theorem on Pole Assignment. IEEE Trans. Aut. Control, AC-25, 1980, 1128-1133.

[7] Jones, E.R.L., A.C.Pugh and G.E.Hayton. Necessary Conditions for the General Pole Assignment via Constant Output Feedback. Int. J. Control, Vol. 51, 1990, 771-784.

[8] Kailath, T. Linear Systems. Prentice-Hall, Inc., Englewood Cliffs, New York, 1980.

[9] Kalman, R.E. Kronecker Invariants and Feedback. Ord. Diff. Equations, ed. by Weis, Proc. Conf. Ord. Diff. Equations, Washington, 1971.

[10] Kučera, V. Discrete Linear Control: The Polynomial Equation Approach. Wiley, Chichester 1979.

[11] Kučera, V. Assigning the invariant factors by feedback. Kybernetika 17, 1981, 118-127.

[12] Kučera, V. and P. Zagalak. Fundamental Theorem of State Feedback for Singular Systems. Automatica 24, 1988, 5, 653-658.

[13] Kučera, V. and P. Zagalak. Constant Solutions of Polynomial Equation. Inf. J. Control, Vol. 53, 1991, No. 2, 495-502.

[14] Kučera, V. Analysis and Design of Discrete Linear Control Systems. Academia, Prague 1992.

[15] Lewis, F.L. and K. Özçaldiran. Reachability and Controllability for Descriptor Systems. Proc. MSCAS(1984), 690-695.

[16] Lewis, F.L. A Survey of Linear Singular Systems. J. Circuits, Systems, Signal. Proc., Special Issue on Singular Systems, Vol. 5, 1986, No 1, 3-36.

[17] Loiseau, J.J. Pole Placement and Connected Problems. Preprints IFAC Workshop on System Structure and Control, Prague 1989, 193-196.

[18] Loiseau, J.J., K. Özçaldiran, M. Malabre and N. Karcanias. A. Feedback classification of Singular Systems. Kybernetika 27, 1991, 289-305.

[19] Malabre, M., Kučera, V. and P. Zagalak. Reachability and Controllability Indices for Linear Descriptor Systems. Systems & Control Letters 15, 1990, 119-123.

[20] Malabre, M. Geometric Characterization of "Complete Controllability Indices" for Singular Systems. Systems & Control Letters 9, 323-327.

[21] Özçaldiran, K. Control of Descriptor Systems. Ph.D. Thesis, Georgia Institute of Technology, Atlanta 1987.

[22] Özçaldiran, K. Fundamental Theorem of Linear State Feedback for Singular Systems. Proc. 29th IEEE Conf. Dec. Control, Hawaii, Honolulu 1990, 67-72.

[23] Özçaldiran, K. and F.L. Lewis. On the Regularizability of Singular Systems. IEEE Trans. Aut. Control, AC-35, 10, 1156-1160.

[24] Rosenbrock, H.H. State-space and Multivariable Theory. Wiley, New York, 1970.

[25] Rosenbrock, H.H. Structural Properties of Linear Dynamical Systems. Int. J. Control 20, 1974, 191-202.

[26] Sá, E. M. Imbedding conditions for λ-matrices. Lin. Alg. Applic. Vol. 24, 1979, 33-50.

[27] Thompson R.C. Interlacing inequalities for invariant factors. Lin. Alg. Applic., Vol. 24, 1979, 1-31.

[28] Thompson, R.C. Invariant Factors of Algebraic Combinations of Matrices. Frequency Domain and State Space Methods for Linear Systems, edited by C.I. byrnes and A. Lindquist, Elsevier Sci. Pub. B.V. (North-Holland), 1986, 73-87.

[29] Vardulakis, A.I.G. Linear Multivariable Control. Wiley, Chichester, 1991.

[30] Wonham, W.M. On Pole Assignment in Multi-input Controllable Linear Systems. IEEE Trans. Aut. Control, AC-12, 1967, 660-665.

[31] Zaballa, I. Matrices with Prescribed Rows and Invariant Factors. Lin. Alg. Applic., Vol. 87, 1987, 113-146.

[32] Zaballa, I. Interlacing Inequalities and Control Theory. Lin. Alg. Applic., Vol. 101, 1988, 9-31.

[33] Zagalak, P. and V. Kučera. Fundamental Theorem of Proportional State Feedback for Descriptor Systems. Kybernetika 28, 1992, 81-89.

[34] Zagalak, P. and V. Kučera. Fundamental Theorem of State Feedback: The Case of Infinite Poles. Kybernetika, Vol. 27, 1991, 1-11.

[35] Zagalak, P. and J. J. Loiseau. Invariant Factors Assignment in Linear Systems. To appear in proc. SINS'92, Texas 1992.

[36] Zagalak, P. and J.J. Loiseau. On the problem of pole assignment by feedback in linear systems. Part I: Constant feedback. Submitted.

Appendix

Some results of matrix theory that are needed for the proofs of the above results and could be at the same time of independent interest are collected here. We begin with some basic concepts.

Let $P(s)$ be a $p \times q$ polynomial matrix having elements in $K[s]$, the ring of polynomials in the indeterminate s over K where K denotes arbitrary field. The matrix $P(s)$ can be written either as

$$P(s) = diag[s^{d_{r1}}, \ldots, s^{d_{rp}}]P_{lr} + \text{terms of lower degrees in } s$$

or as

$$P(s) = P_{lc} diag[s^{d_{c1}}, \ldots, s^{d_{cq}}] + \text{terms of lower degrees in } s$$

where d_{ri} is the degree of the row i of $P(s)$ while d_{ci} is the degree of the column i of $P(s)$. The matrix $P_{lr}(P_{lc})$ is called *the leading row-degree (column-degree) coefficient matrix of* $P(s)$, respectively. We say that $P(s)$ is *row (column) reduced* if $P_{lr}(P_{lc})$ is of rank $p(q)$, respectively. The matrix $P(s)$ is said to be *irreducible* if its Smith form is I_p when $p = q$, $[I_p, 0]$ if $p < q$, and $\begin{bmatrix} I_q \\ 0 \end{bmatrix}$ for $p > q$.

An $m \times m$ polynomial and irreducible matrix $U(s)$ is said to be unimodular if its Smith form is I_m. Such matrices are characterized by the property $detU(s) \in K - \{0\}$.

Let now $K(s)$ denotes the field of rational functions over K, i.e., the field of all fractions $f(s) := \frac{n(s)}{d(s)}, d(s) \neq 0; n(s), d(s) \in K[s]$. The rational function $f(s)$ is said to be *proper, strictly proper, and biproper* if $deg\ n(s) \leq deg\ d(s), deg\ n(s) < deg\ d(s)$, and $deg\ n(s) = deg\ d(s)$, respectively.

Let $H(s)$ be an $m \times n$ matrix having elements in $K(s)$ and let rank $H(s) = r$. Then there exist unimodular matrices $U(s)$ and $V(s)$ such that

$$U(s)H(s)V(s) = \begin{bmatrix} diag\left[\frac{\epsilon_1(s)}{\psi_1(s)}, \ldots, \frac{\epsilon_r(s)}{\psi_r(s)}\right] & 0 \\ 0 & 0 \end{bmatrix}$$

where $\epsilon_1(s) \leq \epsilon_2(s) \leq \ldots \leq \epsilon_r(s)$ and $\psi_1(s) \geq \psi_2(s) \geq \ldots \geq \psi_r(s)$. Here, $n(s) \leq d(s)$ denotes that the polynomial $d(s)$ is divisible by the polynomial $n(s)$. If the polynomials $\epsilon_i(s)'s$ and $\psi_i(s)'s$ are monic, then they are unique for a given matrix $H(s)$ and the above form of $H(s)$ is commonly referred to as *the Smith-McMillan form of* $H(s)$.

An $m \times m$ rational matrix $B(s)$ having all its entries proper and such that $detB(s) \in K - \{0\}$ is said to be *biproper*.

An $m \times n$ rational matrix $H(s)$ is said to be *irreducible at infinity* if its Smith-McMillan form at infinity is I_m when $m = n$, $[I_m, 0]$ if $m < n$, and $\begin{bmatrix} I_n \\ 0 \end{bmatrix}$ for $m > n$.

Any $m \times n$ rational matrix $H(s)$ can be written in the form

$$H(s) = F^{-1}(s)G(s) = N(s)D^{-1}(s)$$

where $F(s), G(s), N(s)$ and $D(s)$ are polynomial matrices of compatible sizes. The matrices $F(s)$ and $G(s)$ ($N(s)$and$D(s)$) are uniquely determined up to their greatest common left (right) divisors and are said to form a *left (right) matrix fraction description (MFD) of* $H(s)$, respectively.

More details about various properties of polynomial and rational matrices can be found in [8], [10], [14], [24], [29] and references therein.

Lemma 5 *Let* $G(s), F(s), N(s), D(s)$ *be polynomial matrices of respective sizes* $n \times m, n \times n, m \times m, n \geq m$ *such that*

(i) $\begin{bmatrix} F(s) & -G(s) \end{bmatrix} \begin{bmatrix} N(s) \\ D(s) \end{bmatrix} = 0,$

(ii) $[F(s) \quad G(s)]$ *and* $\begin{bmatrix} N(s) \\ D(s) \end{bmatrix}$ *are irreducible,*

(iii) rank $G(s) = m$ *and rank*$F(s) = t \leq n$.

Then there exist unimodular matrices $U_i(s)$ *and* $V_i(s), i = 1, 2$ *such that*

$$S_D := V_1(s)D(s)V_2(s) = diag[\phi_1(s), \ldots, \phi_{t+m-n}(s), 0, \ldots, 0]$$

$$S_F := U_1(s)F(s)U_2(s) = \begin{bmatrix} S_D & 0 \\ 0 & I_{n-m} \end{bmatrix}$$

$$S_N := U_2^{-1}(s)N(s)V_2(s) = \begin{bmatrix} diag[\eta_1(s), \ldots, \eta_{t+m-n}(s), 1, \ldots, 1] \\ 0 \end{bmatrix} =$$

$$= U_1(s)G(s)V_1^{-1}(s) =: S_G$$

where $\phi_1(s) \geq \phi_2(s) \geq \ldots \geq \phi_{t+m-n}(s)$ *and* $\eta_1(s) \leq \eta_2(s) \leq \ldots \leq \eta_{t+m-n}(s)$.

Proof.

Let rank $D(s) = r$. As rank $F(s) = t$, there are unimodular matrices $\bar{U}_1(s), \bar{U}_2(s)$ and $V_1(s), V_2(s)$ such that

$$\bar{U}_1(s)F(s)U_2(s) = \begin{bmatrix} F_{11}(s) & 0 \\ 0 & 0 \end{bmatrix}$$

$$V_1(s)D(s)V_2(s) = \begin{bmatrix} D_{11}(s) & 0 \\ 0 & 0 \end{bmatrix}$$

where $F_{11}(s)$ is $t \times t$ and $D_{11}(s)$ is $r \times r$. Then, by (i),

$$\begin{bmatrix} F_{11} & 0 & -G_{11}(s) & G_{12}(s) \\ 0 & 0 & -G_{21}(s) & -G_{22}(s) \end{bmatrix} \begin{bmatrix} N_{11}(s) & N_{12}(s) \\ N_{21}(s) & N_{22}(s) \\ D_{11}(s) & 0 \\ 0 & 0 \end{bmatrix} = 0 \tag{8.54}$$

where

$$\begin{bmatrix} G_{11}(s) & G_{12}(s) \\ G_{21}(s) & G_{22}(s) \end{bmatrix} = \bar{U}_1(s)G(s)\bar{V}_1^{-1}(s)$$

$$\begin{bmatrix} N_{11}(s) & N_{12}(s) \\ N_{21}(s) & N_{22}(s) \end{bmatrix} = \bar{U}_2^{-1}(s)N(s)\bar{V}_2(s)$$

Now (1) implies that $G_{21}(s) = 0$ and $N_{12}(s) = 0$. As both matrices in (8.54) are irreducible, $G_{22}(s)$ and $N_{22}(s)$ must also be irreducible and $n - t = m - r$. Hence

$$\tilde{U}_1(s) = \begin{bmatrix} I_t & -G_{12}(s)G_{22}^{-1}(s) \\ 0 & G_{22}^{-1}(s) \end{bmatrix}$$

$$\tilde{V}_2(s) = \begin{bmatrix} I_r & 0 \\ -N_{22}^{-1}(s)N_{21}(s) & N_{22}^{-1}(s) \end{bmatrix}$$

are unimodular matrices. Premultiplying (8.54) by $\tilde{U}_1(s)$ and postmultiplying by $\tilde{V}_2(s)$, we obtain

$$\begin{bmatrix} F_{11}(s) & 0 & -G_{11}(s) & 0 \\ 0 & 0 & 0 & -I_{n-t} \end{bmatrix} \begin{bmatrix} N_{11}(s) & 0 \\ 0 & I_{m-r} \\ D_{11}(s) & 0 \\ 0 & 0 \end{bmatrix} = 0$$

We define

$$H(s) := F_{11}^{-1}(s)G_{11}(s) = N_{11}(s)D_{11}^{-1}(s).$$

Then there exist unimodular matrices $\hat{U}_2(s)$ and $\hat{V}_1(s)$ that bring $H(s)$ to its Smith-McMillan form. We have

$$F_{11}(s)\hat{U}_2^{-1}(s) = diag[\phi_1(s), \ldots, \phi_{t+m-n}(s), \underbrace{1, \ldots, 1}_{n-m}]$$

$$\hat{V}_{11}^{-1}(s)D_{11}(s) = diag\,[\phi_1(s), \ldots, \phi_{t+m-n}(s)]$$

and

$$G_{11}(s)\hat{V}_1(s) = \hat{U}_2 N_{11}(s) = \begin{bmatrix} diag[\eta_1(s), \ldots, \eta_{t+m-n}(s), \underbrace{1, \ldots, 1}_{n-t}] \\ 0 \end{bmatrix}$$

Defining the $n \times n$ permutation matrix

$$P := \begin{bmatrix} I_{t+m-n} & 0 & 0 \\ 0 & 0 & I_{n-t} \\ 0 & I_{n-m} & 0 \end{bmatrix},$$

the claim follows on putting

$$U_1(s) := P\tilde{U}_1(s)\bar{U}_1(s)$$

$$U_2(s) := \bar{U}_2(s) \begin{bmatrix} \hat{U}_2^{-1} & 0 \\ 0 & I_{n-t} \end{bmatrix} P^{-1}$$

$$V_1(s) := \begin{bmatrix} \hat{V}_1^{-1}(s) & 0 \\ 0 & I_{n-t} \end{bmatrix} \bar{V}_1(s)$$

$$V_2(s) := \bar{V}_2(s)\tilde{V}_2(s).$$

□

From Lemma 1 follows:

- The matrices $G(s)$ and $N(s)$ have the same invariant polynomials.
- The matrices $F(s)$ and $D(s)$ have the same nonunit invariant polynomials.
- The matrices $F(s)$ and $D(s)$ have the same rank deficiency.

If the matrices $[F(s), -G(s)]$ and $\begin{bmatrix} N(s) \\ D(s) \end{bmatrix}$ are also assumed to be irreducible at infinity, then $F(s)$ and $D(s)$ have the same structure of zeros at infinity. The same holds for $G(s)$ and $N(s)$. (The proof of that easily follows using the conformal mapping as in the proof of Theorem 7).

Lemma 6 *Let $P(s)$ be an $m \times m$, column reduced, polynomial matrix and let $\psi_1(s) \geq \psi_2(s) \geq \ldots \geq \psi_m(s)$ be its invariant polynomials. Let $k_1 \geq k_2 \geq \ldots \geq k_m$ be nonnegative integers such that*

$$\{deg\psi_i(s)\}_m \succ \{k_i\}_m$$

Then there exist unimodular matrices $U(s)$ and $V(s)$ such that the matrix

$$P'(s) := U(s)P(s)V(s)$$

is column reduced with column indices equal to $k_i^{'}s$.

P r o o f.
We first recall the lemma of Rosenbrock [24,Ch.5, Lemma 1].

Lemma 7 *Let $C(s)$ be a column reduced, polynomial $m \times m$ matrix and let $deg_{cj}c(s) < deg_{ck}c(s)$ for some j and k. Then $C(s)$ can be transformed by unimodular transformations to a column reduced matrix $C'(s)$ with column degrees*

$$deg_{ci}C^{'}(s) = deg_{ci}C(s), i \neq j, k$$

$$deg_{cj}C^{'}(s) = deg_{cj}C(s) + 1$$

$$deg_{ck}C^{'}(s) = deg_{ck}C(s) - 1.$$

Using this lemma, it is very easy to prove Lemma 2. We first bring the matrix $P(s)$ to the Smith form and then, using repeatedly Lemma 3, we obtain the matrix $P'(s)$.

□

Remark 1 The assertion of Lemma 2 holds also for any $p \times m, p \geq m$, column reduced matrix $P(s)$.

Let $Q(s)$ be an $n \times m$ polynomial matrix such that

$$Q(s) = T blockdiag \left\{ \begin{bmatrix} 1 \\ s \\ \vdots \\ s^{k_i} \end{bmatrix} \right\}_{i=1}^{m}$$

where T is an $n \times n$ nonsingular matrix over K and $k_1 + k_2 + \ldots + k_m + m = n$. Then the matrix $Q(s)$ is said [35] to be a polynomial basis of n-dimensional linear space over K. For instance, let $N(s), D(s)$ be a P-normal external description of (1). Then $N(s)$ has the above property.

Let further $P(s)$ be a $p \times m$ polynomial matrix over $K[s]$ such that $deg_{ci} P(s) \leq k_i + 1$ for $i = 1, 2, \ldots, m$ and let P_{hc} denote the $p \times m$ matrix over K whose entries in the column i are the coefficients of s^{k_i+1} in the column i of $P(s)$. Otherwise there are zeros.

Lemma 8 *Let $M(s)$ be a $p \times m$ polynomial matrix over $K[s]$ such that $deg_{ci} M(s) \leq k_i + 1, i = 1, 2, \ldots, m$. With the notation above, there exist matrices X, Y and Z over K such that X and Z are invertible and*

$$XP(s) + YQ(s) = M(s)Z$$

if and only if

$$rank P_{hc} = rank M_{hc}.$$

Proof.
Let $F(s) := M(s)Z - XP(s)$. Clearly, $deg_{ci} F(s) \leq k_i, i = 1, 2, \ldots, m$, which implies the existence of Y such that $YQ(s) = F(s)$. Hence the claim follows and vice-versa.

□

Lemma 9 *Let $P(s)$ be an $n \times n$ polynomial matrix having the invariant polynomials $\alpha_1(s) \leq \alpha_2(s) \leq \ldots \leq \alpha_n(s)$. Let $\psi_1(s) \leq \psi_2(s) \leq \ldots \leq \psi_{n+1}(s)$ be monic polynomials. Then there exist polynomials $\delta_1(s), \delta_2(s), \ldots, \delta_{n+1}(s)$ such that the matrix*

$$\overline{P} := \left[\begin{array}{c|c} P(s) & \begin{matrix} \delta_1(s) \\ \delta_n(s) \end{matrix} \\ \hline 0 & \delta_{n+1}(s) \end{array} \right]$$

has $\psi_i(s)$'s as its invariant polynomials if and only if

$$\psi_i(s) \leq \alpha_i(s) \leq \psi_{i+1}(s), \qquad i = 1, 2, \ldots, n. \tag{8.55}$$

P r o o f.
Lemma 5 is a special case of the S-Thompson theorem [26], [28] so that we just hint the proof.

Without any loss of generality, we can suppose that $P(s)$ is of the form

$$P(s) = diag\{\alpha_1(s), \ldots, \alpha_n(s)\}.$$

Then,

$$\psi_1(s) = gcd(\alpha_1(s), \delta_1(s), \ldots, \delta_{n+1}(s))$$

and hence $\psi_1(s) \leq \alpha_1(s)$. Next

$$\begin{aligned}\psi_1(s)\psi_2(s) &= gcd(\alpha_1(s)\alpha_2(s), \alpha_1(s)\delta_2(s), \ldots, \alpha_1(s)\delta_{n+1}, \alpha_2(s)\delta_1(s)) \\ &\leq \alpha_2(s) gcd(\alpha_1(s), \delta_1(s), \ldots, \delta_{n+1}(s)),\end{aligned}$$

which implies $\psi_2(s) \leq \alpha_2(s)$.

Continuing in this way, we can prove the first part of the inequalities (8.55), i.e., $\psi_i(s) \leq \alpha_i(s), i = 1, 2, \ldots, n$. The second part is evident.

To prove sufficiency, we shall construct the $(n+1) \times (n+1)$ matrix

$$\begin{bmatrix} \alpha_1(s) & & & \delta_1(s) \\ & \ddots & & \vdots \\ & & \alpha_n(s) & \delta_n(s) \\ & 0 & & \delta_{n+1}(s) \end{bmatrix}$$

where $\delta_1(s) = \psi_1(s), \delta_2(s) = \frac{\psi_1(s)\ \psi_2(s)}{\alpha_1(s)}, \ldots, \delta_{n+1}(s) = \frac{\psi_1(s) \ldots \psi_{n+1}(s)}{\alpha_1(s) \ldots \alpha_n(s)}$. Clearly, this matrix possesses the invariant polynomials $\psi_i(s)$'s.

□

The following lemma is again a special case of the S-Thompson theorem.

Lemma 10 *Let $P(s)$ be an $m \times n$ polynomial matrix with the invariant polynomials $\phi_1(s) \geq \phi_2(s) \geq \ldots \geq \phi_t(s)$. Let $\psi_1(s) \geq \psi_2(s) \geq \ldots \geq \psi_r(s)$ be monic polynomials. Then there exists an $m \times 1$ polynomial vector $p(s)$ such that $\psi_i(s), i = 1, 2, \ldots, r$ are the invariant polynomials of the matrix $[P(s)\ p(s)]$ if and only if*

a) in the case $r = t + 1$:

$$\psi_{i+1}(s) \leq \psi_i(s) \leq \psi_i(s), \ i = 1, 2, \ldots, t$$

b) in the case $r = t$:

$$\psi_i(s) \leq \psi_i(s) \leq \psi_{i-1}(s), \ i = 1, 2, \ldots, t$$

where $\psi_0(s) = 0$.

Lemma 11 *Let $P(s)$ be an $n \times m$ polynomial matrix over $K[s]$, rank $P(s) = m$ having the invariant polynomials $\psi_1(s) \geq \psi_2(s) \geq \ldots \geq \psi_m(s)$. Then*

$$\{k_i\}_m \geq \{deg\psi_i(s)\}_m$$

for any integers $k_i \geq deg_{ci}P(s), i = 1, 2, \ldots, m$ where $k_1 \geq k_2 \geq \ldots \geq k_m$. If $P(s)$ is $m \times m$ and column reduced with column degrees $k_1 \geq k_2 \geq \ldots \geq k_m$, then

$$\{deg\psi_i(s)\} \succ \{k_i\}_m.$$

Using the same notation as in the above lemmas, we obtain the following two corollaries of Lemma 6.

Corollary 3 *Let rank $P(s) = n$ and rank $[P(s) \;\; p(s)] = n + 1$. Denoting $k_i = deg_{ci}P(s), i = 1, 2, \ldots, n$ and $k_{n+1} := degp(s)$, it implies that*

$$\{deg\psi_i(s)\}_{n+1} \leq \left\{k_i'\right\}_{n+1}$$

where $k_1' \geq k_2' \geq \ldots \geq k_{n+1}'$ are the reordered integers $k_1, k_2, \ldots, k_{n+1}$.

Corollary 4 *Let rank $[P(s) \;\; p(s)] = rankP(s) = n$. Then*

$$\{deg\psi_i(s)\}_n \leq \{k_i\}_n$$

where $k_1 \geq k_2 \geq \ldots \geq k_n$ are the reordered column degrees of $P(s)$.

Chapter 9

Polynomial Equations, Conjugacy and Symmetry

J. Ježek

9.1 Introduction

For linear dynamical systems, there are basically two approaches. First, the state space approach, with equations

$$\begin{aligned} Sx &= Ax + Bu \\ y &= Cx + Du \end{aligned} \tag{9.1}$$

where u, y, x are input, output and state signals, A, B, C, D are matrices representing linear mappings and S is the derivative operator

$$Sx = \frac{\mathrm{d}x}{\mathrm{d}t}. \tag{9.2}$$

Problems of control synthesis finally lead to computation of some matrices, say P, Q, R from given matrices A, B, C, D. Almost any control theory book utilizes this approach, originated from [5].

Second, the input-output approach, with equation

$$a(S)y(t) = b(S)u(t) \tag{9.3}$$

where $a(S), b(S)$ are polynomials in S, i.e. expressions of the form

$$a(S) = a_0 + a_1 S + \cdots + a_n S^n. \tag{9.4}$$

Problems of control synthesis finally lead to computation of some polynomials, say $p(S), q(S)$ from given polynomials $a(S), b(S)$. Books like [4], [1] are devoted to this approach.

For the state space approach, the corresponding mathematical tool is algebra of vectors and matrices, matrix equations, more abstractly the theory of linear spaces, linear mappings, scalar products. It is a well established branch of mathematics.

The properties of equations are well known, the algorithms and numerical methods are commonly used.

For the input-output approach, the corresponding mathematical tool is polynomial algebra, polynomial equations, more abstractly the theory of rings and fields. It is also a well established branch of mathematics. However, the properties and solution algorithms for polynomial equations are generally less commonly known than those for matrix and vector equations. This is specially true for equations occurring in LQ optimization control problems which contain a "conjugacy" operation for polynomials, an analogy of "transpose" operation for matrices.

The aim of this chapter is to present the fundamental notions on polynomial algebra and polynomial equations which are used in control theory. Specially mentioned are operations of conjugacy. The structure of the chapter is as follows. In section 9.2, the process of abstraction from system and control theory to algebra is explained. In sections 9.3, 9.4, the fundamental notions of the polynomial algebra are summarized. Based on these, the theory of polynomial equations is presented in sections 9.5, 9.6. For the operation of conjugacy, the fundamental notions are summarized in section 9.7 and the theory of equations is presented in sections 9.8 and 9.9.

The ideas presented here were partly published in author's papers [2], [3]. Some ideas are new here.

9.2 Signals and Operators

We will treat here linear, time-invariant, single-input-single-output continuous-time systems. The basic objects are *signals*, $u(t)$, real functions of time t, which is real number $-\infty < t < \infty$. With (two-sided) *Laplace transform*, a signal $u(t)$ is described by a *spectrum* $U(s)$, function of complex variable s

$$U(s) = \int_{-\infty}^{\infty} u(t)\mathrm{e}^{-st}\,\mathrm{d}t. \tag{9.5}$$

The function $U(s)$ is holomorphic in $\sigma_1 < \mathrm{Re}\ s < \sigma_2$ for some σ_1, σ_2. In the inverse transform

$$u(t) = \frac{1}{2\pi \mathrm{j}} \int_{\sigma-\mathrm{j}\infty}^{\sigma+\mathrm{j}\infty} U(s)\mathrm{e}^{st}\,\mathrm{d}s, \tag{9.6}$$

the integration path lies in this *domain*. For $U(s)$, the domain must always be given: the same $U(s)$ may correspond to different $u(t)$, depending on the domain. For signals u, v exponentially decreasing both for $t \to \infty$ and for $t \to -\infty$, the domain is $-\varepsilon < \mathrm{Re}\ s < \varepsilon$ and the *scalar product* $< u, v >$ is defined:

$$< u, v > = \int_{-\infty}^{\infty} u(t)v(t)\mathrm{d}t = \frac{1}{2\pi \mathrm{j}} \int_{-\mathrm{j}\infty}^{\mathrm{j}\infty} U(-s)V(s)\,\mathrm{d}s. \tag{9.7}$$

The signals are acted on by *linear operators*. The fundamental operator is that of *derivative* S, (9.2). Its action on signals corresponds to multiplication by s, acting on spectra. The polynomial operators $a(S)$, resp. fractional operators $b(S)/a(S)$ correspond to polynomial functions $a(s)$ resp. fractional functions $b(s)/a(s)$, the *transfer functions*.

The operators and transfer functions can be *added* and *multiplied*, the result is that of parallelly or serially connected systems. Besides addition and multiplication, there is one operation more: *conjugacy*. For every operator c, a *conjugated operator* $c^\star$ is defined by

$$< u, cv >=< c^\star u, v > \tag{9.8}$$

for all u, v. For the derivative operator, $S^\star = -S$. For every transfer function, a conjugated transfer function is

$$s^\star = -s, \quad c^\star(s) = c(-s), \tag{9.9}$$

the domains are

$$\begin{aligned} c(s) \quad &\ldots \quad \sigma_1 < \text{Re } s < \sigma_2 \\ c^\star(s) \quad &\ldots \quad -\sigma_2 < \text{Re } s < -\sigma_1. \end{aligned}$$

An abstraction is possible here. We can treat polynomials a and fractions b/a as algebraical entities: elements for which the operations of addition, multiplication and conjugacy are defined. In this approach, we can forget about signals, operators, spectra and transfer functions. Such an abstraction allows us to concentrate on algebraical properties, all other properties being abstracted out. In this approach, polynomial equations are investigated, existence and uniqueness of solution discussed, algorithms for solution constructed.

9.3 Ring of Polynomials

In the *algebraical approach*, a *polynomial* is defined as a (two-sided) sequence of real numbers a_i, *coefficients*, indexed by $i = \ldots -2, -1, 0, 1, 2, \ldots$, having nonzero entries only for nonnegative i, for finite number of i. The polynomial with all $a_i = 0$ is called *zero polynomial.*

Addition of polynomials is defined by

$$c = a + b, \qquad c_i = a_i + b_i, \tag{9.10}$$

multiplication by

$$d = ab, \qquad d_i = \sum_{k=-\infty}^{\infty} a_{i-k} b_k \tag{9.11}$$

the sum having only a finite number of nonzero terms. Using these operations, any nonzero polynomial a can be written

$$a = a_0 + a_1 q + \cdots + a_n q^n \tag{9.12}$$

where q is a polynomial defined $q_1 = 1$, $q_i = 0$ for $i \neq 1$, called *indeterminate.* The n is called *degree* of polynomial, $\deg a$, the a_n is called the *highest coefficient.* For zero polynomial, we define $\deg 0 = -\infty$ and no highest coefficient. The operations have properties

$$\begin{aligned} \deg(a + b) &\leq \max(\deg a, \deg b) \\ \deg ab &= \deg a + \deg b \end{aligned} \tag{9.13}$$

With these operations, the polynomials form a commutative *ring* with identity element. The ring is denoted $R[q]$. It contains a subring of real numbers, polynomials of degree 0. Ring $R[q]$ is *integrity domain*: if $ab = 0$ then $a = 0$ or $b = 0$ (no zero divisors).

If $c = ab$, we say a, b are *divisors* of c, or c is *divisible* by a, b. If a, b are divisible by d then $a + b$ also is. Two polynomials a, a' are called *associate*, $a \sim a'$ if a is divisor of a' and a' is divisor of a. The relation of association is *equivalence*: to any polynomial, a class of its associates exists. Many properties of polynomials are concerned with classes of associates rather than with individual polynomials.

Units of the ring (divisors of 1, invertible elements) are such u that ex. v, $uv = 1$. They are polynomials of degree 0, i.e. nonzero real numbers. The units form a multiplicative group G_u. For every pair of associates a, a' ex. units u, v such that $a = a'u$, $a' = av$.

For any a, b, the *greatest common divisor* $g = \gcd(a, b)$ is defined as a common divisor with the property that any other common divisor h divides g. It is unique up to associates. If $\gcd(a, b) = 1$ then a, b are called *coprime*.

A polynomial p divisible only by 1 and by p (or by its associates) is called *prime*. All polynomials of degree 1 are primes, and some polynomials of degree 2. These are the only primes in $R[q]$. If ab is divisible by prime p, either a or b is. The ring $R[q]$ is *unique factorization domain*: every nonzero a is expressible $a = up_1 \dots p_n$ with u unit, $p_1, \dots, p_n$ primes, uniquely up to associates.

With $\deg a$, the ring $R[q]$ is *Euclidean domain*: given pair a, b with $a \neq 0$, there are unique p, r such that $b = ap + r$, $\deg r < \deg a$. Here p is called *quotient*, r *remainder*. The ring $R[q]$ is also *Bézoutean domain*: for any a, b and their $g = \gcd(a, b)$, exist u, v such that $g = au + bv$.

For the ring $R[q]$, its *quotient field* of *polynomial fractions* can be created, denoted by $R(q)$. The degree of $c = b/a$ with coprime b, a is defined

$$\deg c = \max(\deg a, \deg b). \tag{9.14}$$

A fraction is called *proper* if $\deg b \leq \deg a$, *strictly proper* if $\deg b < \deg a$.

To every polynomial a, a *polynomial function* $a(q)$ exists: we may treat (9.12) as a rule for computing a function value a, given an argument value q. The argument values and function values may be complex even though the coefficients are real. Then the function $a(q)$ satisfies

$$a(\overline{q}) = \overline{a(q)}. \tag{9.15}$$

The values q_i with $a(q_i) = 0$ are called *roots* of a polynomial. Every polynomial a with $\deg a = n$ can be uniquely expressed

$$a(q) = K(q - q_1) \dots (q - q_n) \tag{9.16}$$

with roots q_i. In this sense, it has n roots. For any q_i, the number of factors $(q - q_i)$ is called *multiplicity* of q_i. With every root q_i, also $\overline{q_i}$ is present, with the same multiplicity. Possible combinations of roots are:

pair	$q_i, \overline{q_i}$	
real single	$q_i,$	$\overline{q_i} = q_i$

To every fraction $c = b/a$, *fractional function* $c(q)$ belongs. For this purpose, argument values and function values are taken from the *closed complex plane* $C^\infty = C \cup \{\infty\}$. The values q_i with $c(q_i) = 0$ resp. $c(q_i) = \infty$ are called *zeros* resp. *poles*. Every $c = b/a$, coprime b, a, $\deg c = n$, has n zeros, n poles (accounting for multiplicities). For finite q_i, zeros are roots of b, poles are roots of a. For $q_i = \infty$, the multiplicity is $\deg b - \deg a$, if positive then q_i is a pole, if negative, a zero. In this sense, the polynomials have only finite zeros (roots) and an infinite pole (multiplicity=degree).

9.4 Elementary Operations

The first from elementary operations, serving as building blocks for all algorithms, is that of *reduction*. For given polynomials a, b with $a \neq 0$, $n = \deg a$, $m = \deg b$, $n \leq m$, define b', reduction of b by a:

$$b' = b - q^{m-n} \frac{b_m}{a_n} a, \tag{9.17}$$

It is $\deg b' < \deg b$. With $a' = a$, it is $\gcd(a', b') = \gcd(a, b)$.

The reductions can be used for computing the *remainder*.

Algorithm 9.1 *Given polynomials a, b with $a \neq 0$, $n = \deg a$, let r be the remainder in $b = ap + r$. For $k = 0, 1, 2, \ldots$, define a sequence b^k with $b^0 = b$, by the recurrent rule*

$$\begin{aligned} & m = \deg b^k, \\ & \text{if } m \geq n \text{ then } b^{k+1} = b^k - q^{m-n} \frac{b^k_m}{a_n} a. \end{aligned}$$

Then for some l, it is $\deg b^l < n$, $b^l = r$.

The second elementary operation is *exchange*: replacing the pair a, b by the pair a', b' where $a' = b$, $b' = a$. The third elementary operation is *multiplication by units*: $a' = Ka$, $b' = Lb$ with nonzero real numbers K, L. In both cases, it is $\gcd(a', b') = \gcd(a, b)$.

The reductions are a special case of $a' = a$, $b' = b + pa$, with polynomial $p(q)$. All elementary operations can be written in matrix form

$$\begin{bmatrix} a' & b' \end{bmatrix} = \begin{bmatrix} a & b \end{bmatrix} M \tag{9.18}$$

where the polynomial matrix

$$M = \begin{bmatrix} M_{11} & M_{12} \\ M_{21} & M_{22} \end{bmatrix} \tag{9.19}$$

has one of three forms

$$\begin{bmatrix} 1 & 0 \\ p & 1 \end{bmatrix}, \begin{bmatrix} 0 & 1 \\ 1 & 0 \end{bmatrix}, \begin{bmatrix} K & 0 \\ 0 & L \end{bmatrix}. \tag{9.20}$$

In all three cases, det M is a nonzero real number, matrix M^{-1} being also polynomial matrix. Such polynomial matrices are called *unimodular.* They form a multiplicative group U. The three types of matrices in (9.20) are generators of the group: any unimodular matrix is a product of such matrices. The $\gcd(a,b)$ remains invariant when the pair a, b is multiplied by a unimodular matrix.

The elementary operations can be used for *computing the gcd.* The idea is to replace the pair a, b by a', b' with the same gcd, by means of elementary operations. By repeating the process, we end up with a pair whose gcd is evident. This is the *Euclidean algorithm.*

Algorithm 9.2 *Given polynomials a, b with $a \neq 0$, let $g = \gcd(a,b)$. For $k = 0, 1, 2, \ldots$ define sequences a^k, b^k with $a^0 = a$, $b^0 = b$ by the recurrent rule*

$$\begin{aligned}
& n = \deg a^k, \;\; m = \deg b^k, \\
& \text{if } m \geq n \text{ then} \\
& \quad a^{k+1} = a^k, \\
& \quad b^{k+1} = b^k - q^{m-n}\frac{b^k_m}{a^k_n}a^k, \\
& \text{else if } b^k \neq 0 \text{ then} \\
& \quad a^{k+1} = b^k, \\
& \quad b^{k+1} = a^k.
\end{aligned}$$

It is $\gcd(a^k, b^k) = g$. *Then for some l, it is $b^l = 0$, $a^l = g$.*

9.5 Homogeneous Equations

Polynomial equations are equations in $R[q]$, with some given polynomials, say a, b, and some unknown polynomials, say x, y. As $R[q]$ is a ring only, not a field, the usual rule "the number of equations must be equal to the number of unknowns" does not apply here. There may be less equations than unknowns, the *general solution* and some *special solutions* satisfying further requirements.

Definition 9.1 *For a, b not both zero, a* linear homogeneous equation *is*

$$ax + by = 0 \tag{9.21}$$

A special case is a coprime equation *with coprime $\hat{a}, \hat{b}$*

$$\hat{a}x + \hat{b}y = 0 \tag{9.22}$$

Lemma 9.1 *Denoting*

$$g = \gcd(a,b), \;\; a = \hat{a}g, \;\; b = \hat{b}g, \tag{9.23}$$

equation (9.21) can be cancelled *to (9.22).*

Lemma 9.2 (coprime equation). *The* general solution *of (9.22) is*

$$x = \hat{b}t, \quad y = -\hat{a}t \tag{9.24}$$

where t is an arbitrary polynomial.

Proof Let p be a prime factor of $\hat{a}$. The first term in (9.22) is divisible by p, so the second term must also be. As $\hat{a}, \hat{b}$ are coprime, $\hat{b}$ is not divisible by p, so y must be. This holds for all prime factors of $\hat{a}$, so y is divisible by $\hat{a}$. Similarly, x is divisible by $\hat{b}$. Exist t, u such that $x = \hat{b}t$, $y = \hat{a}u$, from (9.22) follows $u = -t$. □

Lemma 9.3 (general equation). *With (9.23), the* general solution *of equation (9.21) is (9.24).*

Definition 9.2 *A* trivial solution *is $x = 0$, $y = 0$. A* minimal solution *is a non-trivial solution with* $\deg x, \deg y$ *simultaneously minimal.*

Lemma 9.4 (coprime equation). *For equation (9.22), there exists* minimal solution

$$x_{\min} = \hat{b}K, \quad y_{\min} = -\hat{a}K \tag{9.25}$$

unique up to multiplication by an arbitrary real number K. It is $\deg x_{\min} = \deg \hat{b}$, $\deg y_{\min} = \deg \hat{a}$. *The pair* $x_{\min}, y_{\min}$ *is coprime.*

Lemma 9.5 (general equation). *For equation(9.21), there exists* minimal solution *(9.25) with (9.23). It is* $\deg x_{\min} = \deg b - \deg g$, $\deg y_{\min} = \deg a - \deg g$. *The pair* $x_{\min}, y_{\min}$ *is coprime.*

The elementary operations can be used for *computing the minimal solution* of (9.21). First, two preparatory lemmas.

Lemma 9.6 *Given polynomials a, b with $a \neq 0$, $n = \deg a$, $m = \deg b$, $m \geq n$, let x, y be minimal solution of (9.21). Define*

$$\begin{aligned} a' &= a, \\ b' &= b - q^{m-n}\frac{b_m}{a_n}a. \end{aligned} \tag{9.26}$$

Let x', y' be minimal solution of

$$a'x' + b'y' = 0. \tag{9.27}$$

Then

$$\begin{aligned} x &= x' - q^{m-n}\frac{b_m}{a_n}y' \\ y &= y'. \end{aligned} \tag{9.28}$$

Proof By substituting (9.28) into (9.21) we see: if x', y' solve (9.27) with (9.26) then x, y solve (9.21). If x', y' is minimal solution, the pair x', y' is coprime, then the pair x, y is also coprime and minimal. □

Lemma 9.7 *Given polynomials a, b, not both zero, let x, y be minimal solution of (9.21). Define $a' = b$, $b' = a$. Let x', y' be minimal solution of (9.27). Then $x' = y$, $y' = x$.*

Now, a modification of Euclidean algorithm for *computing the minimal solution* x, y of (9.21) simultaneously with g. The idea is to replace (9.21) by equation $a'x' + b'y' = 0$ whose minimal solution x', y' corresponds to x, y. By repeating the process, we end up with an equation whose minimal solution is evident. Then a backward run follows, recovering x, y from x', y'.

Algorithm 9.3 *Given polynomials a, b with $a \neq 0$, let $g = \gcd(a, b)$, let x, y be minimal solution of (9.21). For $k = 0, 1, 2, \ldots$ define sequences a^k, b^k, f^k, n^k, m^k with $a^0 = a$, $b^0 = b$ by the recurrent rule*

$$
\begin{aligned}
& n^k = \deg a^k,\ m^k = \deg b^k, \\
& \text{if } m^k \geq n^k \text{ then} \\
& \quad f^k = 1, \\
& \quad a^{k+1} = a^k, \\
& \quad b^{k+1} = b^k - q^{m^k - n^k} \frac{b^k_{m^k}}{a^k_{n^k}} a^k, \\
& \text{else if } b^k \neq 0 \text{ then} \\
& \quad f^k = 2, \\
& \quad a^{k+1} = b^k, \\
& \quad b^{k+1} = a^k.
\end{aligned}
$$

It is $\gcd(a^k, b^k) = g$. Then for some l, it is $b^l = 0$, $a^l = g$. For $k = l, l-1, \ldots, 0$, let x^k, y^k be minimal solution of $a^k x^k + b^k y^k = 0$. It is $x^l = 0$, $y^l = K$ with nonzero real K. The recurrent rule holds

$$
\begin{aligned}
& \text{if } f^k = 1 \text{ then} \\
& \quad x^k = x^{k+1} - q^{m^k - n^k} \frac{b^k_{m^k}}{a^k_{n^k}} y^{k+1}, \\
& \quad y^k = y^{k+1}, \\
& \text{if } f^k = 2 \text{ then} \\
& \quad x^k = y^{k+1}, \\
& \quad y^k = x^{k+1}.
\end{aligned}
$$

Finally, $x^0 = x$, $y^0 = y$.

The idea of transforming (9.21) by elementary operations can be expressed generally in terms of unimodular matrices.

Lemma 9.8 *For any unimodular matrix M (9.19) define*

$$\begin{bmatrix} a' & b' \end{bmatrix} = \begin{bmatrix} a & b \end{bmatrix} M. \tag{9.29}$$

The solutions of

$$\begin{bmatrix} a & b \end{bmatrix} \begin{bmatrix} x \\ y \end{bmatrix} = 0 \tag{9.30}$$

and

$$\begin{bmatrix} a' & b' \end{bmatrix} \begin{bmatrix} x' \\ y' \end{bmatrix} = 0 \tag{9.31}$$

are connected by bijective mapping

$$\begin{bmatrix} x \\ y \end{bmatrix} = M \begin{bmatrix} x' \\ y' \end{bmatrix}. \tag{9.32}$$

The solution x, y is minimal just if x', y' is.

Proof By substituting (9.32) into (9.30). Bijectivity follows from unimodularity. Minimality follows from the fact that multiplication by M in (9.32) preserves $\gcd(x, y)$. □

9.6 Nonhomogeneous Equations

Definition 9.3 *A* nonhomogeneous linear equation *is*

$$ax + by = c \tag{9.33}$$

with a, b not both zero. A special case is a coprime equation *with $\hat{a}, \hat{b}$ coprime*

$$\hat{a}x + \hat{b}y = \hat{c}. \tag{9.34}$$

Lemma 9.9 *The necessary and sufficient condition for* solvability *of (9.33) is: $g = \gcd(a, b)$ divides c.*

Proof Necessity: both terms in lhs of (9.33) are divisible by g. If x, y exist, the rhs must be also divisible. Sufficiency: by the Bézout property, ex. u, v such that $g = au + bv$. If $c = \hat{c}g$ then $x = \hat{c}u$, $y = \hat{c}v$ is a solution. □

Lemma 9.10 *Denoting*

$$g = \gcd(a, b), \ a = \hat{a}g, \ b = \hat{b}g, \ c = \hat{c}g, \tag{9.35}$$

equation(9.33) can be cancelled *to (9.34).*

Lemma 9.11 *(*coprime equation*). Equation (9.34) has* general solution

$$x = x_p + \hat{b}t, \ y = y_p - \hat{a}t \tag{9.36}$$

where x_p, y_p is some particular solution, t is an arbitrary polynomial.

Proof Let x_p, y_p be given particular solution, x, y any other solution. Then $x - x_p$, $y - y_p$ solve the homogeneous equation (9.22). □

Lemma 9.12 *(*general equation*). Equation (9.33), if solvable, has general solution (9.36) with(9.35).*

Definition 9.4 *A* y-minimal solution *is solution with* $\deg y$ *minimal. Likewise, an* x-minimal solution *is a solution with* $\deg x$ *minimal. A* minimal solution *is a solution with* $\deg x, \deg y$ *simultaneously minimal.*

Lemma 9.13 (coprime equation). *If* $\hat{a} \neq 0$, *equation (9.34) has unique* minimal solution $x_{\min}, y_{\min}$. *It is* $\deg y_{\min} < \deg \hat{a}$. *It is* $\deg x_{\min} = \deg \hat{c} - \deg \hat{a}$ *when* $\deg \hat{c} \geq \deg \hat{a} + \deg \hat{b}$. *It is* $\deg x_{\min} < \deg \hat{b}$ *when* $\deg \hat{c} < \deg \hat{a} + \deg \hat{b}$. *Similarly, if* $\hat{b} \neq 0$, *equation (9.34) has unique* x-minimal solution $x_{\min}, y_{\min}$, *it is* $\deg x_{\min} < \deg \hat{b}$. *It is* $\deg y_{\min} = \deg \hat{c} - \deg \hat{b}$ *when* $\deg \hat{c} \geq \deg \hat{a} + \deg \hat{b}$. *It is* $\deg y_{\min} < \deg \hat{a}$ *when* $\deg \hat{c} < \deg \hat{a} + \deg \hat{b}$. *If* $\hat{a} \neq 0, \hat{b} \neq 0$, $\deg \hat{c} < \deg \hat{a} + \deg \hat{b}$ *then the* x*-minimal and the* y*-minimal solutions coincide, being unique* minimal solution.

Proof Given x_p, y_p, construct y-minimal solution by taking t, y in (9.36) the quotient and remainder in division y_p by $\hat{a}$. Then $\deg y_{\min} < \deg \hat{a}$, uniquely. The degree of $x_{\min}$: when $\deg \hat{c} \geq \deg \hat{a} + \deg \hat{b}$, then in the lhs of (9.34) the second term $\hat{b} y_{\min}$ has $\deg \hat{b} y_{\min} < \deg \hat{c}$, so the first term $\hat{a} x_{\min}$ must have $\deg \hat{a} x_{\min} = \deg \hat{c}$, $\deg x_{\min} = \deg \hat{c} - \deg \hat{a}$. When $\deg \hat{c} < \deg \hat{a} + \deg \hat{b}$, then the second term has $\deg \hat{b} y_{\min} < \deg \hat{a} + \deg \hat{b}$, so the first term must also have $\deg \hat{a} x_{\min} < \deg \hat{a} + \deg \hat{b}$, $\deg x_{\min} < \deg \hat{a}$. □

Lemma 9.14 (general equation). *If the equation (9.33) with (9.35) is solvable and* $a \neq 0$, *it has unique* y-minimal solution *satisfying* $\deg y_{\min} < \deg a - \deg g$, *it is* $\deg x_{\min} = \deg c - \deg a$ *when* $\deg c \geq \deg a + \deg b - \deg g$, *it is* $\deg x_{\min} < \deg b - \deg g$ *when* $\deg c < \deg a + \deg b - \deg g$. *Similarly, for* $b \neq 0$, *there is unique* x-minimal solution *satisfying* $\deg x_{\min} < \deg b - \deg g$, *it is* $\deg y_{\min} = \deg c - \deg b$ *when* $\deg c \geq \deg a + \deg b - \deg g$, *it is* $\deg y_{\min} < \deg a - \deg g$ *when* $\deg c < \deg a + \deg b - \deg g$. *If* $a \neq 0$, $b \neq 0$, $\deg c < \deg a + \deg b - \deg g$, *then* y*- and* x*-minimal solutions coincide, being unique* minimal solution.

The meaning of minimal solutions can be seen by writing (9.34) in the form

$$\frac{x}{\hat{b}} + \frac{y}{\hat{a}} = \frac{\hat{c}}{\hat{a}\hat{b}}. \tag{9.37}$$

Then the problem of finding y-minimal solution is: given fraction $\hat{c}/\hat{a}\hat{b}$, decompose it into two fractions with given denominators, such that $y/\hat{a}$ be strictly proper. The problem of minimal solutions is: given strictly proper $\hat{c}/\hat{a}\hat{b}$, decompose it into two strictly proper fractions with given denominators.

The elementary operations can be used for *computing the* y*-minimal solution* of (9.33). Preparatory lemmas:

Lemma 9.15 *Given polynomials* a, b, c *with* $a \neq 0$, $n = \deg a$, $p = \deg c$, $p \geq n$, *let* x, y *be minimal solution of (9.33). Define*

$$\begin{aligned} a' &= a, \\ b' &= b, \\ c' &= c - q^{p-n} \frac{c_p}{a_n} a. \end{aligned} \tag{9.38}$$

Let x', y' be y'-minimal solution of

$$a'x' + b'y' = c'. \tag{9.39}$$

Then

$$\begin{aligned} x &= x' + q^{p-n}\frac{c_p}{a_n} \\ y &= y'. \end{aligned} \tag{9.40}$$

Proof By substituting (9.40) into (9.33). □

Lemma 9.16 *Given polynomials a, b, c with $a \neq 0$, $n = \deg a$, $m = \deg b$, $m \geq n$, let x, y be y-minimal solution of (9.33). Define*

$$\begin{aligned} a' &= a \\ b' &= b - q^{m-n}\frac{b_m}{a_n}a \\ c' &= c. \end{aligned} \tag{9.41}$$

Let x', y' be y'-minimal solution of (9.39). Then

$$\begin{aligned} x &= x' - q^{m-n}\frac{b_m}{a_n}y' \\ y &= y' \end{aligned} \tag{9.42}$$

Lemma 9.17 *Given polynomials a, b, c with $a \neq 0, b \neq 0$, $n = \deg a, m = \deg b, p = \deg c$, $p < n + m$. Let x, y be minimal solution of (9.33). Define $a' = b$, $b' = a$, $c' = c$. Let x', y' be minimal solution of (9.39). Then $x = y'$, $y = x'$.*

Now, a modification of Euclidean algorithm for computing the *y-minimal solution* of (9.33) simultaneously with g.

Algorithm 9.4 *Given polynomials a, b, c with $a \neq 0$, let $g = \gcd(a, b)$, let x, y be y-minimal solution of (9.33). For $k = 0, 1, 2, \ldots$ define sequences a^k, b^k, c^k, f^k, m^k, n^k, p^k by the recurrent rule*

$n^k = \deg a^k$, $m^k = \deg b^k$, $p^k = \deg c^k$,
if $p^k \geq n^k$ then
  $f^k = 1$,
  $a^{k+1} = a^k$,
  $b^{k+1} = b^k$,
  $c^{k+1} = c^k - q^{p^k - n^k}\dfrac{c^k_{p^k}}{a^k_{n^k}}a^k$,
else if $m^k \geq n^k$ then
  $f^k = 2$,
  $a^{k+1} = a^k$,
  $b^{k+1} = b^k - q^{m^k - n^k}\dfrac{b^k_{m^k}}{a^k_{n^k}}a^k$,

$$c^{k+1} = c^k,$$

else if $b^k \neq 0$ then

$$f^k = 3,$$
$$a^{k+1} = b^k,$$
$$b^{k+1} = a^k,$$
$$c^{k+1} = c^k.$$

It is $\gcd(a^k, b^k) = g$. *Then for some* l, *it is* $b^l = 0$, $a^l = g$. *If* $c^l \neq 0$, *the equation is unsolvable. Otherwise, for* $k = l, l-1, \ldots, 0$, *let* x^k, y^k *be* y^k*-minimal solution of* $a^k x^k + b^k y^k = c^k$. *It is* $x^l = 0$, $y^l = 0$. *The recurrent rule holds*

if $f^k = 1$ then

$$x^k = x^{k+1} + q^{p^k - n^k} \frac{c^k_{p^k}}{a^k_{n^k}},$$
$$y^k = y^{k+1},$$

if $f^k = 2$ then

$$x^k = x^{k+1} - q^{m^k - n^k} \frac{b^k_{m^k}}{a^k_{n^k}} y^{k+1},$$
$$y^k = y^{k+1},$$

if $f^k = 3$ then

$$x^k = y^{k+1},$$
$$y^k = x^{k+1}.$$

Finally, $x^0 = x$, $y^0 = y$.

The idea of transforming (9.33)by elementary operations can be expressed generally in terms of unimodular matrices:

Lemma 9.18 *Unimodular matrices of the form*

$$M = \begin{bmatrix} M_{11} & M_{12} & M_{13} \\ M_{21} & M_{22} & M_{23} \\ 0 & 0 & 1 \end{bmatrix} \tag{9.43}$$

form a multiplicative group V. *For any such* M, *define*

$$\begin{bmatrix} a' & b' & c' \end{bmatrix} = \begin{bmatrix} a & b & c \end{bmatrix} M. \tag{9.44}$$

Equation

$$\begin{bmatrix} a & b & c \end{bmatrix} \begin{bmatrix} x \\ y \\ -1 \end{bmatrix} = 0 \tag{9.45}$$

is solvable just if equation

$$\begin{bmatrix} a' & b' & c' \end{bmatrix} \begin{bmatrix} x' \\ y' \\ -1 \end{bmatrix} = 0 \tag{9.46}$$

is. The solutions are connected by bijective mapping

$$\begin{bmatrix} x \\ y \\ -1 \end{bmatrix} = M \begin{bmatrix} x' \\ y' \\ -1 \end{bmatrix} \tag{9.47}$$

Proof The submatrix (9.19) of (9.43) is also unimodular. Multiplication by M in (9.44) preserves $\gcd(a, b)$ and $\gcd(a, b, c)$. From this, simultaneous solvability of (9.45) and (9.46) follows. □

The modifications of Euclidean algorithm can be used for *numerical computation.* They are very efficient. If the degrees of polynomials are $\sim n$, then number of *computations* is $\sim n^2$, number of *memories* $\sim n$. Of course, as the equations are linear and the number of unknowns is finite, standard techniques can also be used. E.g.

$$\begin{aligned}&(a_0 + a_1 q + a_2 q^2)(x_0 + x_1 q + x_2 q^2) + (b_0 + b_1 q + b_2 q^2 + b_3 q^3)(y_0 + y_1 q) = \\ &c_0 + c_1 q + c_2 q^2 + c_3 q^3 + c_4 q^4\end{aligned} \tag{9.48}$$

leads to

$$\begin{bmatrix} a_0 & 0 & 0 & b_0 & 0 \\ a_1 & a_0 & 0 & b_1 & b_0 \\ a_2 & a_1 & a_0 & b_2 & b_1 \\ 0 & a_2 & a_1 & b_3 & b_2 \\ 0 & 0 & a_2 & 0 & b_3 \end{bmatrix} \begin{bmatrix} x_0 \\ x_1 \\ x_2 \\ y_0 \\ y_1 \end{bmatrix} = \begin{bmatrix} c_0 \\ c_1 \\ c_2 \\ c_3 \\ c_4 \end{bmatrix}, \tag{9.49}$$

the matrix having a special structure. Now the number of computations is $\sim n^3$, number of memories $\sim n^2$. Determinant of the matrix is nonzero just if a, b are coprime. This means that the numerically troublesome cases are those approaching the presence of a common factor. The troubles are the same when Euclidean algorithm is used.

The polynomial elementary operations may be interpreted as special column operations on (9.49) preserving the matrix structure. The general row and column operations, e.g. orthogonal transformations, destroy it.

The only possible orthogonal transformations not destroying the polynomial structure, are for $m = n$. Given polynomials b, a, $\deg b = \deg a = n$, define b', a', *orthogonal reduction* of b, a

$$\begin{aligned} b' &= kb + ha \\ a' &= -hb + ka \end{aligned} \tag{9.50}$$

where numbers k, h are

$$k = \frac{a_n}{\sqrt{a_n^2 + b_n^2}}, \quad h = -\frac{b_n}{\sqrt{a_n^2 + b_n^2}}. \tag{9.51}$$

This can be used in the algorithms whenever $m = n$.

9.7 Conjugacy and Symmetry

To treat the properties of conjugated operators in the algebraical approach, we should equip $R[q]$ with one operation more: *conjugacy* $a \mapsto a^\star$, which would have the property (9.9). For continuous-time systems, it is done by defining

$$(a^\star)_i = (-1)^i a_i. \tag{9.52}$$

It has properties

$$\begin{aligned} (a+b)^\star = a^\star + b^\star, \quad (ab)^\star = a^\star b^\star, \\ a^{\star\star} = a, \quad \deg a^\star = \deg a. \end{aligned} \tag{9.53}$$

For polynomial functions

$$a^\star(q) = a(-q), \tag{9.54}$$

in accord with (9.9).

The quotient field $R(q)$ inherits the conjugacy operation from $R[q]$. For $c = a/b$,

$$\left(\frac{a}{b}\right)^\star = \frac{a^\star}{b^\star}, \quad c^\star(q) = c(-q). \tag{9.55}$$

A polynomial with the property $a^\star = a$ is called *symmetric*, with $a^\star = -a$ *skew-symmetric*. The symmetric polynomial is even function of q. It is of even degree and has the form

$$a = a_0 + a_2 q^2 + \cdots. \tag{9.56}$$

The skew-symmetric polynomial is odd function of q:

$$a = a_1 q + a_3 q^3 + \cdots. \tag{9.57}$$

It is of odd degree and can be expressed

$$a = qa' \tag{9.58}$$

with symmetric a'. Any polynomial can be expressed as a sum of symmetric and skew-symmetric parts:

$$a = a_S + a_W, \quad a^\star = a_S - a_W \tag{9.59}$$

where

$$a_S = \frac{a + a^\star}{2}, \quad a_W = \frac{a - a^\star}{2}. \tag{9.60}$$

Conversely, given a pair a_S, a_W with a_S symmetric, a_W skew-symmetric, it defines polynomial a by (9.59). It is evident that

$$\begin{aligned} \deg a &= \max(\deg a_S, \deg a_W), \\ \gcd(a, a^\star) &= \gcd(a_S, a_W). \end{aligned} \tag{9.61}$$

One thing more is taken from system theory to algebra: *stability*. In complex q-plane, the domain $\operatorname{Re} q \leq 0$ is called *stable*, $\operatorname{Re} q \geq 0$ *antistable*, $\operatorname{Re} q < 0$ *strictly stable*, $\operatorname{Re} q > 0$ *strictly antistable*. The line $\operatorname{Re} q = 0$ is called *boundary* of stability. A fraction is called (strictly) (anti-) stable if all its poles lie in such a domain. A polynomial is called (strictly) (anti-) Hurwitz if all its roots lie in (strictly) (anti-) stable domain. If polynomial a is Hurwitz then $a^\star$ is anti-Hurwitz. If strictly then $a, a^\star$ are coprime.

With every root q_i, the (skew-) symmetric polynomial has also root $-q_i$ of the same multiplicity. As this is combined with the complex conjugacy of roots, we have the following combination :

quadruple	$q_i, -q_i, \overline{q_i}, -\overline{q_i}$	
real couple	$q_i, -q_i,$	$\overline{q_i} = q_i, \; -\overline{q_i} = -q_i$
boundary couple	$q_i, -q_i,$	$\overline{q_i} = -q_i, \; -\overline{q_i} = q_i$
zero	$0,$	$\overline{0} = -0 = -\overline{0} = 0$

If a symmetric polynomial has root 0, it is of even multiplicity. Every skew-symmetric polynomial has root 0 with odd multiplicity.

A symmetric polynomial as a function acquries only real values on the stability boundary. The polynomial is called *symmetric positive* (*nonnegative*) if all these values are such. Symmetric positive polynomials have no roots in the boundary. Symmetric nonnegative may have such roots but only with even multiplicity.

Consider the *greatest common divisors* $\gcd(a, a^\star)$. If g is gcd, $g^\star$ is also gcd. The $g, g^\star$ are associates, ex. unit u such that $g = ug^\star$. From $g = uu^\star g$, it is $uu^\star = 1$. Generally, the unit u satisfying $uu^\star = 1$, is called *orthogonal.* Such elements form a multiplicative group R_o, subgroup of the group of units R_u. In $R[q]$, the only orthogonals are ± 1, so there are two cases of g: symmetric $g^\star = g$, skew-symmetric $g^\star = -g$. The Bézout expression can be written $g = \frac{1}{2}(au^\star + a^\star u)$ in the first case, $g = \frac{1}{2}(au^\star - a^\star u)$ in the second.

Given a_S, a_W with a_S symmetric, a_W skew-symmetric, their $g = \gcd(a_S, a_W)$ is also of two cases: either $g^\star = g$ or $g^\star = -g$. The Bézout expression can be written $g = a_S u_S - a_W u_W$ in the first case, $g = a_S u_W - a_W u_S$ in the second.

The Algorithm 9.2 for computing *greatest common divisor* can be modified to the case of a_S, a_W, an equivalent of the case $a, a^\star$. Note that now a_S is of different type from a_W (the former one is symmetric, the latter one skew-symmetric). This is why the exchange operation cannot be used.

Algorithm 9.5 *Given polynomials a_S, a_W, not both zero, $a_S^\star = a_S$, $a_W^\star = -a_W$, let $g = \gcd(a_S, a_W)$. For $k = 0, 1, 2, \ldots$ construct sequences a_S^k, a_W^k with $a_S^0 = a_S, a_W^0 = a_W$ by the recurrent rule*

$$
\begin{aligned}
& n = \deg a_S^k, \quad m = \deg a_W^k, \\
& \text{if } n > m \text{ and } a_W^k \neq 0 \text{ then} \\
& \qquad a_S^{k+1} = a_S^k - q^{n-m} \frac{a_{Sn}^k}{a_{Wm}^k} a_W^k, \\
& \qquad a_W^{k+1} = a_W^k, \\
& \text{else if } m > n \text{ and } a_S^k \neq 0 \text{ then} \\
& \qquad a_S^{k+1} = a_S^k \\
& \qquad a_W^{k+1} = a_W^k - q^{m-n} \frac{a_{Wm}^k}{a_{Sn}^k} a_S^k.
\end{aligned}
$$

It is $(a_S^k)^\star = a_S^k$, $(a_W^k)^\star = -a_W^k$, $\gcd(a_S^k, a_W^k) = g$. Then for some l, it is either $a_W^l = 0$, $a_S^l = g$, $g^\star = g$ or $a_S^l = 0$, $a_W^l = g$, $g^\star = -g$.

If a is strictly Hurwitz then $a, a^\star$ are coprime and so are a_S, a_W. The a^k correspodning to a_S^k, a_W^k is also strictly Hurwitz as the recurrent rule is the same as that for Routh-Hurwitz stability test.

9.8 Symmetric Homogeneous Equations

Definition 9.5 *For $a \neq 0$, a* linear symmetric homogeneous equation *is*

$$ax^\star + a^\star x = 0, \tag{9.62}$$

a linear skew-symmetric homogeneous equation *is*

$$ax^\star - a^\star x = 0. \tag{9.63}$$

A special case is coprime equation *with $\hat{a}, \hat{a}^\star$ coprime, either*

$$\hat{a}x^\star + \hat{a}^\star x = 0 \tag{9.64}$$

or

$$\hat{a}x^\star - \hat{a}^\star x = 0. \tag{9.65}$$

These two types of equation should be investigated together. The behaviour depends on properties of $\gcd(a, a^\star)$, whether it is symmetric or skew-symmetric.

Lemma 9.19 *Denote*

$$g = \gcd(a, a^\star), \quad a = \hat{a}g. \tag{9.66}$$

When $g^\star = g$, equation (9.62) can be cancelled *to (9.64), equation (9.63) to (9.65). When $g^\star = -g$, equation (9.62) to (9.65), equation (9.63) to (9.64).*

Lemma 9.20 *(coprime equation) For (9.64), the* general solution *is*

$$x = \hat{a}q\tau \tag{9.67}$$

where τ is an arbitrary symmetric polynomial. For (9.65), it is

$$x = \hat{a}\tau. \tag{9.68}$$

Proof Equation (9.64) can be thought of as equation $\hat{a}y + \hat{a}^\star x = 0$ with requirement $y = x^\star$. By Lemma 9.2, the general solution is $x = \hat{a}t$, $y = -\hat{a}^\star t$, i.e. $x^\star = -\hat{a}^\star t$. From that, $t^\star = -t$, t skew-symmetric, $t = q\tau$ with τ symmetric. For equation (9.65), $\hat{a}y - \hat{a}^\star x = 0$, $y = x^\star$, general solution $x = \hat{a}t$, $y = \hat{a}^\star t$, i.e. $x^\star = \hat{a}^\star t$. From that, $t^\star = t$ symmetric. □

Lemma 9.21 *(general equation). With (9.66), the* general solution *of (9.62), (9.63) is (9.67) when the cancellation leads to (9.64), or (9.68) when to (9.65).*

Definition 9.6 *A* trivial solution *is $x = 0$. A* minimal solution *is a nontrivial solution with* $\deg x$ *minimal.*

Lemma 9.22 *(coprime equation). For (9.64), there exists* minimal solution

$$x_{\min} = \hat{a}qK, \tag{9.69}$$

unique up to multiplication by an arbitrary nonzero real number K. It is $\deg x_{\min} = \deg \hat{a} + 1$. *For (9.65), there exists minimal solution*

$$x_{\min} = \hat{a}K, \tag{9.70}$$

$\deg x_{\min} = \deg \hat{a}$.

Lemma 9.23 *(general equation). For (9.62), (9.63) with (9.66), there exists a minimal solution. It is (9.69) with* $\deg x_{\min} = \deg a - \deg g + 1$ *when the cancellation leads to (9.64), or (9.70) with* $\deg x_{\min} = \deg a - \deg g$ *when to (9.65).*

The (skew-) symmetric equation can be also investigated by converting it to a special type of a *polynomial equation with two unknowns.* We develop first a theory of such equations, then (Lemma 9.29) we show the conversion.

Definition 9.7 *Given symmetric a_S, skew-symmetric a_W, not both zero,* symmetric equation *is*

$$a_S x_S - a_W x_W = 0, \tag{9.71}$$

skew-symmetric equation *is*

$$a_W x_S - a_S x_W = 0 \tag{9.72}$$

with reqirement x_S to be symmetric, x_W to be skew-symmetric. For coprime $\hat{a}_S, \hat{a}_W$, there is a special case of coprime equation

$$\hat{a}_S x_S - \hat{a}_W x_W = 0 \tag{9.73}$$

and

$$\hat{a}_W x_S - \hat{a}_S x_W = 0. \tag{9.74}$$

Lemma 9.24 *Denote*

$$g = \gcd(a_S, a_W). \tag{9.75}$$

When $g^\star = g$, denoting

$$a_S = \hat{a}_S g, \ a_W = \hat{a}_W g, \tag{9.76}$$

equation (9.71) can be cancelled *to (9.73), equation (9.72) to (9.74). When $g^\star = -g$, denoting*

$$a_S = \hat{a}_W g, \ a_W = \hat{a}_S g, \tag{9.77}$$

equation(9.71) can be cancelled to (9.74), equation(9.72) to (9.73).

Lemma 9.25 *(coprime equation). For (9.73), the* general solution *is*

$$x_S = \hat{a}_W q \tau, \ x_W = \hat{a}_S q \tau \tag{9.78}$$

where τ is an arbitrary symmetric polynomial. For (9.74),

$$x_S = \hat{a}_S \tau, \ x_W = \hat{a}_W \tau. \tag{9.79}$$

Proof The general solution of (9.73) is $x_S = \hat{a}_W t$, $x_W = \hat{a}_S t$ with arbitrary polynomial t. The requirement $x_S^\star = x_S$, $x_W^\star = -x_W$ leads to $t^\star = -t$, $t = q\tau$ with $\tau^\star = \tau$. The general solution of (9.74) is $x_S = \hat{a}_S t$, $x_W = \hat{a}_W t$, $t^\star = t$. □

Lemma 9.26 *(general equation). With (9.66), (9.76), (9.77), the* general solution *of (9.71), (9.72) is (9.78) when the cancellation leads to (9.73), or (9.79) when to (9.74).*

Lemma 9.27 *(coprime equation). For (9.73), there exists minimal solution*

$$x_{S\min} = \hat{a}_W q K, \quad x_{W\min} = \hat{a}_S q K, \tag{9.80}$$

unique up to multiplication by an arbitrary nonzero real number K. It is $\deg x_{S\min} = \deg \hat{a}_W + 1$, $\deg x_{W\min} = \deg \hat{a}_S + 1$. *For (9.74), there exists minimal solution*

$$x_{S\min} = \hat{a}_S K, \quad x_{W\min} = \hat{a}_W K. \tag{9.81}$$

It is $\deg x_{S\min} = \deg \hat{a}_S$, $\deg x_{W\min} = \deg \hat{a}_W$.

Lemma 9.28 *(general equation). With (9.75), (9.76), (9.77), a* minimal solution *of equation(9.71) exists: if $g^\star = g$ then (9.80) with* $\deg x_{S\min} = \deg a_W - \deg g + 1$, $\deg x_{W\min} = \deg a_S - \deg g + 1$, *if $g^\star = -g$ then (9.81) with* $\deg x_{S\min} = \deg a_W - \deg g$, $\deg x_{W\min} = \deg a_S - \deg g$. *For equation (9.72), a minimal solution exists: if $g^\star = g$ then (9.81) with* $\deg x_{S\min} = \deg a_S - \deg g$, $\deg x_{W\min} = \deg a_W - \deg g$, *if $g^\star = -g$ then (9.80) with* $\deg x_{S\min} = \deg a_S - \deg g + 1$, $\deg x_{W\min} = \deg a_W - \deg g + 1$.

Lemma 9.29 *By (9.59), (9.60) for a, and similarly for x, equation (9.62) can be converted to (9.71), equation (9.63) to (9.72). The conversion has properties (9.61), similarly for x. The minimal solutions correspond each to other.*

The Algorithm 9.3 can be modified for the case of a_S, a_W, it computes the *minimal solution* simultaneously with g. For homogeneous symmetric equation we have:

Algorithm 9.6 *Given polynomials a_S, a_W, not both zero, $a_S^\star = a_S, a_W^\star = -a_W$, let $g = \gcd(a_S, a_W)$, let x_S, x_W be minimal solution of $a_S x_S - a_W x_W = 0$. For $k = 0, 1, 2, \ldots$ construct sequences a_S^k, a_W^k, f^k, n^k, m^k with $a_S^0 = a_S$, $a_W^0 = a_W$ by the recurrent rule*

$$\begin{aligned}
&n^k = \deg a_S^k, \ m^k = \deg a_W^k, \\
&\text{if } n^k > m^k \text{ and } a_W^k \neq 0 \text{ then} \\
&\quad f^k = 1, \\
&\quad a_S^{k+1} = a_S^k - q^{n^k - m^k} \frac{a_{Sn^k}^k}{a_{Wm^k}^k} a_W^k, \\
&\quad a_W^{k+1} = a_W^k, \\
&\text{else if } m^k > n^k \text{ and } a_S^k \neq 0 \text{ then} \\
&\quad f^k = 2, \\
&\quad a_S^{k+1} = a_S^k, \\
&\quad a_W^{k+1} = a_W^k - q^{m^k - n^k} \frac{a_{Wm^k}^k}{a_{Sn^k}^k} a_S^k.
\end{aligned}$$

It is $(a_S^k)^\star = a_S^k$, $(a_W^k)^\star = -a_W^k$, $\gcd(a_S^k, a_W^k) = g$. *Then for some* l, *it is either* $a_W^l = 0$, $a_S^l = g$, $g^\star = g$ *or* $a_S^l = 0$, $a_W^l = g$, $g^\star = -g$.

For $k = l, l-1, \ldots, 0$, *let* x_S^k, x_W^k *be minimal solution of* $a_S^k x_S^k - a_W^k x_W^k = 0$. *When* $a_W^l = 0$, *it is* $x_S^l = 0$, $x_W^l = Kq$ *where* K *is arbitrary real number. When* $a_S^l = 0$, *it is* $x_S^l = K$, $x_W^l = 0$. *The recurrent rule holds*

$$\begin{aligned}
&\text{if } f^k = 1 \text{ then}\\
&\quad x_S^k = x_S^{k+1},\\
&\quad x_W^k = x_W^{k+1} + q^{n^k - m^k} \frac{a_{Sn^k}^k}{a_{Wm^k}^k} x_S^{k+1},\\
&\text{if } f^k = 2 \text{ then}\\
&\quad x_S^k = x_S^{k+1} + q^{m^k - n^k} \frac{a_{Wm^k}^k}{a_{Sn^k}^k} x_W^{k+1},\\
&\quad x_W^k = x_W^{k+1}
\end{aligned}$$

Finally, $x_S^0 = x_S$, $x_W^0 = x_W$.

For homogeneous skew-symmetric equation we have:

Algorithm 9.7 *Given polynomials* a_S, a_W, *not both zero,* $a_S^\star = a_S, a_W^\star = -a_W$, *let* $g = \gcd(a_S, a_W)$, *let* x_S, x_W *be minimal solution of* $a_W x_S - a_S x_W = 0$. *For* $k = 0, 1, 2, \ldots$ *construct sequences* a_S^k, a_W^k, f^k, n^k, m^k *with* $a_S^0 = a_S$, $a_W^0 = a_W$ *by the recurrent rule*

$$\begin{aligned}
&n^k = \deg a_S^k,\ m^k = \deg a_W^k,\\
&\text{if } n^k > m^k \text{ and } a_W^k \neq 0 \text{ then}\\
&\quad f^k = 1,\\
&\quad a_S^{k+1} = a_S^k - q^{n^k - m^k} \frac{a_{Sn^k}^k}{a_{Wm^k}^k} a_W^k,\\
&\quad a_W^{k+1} = a_W^k,\\
&\text{else if } m^k > n^k \text{ and } a_S^k \neq 0 \text{ then}\\
&\quad f^k = 2,\\
&\quad a_S^{k+1} = a_S^k,\\
&\quad a_W^{k+1} = a_W^k - q^{m^k - n^k} \frac{a_{Wm^k}^k}{a_{Sn^k}^k} a_S^k.
\end{aligned}$$

It is $(a_S^k)^\star = a_S^k$, $(a_W^k)^\star = -a_W^k$, $\gcd(a_S^k, a_W^k) = g$. *Then for some* l, *it is either* $a_W^l = 0$, $a_S^l = g$, $g^\star = g$ *or* $a_S^l = 0$, $a_W^l = g$, $g^\star = -g$.

For $k = l, l-1, \ldots, 0$, *let* x_S^k, x_W^k *be minimal solution of* $a_W^k x_S^k - a_S^k x_W^k = 0$. *When* $a_W^l = 0$, *it is* $x_S^l = K$, $x_W^l = 0$, *where* K *is arbitrary real number. When* $a_S^l = 0$, *it is* $x_S^l = 0$, $x_W^l = Kq$. *The recurrent rule holds*

$$\begin{aligned}
&\text{if } f^k = 1 \text{ then}\\
&\quad x_S^k = x_S^{k+1} + q^{n^k - m^k} \frac{a_{Sn^k}^k}{a_{Wm^k}^k} x_W^{k+1},\\
&\quad x_W^k = x_W^{k+1},
\end{aligned}$$

$$\text{if } f^k = 2 \text{ then}$$
$$x_S^k = x_S^{k+1},$$
$$x_W^k = x_W^{k+1} + q^{m^k - n^k} \frac{a^k_{Wm^k}}{a^k_{Sn^k}} x_S^{k+1}.$$

Finally, $x_S^0 = x_S$, $x_W^0 = x_W$.

The transformation of (skew-) symmetric homogeneous equations can be also written by means of unimodular matrices. It is easily seen that matrices M, see (9.19), preserving the form of (9.62), (9.63) are those satisfying

$$M_{11}^\star = M_{22}, \quad M_{12}^\star = M_{21}. \tag{9.82}$$

They form a multiplicative group U_S, subgroup of U. Matrices N preserving the form (9.71), (9.72), i.e.

$$\begin{bmatrix} a'_S & a'_W \end{bmatrix} = \begin{bmatrix} a_S & a_W \end{bmatrix} N \tag{9.83}$$

are those satisfying

$$N_{11}^\star = N_{11}, \quad N_{12}^\star = -N_{12}, \quad N_{21}^\star = -N_{21}, \quad N_{22}^\star = N_{22}. \tag{9.84}$$

They form a group isomorphic to U_S. From (9.59), (9.60), the isomorphism is

$$N = \frac{1}{2}\begin{bmatrix} 1 & 1 \\ 1 & -1 \end{bmatrix} M \begin{bmatrix} 1 & 1 \\ 1 & -1 \end{bmatrix}. \tag{9.85}$$

9.9 Symmetric Nonhomogeneous Equations

Definition 9.8 *For* $a \neq 0$, $b^\star = b$, *a* linear symmetric nonhomogeneous equation *is*

$$\frac{1}{2}(ax^\star + a^\star x) = b, \tag{9.86}$$

for $a \neq 0$, $b^\star = -b$, *a* linear skew-symmetric nonhomogeneous equation *is*

$$\frac{1}{2}(ax^\star - a^\star x) = b. \tag{9.87}$$

A special case of coprime equations *is with coprime* $\hat{a}, \hat{a}^\star$

$$\frac{1}{2}(\hat{a}x^\star + \hat{a}^\star x) = \hat{b} \tag{9.88}$$

and

$$\frac{1}{2}(\hat{a}x^\star - \hat{a}^\star x) = \hat{b}. \tag{9.89}$$

Lemma 9.30 *The necessary and sufficient condition for* solvability *of (9.86), (9.87) is:* $\gcd(a, a^\star)$ *divides* b.

Proof Necessity: both terms in lhs of the equation are divisible by g. If x exists, the rhs must be also divisible. Sufficiency: when $g^\star = g$, the Bézoutean expression is $g = \frac{1}{2}(aw^\star + a^\star w)$, when $g^\star = -g$, $g = \frac{1}{2}(aw^\star - a^\star w)$. For both cases and both equations, if $b = \hat{b}g$, then $x = w\hat{b}^\star$ is a solution. □

Lemma 9.31 *Denote*

$$g = \gcd(a, a^\star), \quad a = \hat{a}g, \quad b = \hat{b}g. \tag{9.90}$$

When $g^\star = g$, *equation (9.86) can be* cancelled *to (9.88), equation(9.87) to (9.89). When* $g^\star = -g$, *equation (9.86) to (9.89), equation (9.87) to (9.88).*

Definition 9.9 *A* minimal solution *is solution with* $\deg x$ *minimal.*

Lemma 9.32 *(coprime equation). Equation(9.88) with* $\deg \hat{b} < 2\deg \hat{a}$, *has unique* minimal solution $x_{\min}$ *satisfying* $\deg x_{\min} < \deg \hat{a}$. *Equation(9.89) with* $\deg \hat{b} < 2\deg \hat{a} - 1$ *has unique minimal solution* $x_{\min}$ *satisfying* $\deg x_{\min} < \deg \hat{a} - 1$.

Proof Equation (9.88) can be thought of as equation

$$\frac{1}{2}(\hat{a}y + \hat{a}^\star x) = \hat{b} \tag{9.91}$$

with requirement $y = x^\star$. By Lemma 9.10, unique minimal solution $x_{\min}, y_{\min}$ exists with $\deg x_{\min} < \deg \hat{a}$, $\deg y_{\min} < \deg \hat{a}$. Taking the conjugate of (9.91), we see that the pair $y^\star_{\min}, x^\star_{\min}$ is also minimal solution of (9.91). As the minimal solution is unique, it is $y_{\min} = x^\star_{\min}$. So $x_{\min}$ is the required minimal solution of (9.88).

For equation (9.89), the same construction holds. We must now prove that $x_{\min}$ satisfies more: $\deg x_{\min} < \deg \hat{a} - 1$. Assumption $\deg x_{\min} = \deg \hat{a} - 1$ implies $\deg \hat{a}x^\star_{\min} = 2\deg \hat{a} - 1$. As this degree is odd, it is equal to $\deg(\hat{a}x^\star_{\min} - \hat{a}^\star x_{\min}) = \deg \hat{b}$, contradictory with assumption about $\deg \hat{b}$. So the assumption $\deg x_{\min} = \deg \hat{a} - 1$ is false, it must be $\deg x_{\min} < \deg \hat{a} - 1$. □

Lemma 9.33 *(general equation). Equations (9.86), (9.87), if solvable, have unique* minimal solution*: If the cancellation leads to (9.88) then it is: assumption* $\deg b < 2\deg a - \deg g$, *property* $\deg x_{\min} < \deg a - \deg g$. *If the cancellation leads to (9.89) then the minimal solution is: assumption* $\deg b < 2\deg a - \deg g - 1$, *property* $\deg x_{\min} < \deg a - \deg g - 1$.

The meaning of minimal solutions can be seen by writing (9.88) in the form

$$\frac{x^\star}{\hat{a}^\star} + \frac{x}{\hat{a}} = \frac{2\hat{b}}{\hat{a}^\star \hat{a}}. \tag{9.92}$$

The problem of finding the minimal solution is: given symmetric fraction $2\hat{b}/\hat{a}^\star\hat{a}$, strictly proper (for $q \to \infty$, $\sim 1/q^2$), decompose it into a sum of two mutually conjugated fractions with given denominator, strictly proper ($\sim 1/q$).

For (9.89),

$$\frac{x^\star}{\hat{a}^\star} - \frac{x}{\hat{a}} = \frac{2\hat{b}}{\hat{a}^\star \hat{a}}. \tag{9.93}$$

Given skew-symmetric fraction with $\sim 1/q^3$, decompose it into a difference of two mutually conjugated fractions with given denominator, with $\sim 1/q^2$.

The (skew-) symmetric nonhomogeneous equation can be also investigated by converting it to a *polynomial equation with two unknowns.*

Definition 9.10 *With a_S symmetric, a_W skew-symmetric, not both zero, the* symmetric equation *is*

$$a_S x_S - a_W x_W = b \tag{9.94}$$

with symmetric b; the skew-symmetric equation *is*

$$a_W x_S - a_S x_W = b \tag{9.95}$$

with skew-symmetric b, both equations with requirement x_S symmetric, x_W skew-symmetric. Coprime equations *are with coprime $\hat{a}_S, \hat{a}_W$*

$$\hat{a}_S x_S - \hat{a}_W x_W = \hat{b}, \tag{9.96}$$

and

$$\hat{a}_W x_S - \hat{a}_S x_W = \hat{b}. \tag{9.97}$$

Lemma 9.34 *The necessary and sufficient contition for* solvability *of (9.86), (9.87) is:* $\gcd(a_S, a_W)$ *divides b.*

Proof Necessity is clear from divisibility. Sufficiency: If $g^\star = g$, the Bézout expression is $g = a_S u_S - a_W u_W$, if $g^\star = -g$, $g = a_W u_S - a_S u_W$. If $b = \hat{b}g$, either $x_S = u_S\hat{b}$, $x_W = u_W\hat{b}$ or $x_S = -u_W\hat{b}$, $x_W = -u_S\hat{b}$ is a solution. □

Lemma 9.35 *When $g^\star = g$, equation (9.94) is* cancelled *to (9.96), equation (9.95) to (9.97). When $g^\star = -g$, equation (9.94) to (9.97), equation(9.95) to (9.96).*

Definition 9.11 *A* minimal solution *is solution with* $\max(\deg x_S, \deg x_W)$ *minimal.*

Lemma 9.36 *(coprime equation). Equation (9.96) with assumption*

$$\deg \hat{b} < 2\max(\deg \hat{a}_S, \deg \hat{a}_W)$$

has unique minimal solution *with property*

$$\max(\deg x_S, \deg x_W) < \max(\deg \hat{a}_S, \deg \hat{a}_W).$$

Equation (9.97) with assumption

$$\deg \hat{b} < 2\max(\deg \hat{a}_S, \deg \hat{a}_W) - 1$$

has unique minimal solution with property

$$\max(\deg x_S, \deg x_W) < \max(\deg \hat{a}_S, \deg \hat{a}_W) - 1.$$

Proof For equation (9.96), distinguish two cases. The first, suppose $\deg \hat{a}_W < \deg \hat{a}_S$, it is $\deg \hat{b} < 2\deg \hat{a}_S$. By Lemma 9.10, take the x_W-minimal solution. It is $\deg x_W < \deg \hat{a}_S$, the unique solution with this property. It is either $\deg x_S < \deg \hat{a}_W < \deg \hat{a}_S$, or $\deg x_S = \deg \hat{b} - \deg \hat{a}_S < \deg \hat{a}_S$. So we have proved $\max(\deg x_S, \deg x_W) < \deg \hat{a}_S = \max(\deg \hat{a}_S, \deg \hat{a}_W)$. The second case, suppose $\deg \hat{a}_S < \deg \hat{a}_W$, it is $\deg \hat{b} < 2\deg \hat{a}_W$. Take the x_S-minimal solution, it is $\deg x_S < \deg \hat{a}_W$, the unique solution with this property. It is either $\deg x_W <$

$\deg \hat{a}_S < \deg \hat{a}_W$ or $\deg x_W = \deg \hat{b} - \deg \hat{a}_W < \deg \hat{a}_W$. So we have proved $\max(\deg x_S, \deg x_W) < \deg \hat{a}_W = \max(\deg \hat{a}_S, \deg \hat{a}_W)$.

For equation (9.97), first, suppose $\deg \hat{a}_W < \deg \hat{a}_S$, it is $\deg \hat{b} < 2\deg \hat{a}_S - 1$. Take x_S-minimal solution. It is $\deg x_S < \deg \hat{a}_S$, uniquely. As both these numbers are even, it is $\deg x_S < \deg \hat{a}_S - 1$. It is either $\deg x_W < \deg \hat{a}_W \leq \deg \hat{a}_S - 1$, or $\deg x_W = \deg \hat{b} - \deg \hat{a}_S < \deg \hat{a}_S - 1$. So we have proved $\max(\deg x_S, \deg x_W) < \deg \hat{a}_S - 1 = \max(\deg \hat{a}_S, \deg \hat{a}_W) - 1$. Second, suppose $\deg \hat{a}_S < \deg \hat{a}_W$, it is $\deg \hat{b} < 2\deg \hat{a}_W - 1$. Take x_W-minimal solution: it is $\deg x_W < \deg \hat{a}_W$, uniquely. By oddness, $\deg x_W < \deg \hat{a}_W - 1$. It is either $\deg x_S < \deg \hat{a}_S \leq \deg \hat{a}_W - 1$, or $\deg x_S = \deg \hat{b} - \deg \hat{a}_W < \deg \hat{a}_W - 1$. So we have proved $\max(\deg x_S, \deg x_W) < \deg \hat{a}_W - 1 = \max(\deg \hat{a}_S, \deg \hat{a}_W) - 1$.

To prove that our solution satisfies $x_S^\star = x_S$, $x_W^\star = -x_W$: it is x_S- or x_W-minimal solution. Conjugating the equations, we see that the pair $x_S^\star, -x_W^\star$ is also x_S- or x_W-minimal solution. As such a solution is unique, it is $x_S^\star = x_S$, $x_W^\star = -x_W$. □

Lemma 9.37 *(general equation). Equations (9.94), (9.95), if solvable, have unique* minimal solution. *If the cancellation leads to (9.96) then it is: assumption*

$$\deg b < 2\max(\deg a_S, \deg a_W) - \deg g,$$

property

$$\max(\deg x_S, \deg x_W) < \max(\deg a_S, \deg a_W) - \deg g.$$

If the cancellation leads to (9.97), then the minimal solution is: assumption

$$\deg b < 2\max(\deg a_S, \deg a_W) - \deg g - 1,$$

property

$$\max(\deg x_S, \deg x_W) < \max(\deg a_S, \deg a_W) - \deg g - 1.$$

Lemma 9.38 *By (9.59), (9.60) and similarly for x, equation (9.86) can be converted to (9.94), equation (9.87) to (9.95). The conversion has properties (9.61), similarly for x. The minimal solutions correspond each to other.*

Note that the unique minimal solution of (9.96), (9.97) exists not always but only when a condition for $\deg \hat{b}$ is satisfied. This situation is like that for equation (9.34), Lemma 9.13, where the unique minimal solution existed only when a condition for $\deg \hat{c}$ was satisfied, whereas the y-minimal and x-minimal solution existed always. For (9.96), (9.97), we also want to define a sort of special solution existing always. That is why we define:

Definition 9.12 *A* doubly minimal solution *of (9.96), (9.97) is a solution satisfying:*
1) $\max(\deg x_S, \deg x_W)$ *is minimal,*
2) from all such solutions, $\min(\deg x_S, \deg x_W)$ *is minimal.*

Lemma 9.39 *(coprime equation). For equations (9.96), (9.97), unique doubly minimal solution exists. If the assumptions of Lemma 9.36 are satisfied then it coincides with the minimal solution. Otherwise, for equation (9.96), it is: assumption*

$$\deg \hat{b} \geq 2\max(\deg \hat{a}_S, \deg \hat{a}_W),$$

properties

$$\max(\deg x_S, \deg x_W) = \deg \hat{b} - \max(\deg \hat{a}_S, \deg \hat{a}_W),$$
$$\min(\deg x_S, \deg x_W) < \max(\deg \hat{a}_S, \deg \hat{a}_W).$$

For equation (9.97), it is: assumption

$$deg \hat{b} \geq 2\max(\deg \hat{a}_S, \deg \hat{a}_W) - 1,$$

properties

$$\max(\deg x_S, \deg x_W) = \deg \hat{b} - \max(\deg \hat{a}_S, \deg \hat{a}_W)$$
$$\min(\deg x_S, \deg x_W) < \max(\deg \hat{a}_S, \deg \hat{a}_W) - 1.$$

Proof In the first case, the coincidence with the minimal solution is clear. Consider the second case. For equation (9.96), the rhs has $\deg \hat{b}$, so for any solution it must be either

$$\deg x_S = \deg \hat{b} - \deg \hat{a}_S \geq \deg \hat{b} - \max(\deg \hat{a}_S, \deg \hat{a}_W)$$

or

$$\deg x_W = \deg \hat{b} - \deg \hat{a}_W \geq \deg \hat{b} - \max(\deg \hat{a}_S, \deg \hat{a}_W).$$

From this,

$$\max(\deg x_S, \deg x_W) \geq \deg \hat{b} - \max(\deg \hat{a}_S, \deg \hat{a}_W).$$

So $\deg \hat{b} - \max(\deg \hat{a}_S, \deg \hat{a}_W)$ is minimal possible value of $\max(\deg x_S, \deg x_W)$. If $\deg \hat{a}_S > \deg \hat{a}_W$, take the x_W-minimal solution. It has

$$\deg x_W < \deg \hat{a}_S = \max(\deg \hat{a}_S, \deg \hat{a}_W) \leq \deg \hat{b} - \max(\deg \hat{a}_S, \deg \hat{a}_W),$$
$$\deg x_S = \deg \hat{b} - \deg \hat{a}_S = \deg \hat{b} - \max(\deg \hat{a}_S, \deg \hat{a}_W).$$

So, this solution has the minimal possible value of $\max(\deg x_S, \deg x_W)$ as well as of $\deg x_W$, i.e. of $\min(\deg x_S, \deg x_W)$, so it is doubly minimal. Similarly, if $\deg \hat{a}_W > \deg \hat{a}_S$, the x_S-minimal solution has this property. For equation (9.97), the minimal possible value of $\max(\deg x_S, \deg x_W)$ is the same. If $\deg \hat{a}_S > \deg \hat{a}_W$ we take the x_S-minimal solution, if $\deg \hat{a}_W > \deg \hat{a}_S$ we take the x_W-minimal solution. □

Lemma 9.40 *(general equation). Equations (9.94), (9.95), if solvable, have unique* doubly minimal solution. *If the assumptions of Lemma 9.37 are satisfied, it coincides with the minimal solution. Otherwise, if the cancellation leads to (9.96) then the doubly minimal solution is: assumption*

$$\deg b > 2\max(\deg a_S, \deg a_W) - \deg g,$$

properties

$$\max(\deg x_S, \deg x_W) = \deg b - \max(\deg a_S, \deg a_W),$$
$$\min(\deg x_S, \deg x_W) < \max(\deg a_S, \deg a_W) - \deg g.$$

If the cancellation leads to (9.97), then the doubly minimal solution is: assumption

$$\deg b > 2\max(\deg a_S, \deg a_W) - \deg g - 1,$$

properties

$$\max(\deg x_S, \deg x_W) = \deg b - \max(\deg a_S, \deg a_W),$$
$$\min(\deg x_S, \deg x_W) < \max(\deg a_S, \deg a_W) - \deg g - 1.$$

The algorithm 9.4 can be modified for the case of a_S, a_W, b, it computes the doubly minimal solution simultaneously with g. For nonhomogeneous symmetric equation, we have

Algorithm 9.8 *Given polynomials a_S, a_W, b with a_S, a_W not both zero, $a_S^\star = a_S$, $a_W^\star = -a_W$, $b^\star = b$, let $g = \gcd(a_S, a_W)$, let x_S, x_W be doubly minimal solution of $a_S x_S - a_W x_W = b$. For $k = 0, 1, 2, \ldots$ construct sequences a_S^k, a_W^k, b^k, f^k, n^k, m^k, p^k with $a_S^0 = a_S$, $a_W^0 = a_W$, $b^0 = b$ by the recurrent rule*

$$
\begin{aligned}
& n^k = \deg a_S^k,\ m^k = \deg a_W^k,\ p = \deg b^k, \\
& \text{if } p^k \geq \max(n^k, m^k) \text{ and } n^k > m^k \text{ then} \\
& \quad f^k = 1, \\
& \quad a_S^{k+1} = a_S^k, \\
& \quad a_W^{k+1} = a_W^k, \\
& \quad b^{k+1} = b^k - q^{p^k - n^k} \frac{b^k_{p^k}}{a^k_{Sn^k}} a_S^k, \\
& \text{else if } p^k \geq \max(n^k, m^k) \text{ and } m^k > n^k \text{ then} \\
& \quad f^k = 2, \\
& \quad a_S^{k+1} = a_S^k, \\
& \quad a_W^{k+1} = a_W^k, \\
& \quad b^{k+1} = b^k - q^{p^k - m^k} \frac{b^k_{p^k}}{a^k_{Wm^k}} a_W^k, \\
& \text{else if } n^k > m^k \text{ and } a_W^k \neq 0 \text{ then} \\
& \quad f^k = 3, \\
& \quad a_S^{k+1} = a_S^k - q^{n^k - m^k} \frac{a^k_{Sn^k}}{a^k_{Wm^k}} a_W^k, \\
& \quad a_W^{k+1} = a_W^k, \\
& \quad b^{k+1} = b^k, \\
& \text{else if } m^k > n^k \text{ and } a_S^k \neq 0 \text{ then} \\
& \quad f^k = 4, \\
& \quad a_S^{k+1} = a_S^k, \\
& \quad a_W^{k+1} = a_W^k - q^{m^k - n^k} \frac{a^k_{Wm^k}}{a^k_{Sn^k}} a_S^k, \\
& \quad b^{k+1} = b^k.
\end{aligned}
$$

It is $(a_S^k)^\star = a_S^k$, $(a_W^k)^\star = -a_W^k$, $(b^k)^\star = b^k$, $\gcd(a_S^k, a_W^k) = g$. Then for some l, it is either $a_W^l = 0$, $a_S^l = g$, $g^\star = g$ or $a_S^l = 0$, $a_W^l = g$, $g^\star = -g$. If $b^l \neq 0$, the equation is unsolvable. Otherwise, for $k = l, l-1, \ldots, 0$, let x_S^k, x_W^k be doubly minimal solution of $a_S^k x_S^k - a_W^k x_W^k = b^k$. It is $x_S^l = 0$, $x_W^l = 0$. The recurrent rule holds

$$
\begin{aligned}
& \text{if } f^k = 1 \text{ then} \\
& \quad x_S^k = x_S^{k+1} + q^{p^k - n^k} \frac{b^k_{p^k}}{a^k_{Sn^k}},
\end{aligned}
$$

$$
\begin{aligned}
&\quad x_W^k = x_W^{k+1}, \\
&\text{if } f^k = 2 \text{ then} \\
&\quad x_S^k = x_S^{k+1}, \\
&\quad x_W^k = x_W^{k+1} - q^{p^k - m^k} \frac{b_{p^k}^k}{a_{Wm^k}^k}, \\
&\text{if } f^k = 3 \text{ then} \\
&\quad x_S^k = x_S^{k+1}, \\
&\quad x_W^k = x_W^{k+1} + q^{n^k - m^k} \frac{a_{Sn^k}^k}{a_{Wm^k}^k} x_S^{k+1}, \\
&\text{if } f^k = 4 \text{ then} \\
&\quad x_S^k = x_S^{k+1} + q^{m^k - n^k} \frac{a_{Wm^k}^k}{a_{Sn^k}^k} x_W^{k+1}, \\
&\quad x_W^k = x_W^{k+1}.
\end{aligned}
$$

Finally, $x_S^0 = x_S$, $x_W^0 = x_W$.

For nonhomogeneous skew-symmetric equation we have:

Algorithm 9.9 *Given polynomials* a_S, a_W, b *with* a_S, a_W *not both zero,* $a_S^\star = a_S$, $a_W^\star = -a_W$, $b^\star = -b$, *let* $g = \gcd(a_S, a_W)$, *let* x_S, x_W *be doubly minimal solution of* $a_W x_S - a_S x_W = b$. *For* $k = 0, 1, 2, \ldots$ *construct sequences* a_S^k, a_W^k, b^k, f^k, n^k, m^k, p^k *with* $a_S^0 = a_S$, $a_W^0 = a_W$, $b^0 = b$ *by the recurrent rule*

$$
\begin{aligned}
&n^k = \deg a_S^k, \ m^k = \deg a_W^k, \ p = \deg b^k, \\
&\text{if } p^k \geq \max(n^k, m^k) \text{ and } n^k > m^k \text{ then} \\
&\quad f^k = 1, \\
&\quad a_S^{k+1} = a_S^k, \\
&\quad a_W^{k+1} = a_W^k, \\
&\quad b^{k+1} = b^k - q^{p^k - n^k} \frac{b_{p^k}^k}{a_{Sn^k}^k} a_S^k, \\
&\text{else if } p^k \geq \max(n^k, m^k) \text{ and } m^k > n^k \text{ then} \\
&\quad f^k = 2, \\
&\quad a_S^{k+1} = a_S^k, \\
&\quad a_W^{k+1} = a_W^k, \\
&\quad b^{k+1} = b^k - q^{p^k - m^k} \frac{b_{p^k}^k}{a_{Wm^k}^k} a_W^k, \\
&\text{else if } n^k > m^k \text{ and } a_W^k \neq 0 \text{ then} \\
&\quad f^k = 3, \\
&\quad a_S^{k+1} = a_S^k - q^{n^k - m^k} \frac{a_{Sn^k}^k}{a_{Wm^k}^k} a_W^k, \\
&\quad a_W^{k+1} = a_W^k, \\
&\quad b^{k+1} = b^k,
\end{aligned}
$$

$$\begin{aligned}
&\text{else if } \ m^k > n^k \text{ and } a_S^k \neq 0 \text{ then} \\
&\quad f^k = 4, \\
&\quad a_S^{k+1} = a_S^k, \\
&\quad a_W^{k+1} = a_W^k - q^{m^k - n^k} \frac{a^k_{Wm^k}}{a^k_{Sn^k}} a_S^k, \\
&\quad b^{k+1} = b^k.
\end{aligned}$$

It is $(a_S^k)^\star = a_S^k$, $(a_W^k)^\star = -a_W^k$, $(b^k)^\star = -b^k$, $\gcd(a_S^k, a_W^k) = g$. *Then for some* l, *it is either* $a_W^l = 0$, $a_S^l = g$, $g^\star = g$ *or* $a_S^l = 0$, $a_W^l = g$, $g^\star = -g$. *If* $b^l \neq 0$, *the equation is unsolvable. Otherwise, for* $k = l, l-1, \ldots, 0$, *let* x_S^k, x_W^k *be doubly minimal solution of* $a_W^k x_S^k - a_S^k x_W^k = b^k$. *It is* $x_S^l = 0$, $x_W^l = 0$. *The recurrent rule holds*

$$\begin{aligned}
&\text{if } f^k = 1 \text{ then} \\
&\quad x_S^k = x_S^{k+1}, \\
&\quad x_W^k = x_W^{k+1} - q^{p^k - n^k} \frac{b^k_{p^k}}{a^k_{Sn^k}}, \\
&\text{if } f^k = 2 \text{ then} \\
&\quad x_S^k = x_S^{k+1} + q^{p^k - m^k} \frac{b^k_{p^k}}{a^k_{Wm^k}}, \\
&\quad x_W^k = x_W^{k+1}, \\
&\text{if } f^k = 3 \text{ then} \\
&\quad x_S^k = x_S^{k+1} + q^{n^k - m^k} \frac{a^k_{Sn^k}}{a^k_{Wm^k}} x_W^{k+1}, \\
&\quad x_W^k = x_W^{k+1}, \\
&\text{if } f^k = 4 \text{ then} \\
&\quad x_S^k = x_S^{k+1}, \\
&\quad x_W^k = x_W^{k+1} + q^{m^k - n^k} \frac{a^k_{Wm^k}}{a^k_{Sn^k}} x_S^{k+1}.
\end{aligned}$$

Finally, $x_S^0 = x_S$, $x_W^0 = x_W$.

The transformations of (skew-) symmetric nonhomogeneous equations can be also written by means of unimodular matrices. It is easily seen that matrices of type (9.43) preserving the form of (9.86), (9.87) are those satisfying (9.82) and $M_{13}^\star = M_{23}$. They form group V_S, subgroup of V. Matrices N of the form (9.43) preserving the form (9.94), (9.95), i.e.

$$\begin{bmatrix} a'_S & a'_W & b' \end{bmatrix} = \begin{bmatrix} a_S & a_W & b \end{bmatrix} N, \tag{9.98}$$

$$b^\star = b, \quad b'^\star = b'$$

and

$$\begin{bmatrix} a'_W & a'_S & b' \end{bmatrix} = \begin{bmatrix} a_W & a_S & b \end{bmatrix} N, \tag{9.99}$$

$$b^\star = -b, \quad b'^\star = -b'$$

are those satisfying (9.84) and $N^{\star}_{13} = N_{13}$, $N^{\star}_{23} = -N_{23}$. They form a group isomorphic to V_S. From (9.59), (9.60), the isomorphism is

$$N = \frac{1}{2}\begin{bmatrix} 1 & 1 & 0 \\ 1 & -1 & 0 \\ 0 & 0 & 1 \end{bmatrix} M \begin{bmatrix} 1 & 1 & 0 \\ 1 & -1 & 0 \\ 0 & 0 & 1 \end{bmatrix} \tag{9.100}$$

Bibliography

[1] K. Hunt, *Stochastic Optimal Control Theory with Applications in Self-tuning Control.* Springer-Verlag: Berlin, 1989.

[2] J. Ježek, *New Algorithm for Minimal Solution of Linear Polynomial Equations.* Kybernetika 18(1982), pp. 505-516.

[3] J. Ježek, *Conjugated and Symmetric Polynomial Equations I. Continuous Time Systems.* Kybernetika 19(1983), pp.121-130.

[4] V. Kučera, *Discrete Linear Control: the Polynomial Approach.* Chichester: Wiley, 1979.

[5] L.A. Zadeh and C.A.Desoer, *Linear System Theory - the State Space Approach.* Mc Graw-Hill Book Co: New York 1963.

Chapter 10

J-Spectral Factorisation

M. Šebek

10.1 Prologue

This chapter surveys several existing methods for the J-spectral factorisation of a polynomial para-Hermitian matrix Z with real coefficients. A polynomial matrix Z is said to be *para-Hermitian* if

$$Z^* = Z,$$

where the polynomial matrix Z^* is the *adjoint* of Z, defined by $Z(s) = Z^{\mathrm{T}}(-s)$. We call

$$Z = P^* J P, \tag{10.1}$$

a *spectral factorisation* if J is a signature matrix, and P a square polynomial matrix with real coefficients such that $\det P$ has all its roots in the closed left-half plane. The *signature matrix* J is of the form

$$J = \begin{bmatrix} I_1 & 0 & 0 \\ 0 & -I_2 & 0 \\ 0 & 0 & 0 \end{bmatrix} \tag{10.2}$$

where I_1 and I_2 are unit matrices of not necessarily the same dimensions.

A factorisation of the form $Z = PJP^*$, with P square such that $\det P$ is Hurwitz, is called a *J-spectral cofactorisation.* A spectral cofactorisation may be obtained by transposing a spectral factorisation of the transpose Z^{T} of Z.

The best known spectral factorisation problem is that of factoring a para-Hermitian polynomial matrix Z that is *positive-definite* on the imaginary axis, that is, $Z(j\omega) > 0$ for all $\omega \in \mathcal{R}$. Within the systems and control area, this problem arises in multivariable Wiener filtering and in frequency domain versions of LQG theory. In this case the signature matrix J is the unit matrix, and the desired factorisation is of the form $Z = P^*P$. There are numerous references on this problem (see for instance [1, 2, 3]).

The more general case, with Z possibly *indefinite* on the imaginary axis, is encountered in LQG game theory and $\mathcal{H}_\infty$-optimization theory (see [4]).

The four methods we review are factorisation by

1. diagonalization,
2. successive factor extraction,
3. interpolation, and
4. solution of an algebraic Riccati equation.

The methods above were originally published in the series of papers by H. Kwakernaak and M. Šebek: [5, 6, 7], where also proofs and further details can be found.

10.2 Motivation

Let us first demonstrate that J-spectral factorisation arises naturally during the solution of *Standard $\mathcal{H}_\infty$ Regulation Problem* which itself covers various different control tasks. This problem involves the *reduction* or directly the *minimization* of the ∞-norm of the closed-loop transfer function of a general linear feedback configuration.

The ∞-norm of a rational transfer matrix H is defined as

$$\|H\|_\infty = \sup_{\omega\in\mathcal{R}} \lambda_{\max}^{1/2}\left(H^*(j\omega)H(j\omega)\right), \tag{10.3}$$

where $\lambda_{\max}(A)$ denotes the largest eigenvalue of the (constant) matrix A. The number $\lambda_{\max}^{1/2}(A^{\mathrm{H}}A)$ is the largest of the *singular values*

$$\sigma_i(A) = \lambda_i^{1/2}(A^{\mathrm{H}}A), \quad i = 1,\ 2,\ \cdots, n. \tag{10.4}$$

The superscript H denotes the Hermitian, and λ_i is the ith eigenvalue.

When employing now very popular *Standard Regulation Structure*, the plant is represented by the rational transfer matrix G such that

$$\begin{bmatrix} z \\ y \end{bmatrix} = \begin{bmatrix} G_{11} & G_{12} \\ G_{21} & G_{22} \end{bmatrix} \begin{bmatrix} w \\ u \end{bmatrix}, \tag{10.5}$$

where the external signals are to be understood as follows: The signal w represents *external,* uncontrollable inputs such as disturbances, measurement noise, and reference inputs. The signal u is the *control* input. The output z has the meaning of *control error*, which ideally should be zero. The signal y, finally, is the *measured* output, available for feedback via the compensator

$$u = Ky \tag{10.6}$$

In such a configuration, the closed-loop transfer function H from the external input w to the control error z is easily found to be given by

$$H = G_{11} + G_{12}K(I - G_{22}K)^{-1}G_{21}, \tag{10.7}$$

The standard $\mathcal{H}_\infty$- *optimal* regulation problem is now defined as follows:

Determine the compensator K such that

1. the closed-loop system is stable,

2. the ∞-norm $\|H\|_\infty$ of the closed-loop transfer matrix H is minimal, i.e.,

$$||H||_\infty = \lambda_{\text{min}}.$$

On the other hand, the problem of reducing the norm below a given bound, sometimes called the *sub-optimal* regulation problem is specified as follows:
Determine the compensator K such that

1. the closed-loop system is stable,

2. the ∞-norm $\|H\|_\infty$ of the closed-loop transfer matrix H is less than a given bound $\lambda_{\text{given}} > 0$, i.e.,

$$||H||_\infty \leq \lambda_{\text{given}} > \lambda_{\text{min}}.$$

This problem is important in itself and find its use in many tasks of robust control [4]. In addition, during the search of λ_{min} in the optimal problem solution, the suboptimal solution is to be found in every step of iteration [4].

To solve the sub-optimal problem via polynomial techniques [4], we represent the plant G and K in polynomial matrix fraction form, although on one occasion it is convenient to resort to rational matrix fraction form. In particular, we assume that the plant is represented in left coprime matrix fraction form

$$G = D^{-1}N, \tag{10.8}$$

where D is an invertible square polynomial matrix. This representation is equivalent to characterizing the plant by the set of differential equations

$$D\begin{bmatrix} z \\ y \end{bmatrix} = N\begin{bmatrix} w \\ u \end{bmatrix}. \tag{10.9}$$

By partitioning

$$D = [D_1 \quad D_2], \qquad N = [N_1 \quad N_2], \tag{10.10}$$

we have

$$D_1 z + D_2 y = N_1 w + N_2 u. \tag{10.11}$$

Similarly, we represent the compensator K in *right* matrix fraction form as

$$K = YX^{-1}. \tag{10.12}$$

It was shown in [4] that then $\|H_K\|_\infty \leq \lambda$ for the compensator $K = YX^{-1}$ if and only if

$$[X^* \quad Y^*]\,\Pi_\lambda \begin{bmatrix} X \\ Y \end{bmatrix} \geq 0 \quad \text{on the imaginary axis,} \tag{10.13}$$

where

$$\Pi_\lambda = \begin{bmatrix} D_2^* \\ -N_2^* \end{bmatrix} (N_1 N_1^* - \lambda^2 D_1 D_1^*)^{-1} [D_2 \quad -N_2]. \tag{10.14}$$

This rational inequality (10.13) can be solved by means of two polynomial matrix J-spectral factorisation, one for the denominator, the other for the numerator. The former is the polynomial J-spectral cofactorisation

$$N_1 N_1^* - \lambda^2 D_1 D_1^* = Q_\lambda J' Q_\lambda^*, \tag{10.15}$$

with Q_λ square such that its determinant is Hurwitz. Once Q_λ has been determined, we may obtain polynomial matrices Δ_λ and Λ_λ by the left-to-right fraction conversion

$$Q_\lambda^{-1}[D_2 \quad -N_2] = \Delta_\lambda \Lambda_\lambda^{-1}. \tag{10.16}$$

With this we have

$$\Pi_\lambda = \begin{bmatrix} D_2^* \\ -N_2^* \end{bmatrix} (Q_\lambda^{-1})^* J' Q_\lambda^{-1} [D_2 \quad -N_2] = (\Lambda_\lambda^{-1})^* \Delta_\lambda^* J' \Delta_\lambda \Lambda_\lambda^{-1}. \tag{10.17}$$

By the second polynomial J-spectral factorisation

$$\Delta_\lambda^* J' \Delta_\lambda = \Gamma_\lambda^* J \Gamma_\lambda, \tag{10.18}$$

with Γ_λ square such that its determinant is Hurwitz, we obtain the rational J-spectral factor Z_λ as

$$Z_\lambda = \Gamma_\lambda \Lambda_\lambda^{-1}. \tag{10.19}$$

for which

$$\Pi_\lambda = Z_\lambda^* J Z_\lambda. \tag{10.20}$$

Now any compensator K such that $\|H_K\|_\infty \le \lambda_{\text{given}}$ may be represented as

$$K = \tilde{Y}\tilde{X}^{-1}, \tag{10.21}$$

where

$$\begin{bmatrix} \tilde{X} \\ \tilde{Y} \end{bmatrix} = Z_{\lambda_{\text{given}}}^{-1} \begin{bmatrix} A \\ B \end{bmatrix}, \tag{10.22}$$

with A and B rational stable and A square such that

$$\|BA^{-1}\|_\infty \le 1. \tag{10.23}$$

Finally, the class of all *stabilizing* compensators assuring $\|H_K\|_\infty \le \lambda_{\text{given}}$, i.e., the *class of all solutions to the sub-optimal problem*, is again parameterized by (10.21) and (10.22) with A, B as above and, in addition, A^{-1} *stable*. Thus, we have illustrated that the operation of J-spectral factorisation is crucial for computing the sub-optimal regulator.

10.3 Existence and Uniqueness

In contrast to the standard positive definite case, no necessary and sufficient existence conditions appear to be known for J-spectral factorisation. The following *sufficient* condition is known from the work of Jakubovič [8].

10.3.1 Theorem (Existence of J-spectral factorisation.) Suppose that the multiplicity of the zeros on the imaginary axis of each of the invariant polynomials of the para-Hermitian polynomial matrix Z is even. Then Z has a spectral factorisation $Z = P^*JP$. ◇

The condition of the theorem is violated if and only if any of the invariant factors is not factorable by itself. An example of a non-factorable polynomial is $1 + s^2$. A sufficient condition that implies that of the theorem is that $\det Z$ has no roots on the imaginary axis. This assumption is often invoked in what follows.

The signature matrix J of (10.1) is nonsingular if and only if $\det Z$ is not identical to zero. J is a unit matrix (that is, the block $-I_2$ and the zero diagonal block are absent) if and only if Z is nonnegative-definite on the imaginary axis (that is, $Z(j\omega) \geq 0$ for $\omega \in \mathcal{R}$.) More generally, if $\det Z$ is nonzero, the number of positive and negative eigenvalues of Z on the imaginary axis (that is, the eigenvalues of the Hermitian matrix $Z(j\omega)$, $\omega \in \mathcal{R}$) are constant, and equal the dimensions of I_1 and I_2, respectively.

In the sequel it is often assumed that Z is *diagonally reduced* [9, 2].

10.3.2 Definition (Diagonal reducedness.) Define the *half virtual diagonal degrees* of the $n \times n$ para-Hermitian matrix Z as the least integers $\delta_1, \delta_2, \cdots, \delta_n$ such that the matrix

$$Z_{\mathrm{L}} = \lim_{|s|\to\infty} E^*(s)^{-1} Z(s) E(s)^{-1}, \tag{10.24}$$

where E is the polynomial matrix defined by

$$E(s) = \begin{bmatrix} s^{\delta_1} & 0 & \cdots & \cdots & 0 \\ 0 & s^{\delta_2} & 0 & \cdots & 0 \\ \cdots & \cdots & \cdots & \cdots & \cdots \\ 0 & \cdots & \cdots & 0 & s^{\delta_n} \end{bmatrix} \tag{10.25}$$

is a finite real constant matrix. The matrix Z_{L} is called the *diagonal leading coefficient matrix* and if it is nonsingular, then Z is called *diagonally reduced*. ◇

Notice that the virtual diagonal degrees may equal the *real diagonal degrees* (i.e., the degrees of the diagonal entries), such as in the positive definite case, but they are different in general.

10.3.3 Example (Diagonal reducedness.)[1] The polynomial matrix

$$Z(s) = \begin{bmatrix} 0 & s \\ -s & s^2 \end{bmatrix} \tag{10.26}$$

has half virtual diagonal degrees $\delta_1 = 0$ and $\delta_2 = 1$, and its leading coefficient matrix reads

$$Z_{\mathrm{L}} = \begin{bmatrix} 0 & 1 \\ 1 & -1 \end{bmatrix}. \tag{10.27}$$

[1] This and all other examples are taken from [6].

Z_{L} is nonsingular and, hence, Z is diagonally reduced. Notice that here $\delta_1 = 0$ although the degree of the $(1,1)$ element is deg $0 = -\infty$. The polynomial matrix

$$Z(s) = \begin{bmatrix} 0 & s \\ -s & 0 \end{bmatrix} \tag{10.28}$$

has half virtual diagonal degrees 0 and 1 and

$$Z_{\mathrm{L}} = \begin{bmatrix} 0 & 1 \\ 1 & 0 \end{bmatrix}. \tag{10.29}$$

◇

If a para-Hermitian polynomial matrix Z with finite diagonal leading coefficient matrix Z_{L} is not diagonally reduced, there always exists a unimodular polynomial matrix U such that U^*ZU is diagonally reduced. For the construction of U see Section 10.4.

Factorisations of diagonally reduced polynomial matrices may or may not be *canonical* (see for instance [10]).

10.3.4 Definition (Canonical factorisation.) The spectral factorisation of a diagonally reduced para-Hermitian polynomial matrix Z is *canonical* if there exists a spectral factor P that is column reduced with column degrees equal to the half virtual diagonal degrees of Z. ◇

Factorisations of polynomial matrices that are positive-definite on the imaginary axis are always canonical [2], but this is not true in the indefinite case. Non-canonical factorisations arise in $\mathcal{H}_\infty$ optimization when optimal solutions (as opposed to *sub*optimal) are computed.

10.3.5 Example (Canonical and non-canonical factorisations.) We consider the polynomial matrix

$$Z(s) = \begin{bmatrix} \varepsilon & 1-s \\ 1+s & 1-s^2 \end{bmatrix}, \tag{10.30}$$

with ε real such that $|\varepsilon| < 1$. For $\varepsilon \neq 0$ the matrix Z has the spectral factor and signature matrix

$$P(s) = \begin{bmatrix} -1 & -\frac{2}{\varepsilon}+1+s \\ \sqrt{1-\varepsilon} & \frac{2}{\varepsilon}\sqrt{1-\varepsilon} \end{bmatrix}, \quad J = \begin{bmatrix} 1 & 0 \\ 0 & -1 \end{bmatrix} \tag{10.31}$$

(compare Example 10.6.2.) P is column reduced with column degrees equal to the half diagonal degrees of Z. Hence, for $\varepsilon \neq 0$ the factorisation of Z is canonical. We note that as ε approaches 0, some of the coefficients of P grow without bound.

For $\varepsilon = 0$ the matrix Z reduces to

$$Z(s) = \begin{bmatrix} 0 & 1-s \\ 1+s & 1-s^2 \end{bmatrix}, \tag{10.32}$$

which has the spectral factor and signature matrix

$$P(s) = \tfrac{1}{2}\sqrt{2}\begin{bmatrix} 1+s & \frac{3-s^2}{2} \\ 1+s & \frac{-1-s^2}{2} \end{bmatrix}, \qquad J = \begin{bmatrix} 1 & 0 \\ 0 & -1 \end{bmatrix} \tag{10.33}$$

(again compare Example 10.6.2.) The spectral factor P is not column reduced and its column degrees 1 and 2 do not equal the half virtual diagonal degrees $\delta_1 = 0$ and $\delta_2 = 1$ of Z. By direct computation it may be verified that there exists *no* spectral factor of Z with column degrees 0 and 1. Hence, for $\varepsilon = 0$ the factorisation of Z is non-canonical. Apparently, the spectral factorisation of Z has an essential discontinuity at $\varepsilon = 0$.

On the other hand, there does exists a canonical *co*-factorisation[2] of (10.32) with (left) spectral factor and signature matrix given by

$$P(s) = \begin{bmatrix} 1 & 1 \\ 1+s & 0 \end{bmatrix}, \qquad J = \begin{bmatrix} 1 & 0 \\ 0 & -1 \end{bmatrix}. \tag{10.34}$$

◇

Of the various algorithms that are discussed in this paper only that based on diagonalization (Section 10.5) can handle factorisations that are non-canonical, as well as factorisations of singular polynomial matrices and of polynomial matrices Z such that $\det Z$ has roots on the imaginary axis. The algorithm based on symmetric factor extraction (Section 10.6) can deal with non-canonical factorisations. The two other algorithms (interpolation, Section 10.7, and the one based on solution of a Riccati equation, Section 10.8) can deal with *nearly* non-canonical factorisations. Such factorisations are the subject of Section 10.9.

It remains to discuss the uniqueness of spectral factorisations.

10.3.6 Theorem (Nonuniqueness of J-spectral factorisations.) Let the polynomial matrix P be a spectral factor of the para-Hermitian polynomial matrix Z with corresponding signature matrix J. Then UP is also a spectral factor for any unimodular U such that

$$U^* J U = J. \tag{10.35}$$

U is said to be a *J-unitary matrix*. ◇

Given J there are many J-unitary matrices. In the 2×2 case, with $J = \operatorname{diag}(1, \; -1)$, an example of a non-constant J-unitary matrix is

$$U(s) = \begin{bmatrix} 1+s & s \\ s & -1+s \end{bmatrix}. \tag{10.36}$$

Recall that no nonconstant J-unitary matrices exist for $J = I$ (the positive definite case) and only real orthogonal matrices satisfy (10.35).

[2] A co-factorisation is canonical if there exists a left spectral factor whose row degrees equal the half virtual diagonal degrees.

We conclude this section with a few comments about the factorisation of *constant* square symmetric matrices. Such a factorisation is needed in most of the algorithms that we discuss for polynomial matrices. Positive-definite symmetric constant matrices S may conveniently be factored by Choleski decomposition in the form $S = C^T C$, with C lower- or upper-triangular. The Choleski algorithm, based on successively clearing the nondiagonal entries of each row and corresponding column by using the diagonal entry, may straightforwardly be generalized to non-definite matrices, but fails whenever all diagonal entries are zero, such as in

$$S = \begin{bmatrix} 0 & 1 \\ 1 & 0 \end{bmatrix}. \tag{10.37}$$

Although this difficulty may be remedied, the most reliable method to factor S is to find its Schur decomposition [11, p. 192].

$$S = QTQ^H, \tag{10.38}$$

where the superscript H indicates the complex conjugate transpose. Q is unitary, that is, $Q^H Q = QQ^H = I$, and, in general, T is upper-triangular, with the eigenvalues of S on its diagonal. The eigenvalues may be arranged in any order. If S is symmetric, T is diagonal and real, and Q may be arranged to be orthogonal, that is, real such that $Q^T Q = QQ^T = I$ [11, p. 268]. There exist reliable algorithms for computing the Schur decomposition [11, Chapter 7.]. Once the Schur decomposition (10.38) of S is available it is simple to bring S into the form $S = P^T JP$.

10.4 Factorisation of a Unimodular Matrix

In this section we consider the factorisation of a unimodular para-Hermitian matrix. The algorithm we present consists of the successive application of three types of steps. The factorisation of a unimodular polynomial matrix arises in factorisation of more general para-Hermitian polynomial matrices in the algorithm based on diagonalization (Section 10.5) and that based on successive extraction (Section 10.6).

10.4.1 Algorithm (Factorisation of a unimodular polynomial matrix.) For a given nonsingular para-Hermitian $n \times n$ polynomial matrix Z with finite diagonal leading coefficient matrix we calculate the spectral factor P and the signature matrix J of the factorisation $Z = P^* JP$ by the following steps.

STEP 1. *Deflation* [2]. Find the half virtual diagonal degrees $\delta_1, \delta_2, \cdots, \delta_n$ of Z, and let $P := I$.

(a) Determine the diagonal leading coefficient matrix Z_L of Z corresponding to the half virtual diagonal degrees $\delta_1, \delta_2, \cdots, \delta_n$. If Z_L is nonsingular (which only happens if Z is a constant matrix), go to Step 2. Else, compute a real *null vector* $e = \text{col}(e_1, e_2, \cdots, e_n)$ such that $Z_L e = 0$.

(b) Determine the *active index set* $\mathcal{A}$ as $\mathcal{A} = \{i : e_i \neq 0\}$ and the *highest degree active index set* $\mathcal{M} \subset \mathcal{A}$ as $\mathcal{M} = \{i \in \mathcal{A} : \delta_i \geq \delta_j \text{ for all } j \in \mathcal{A}\}$. Choose $k \in \mathcal{M}$. If $\mathcal{M}$ has several elements, for numerical reasons it is best to select k such that $|e_k|$ is maximal on $\mathcal{M}$.

(c) If $\delta_k = 0$ for every null vector e, go to Step 2. Else, let $a = e/e_k$, and construct the unimodular matrix U and its inverse $V = U^{-1}$ as follows. V and U are obtained by replacing the kth column of the $n \times n$ unit matrix with the polynomial column vectors

$$\begin{bmatrix} a_1 s^{\delta_k - \delta_1} \\ a_2 s^{\delta_k - \delta_2} \\ \cdots \\ a_{k-1} s^{\delta_k - \delta_{k-1}} \\ 1 \\ a_{k+1} s^{\delta_k - \delta_{k+1}} \\ \cdots \\ a_n s^{\delta_k - \delta_n} \end{bmatrix} \quad \text{and} \quad \begin{bmatrix} -a_1 s^{\delta_k - \delta_1} \\ -a_2 s^{\delta_k - \delta_2} \\ \cdots \\ -a_{k-1} s^{\delta_k - \delta_{k-1}} \\ 1 \\ -a_{k+1} s^{\delta_k - \delta_{k+1}} \\ \cdots \\ -a_n s^{\delta_k - \delta_n} \end{bmatrix}, \tag{10.39}$$

respectively.

(d) We now have the factorisation $Z = U^*(V^*ZV)U$, where the kth half virtual diagonal degree of V^*ZV is one less than the corresponding half virtual diagonal degree of Z. Let $Z := V^*ZV$, $P := UP$, $\delta_k := \delta_k - 1$ and return to (a).

STEP 2. *Clearing.* Use all nonzero nonpolynomial diagonal entries of Z to clear the corresponding row and column. That is, suppose that the kth diagonal entry Z_{kk} of Z is a nonzero constant. Then construct the unimodular polynomial matrix V and its inverse $U = V^{-1}$ by replacing the kth row of the $n \times n$ unit matrix with

$$\frac{1}{Z_{kk}} \begin{pmatrix} Z_{k1} & Z_{k2} & \cdots & Z_{k,k-1} & Z_{kk} & Z_{k,k+1} & \cdots & Z_{kn} \end{pmatrix} \tag{10.40}$$

and

$$\frac{1}{Z_{kk}} \begin{pmatrix} -Z_{k1} & -Z_{k2} & \cdots & -Z_{k,k-1} & Z_{kk} & -Z_{k,k+1} & \cdots & -Z_{kn} \end{pmatrix}, \tag{10.41}$$

respectively. Then $Z = U^*(V^*ZV)U$, where the kth row and column of V^*ZV except the diagonal entry consist of zeros. Let $Z := V^*ZV$, $P := PU$ and repeat this step for all nonzero nonpolynomial diagonal entries.

STEP 3. *Finalization.* If at this point Z has been reduced to a constant diagonal matrix, perform a J-factorisation and terminate the algorithm.

(a) Else, bring the polynomial matrix Z by suitable symmetric row and column permutations (possibly nonunique) into the block matrix form

$$U^*ZU = \begin{bmatrix} Z_{00} & 0 & 0 \\ 0 & 0 & Z_{12} \\ 0 & Z_{21} & Z_{22} \end{bmatrix}, \tag{10.42}$$

where Z_{00} is constant diagonal, the diagonal entries of Z_{22} are either zero or strictly polynomial (that is, polynomial with nonzero degree), and U a suitable permutation matrix. Let $Z := U^*ZU$ and $P := U^*P$.

(b) The constant matrix Z_{00} may be factored straightforwardly, while for the factorisation of

$$\begin{bmatrix} 0 & Z_{12} \\ Z_{21} & Z_{22} \end{bmatrix} \tag{10.43}$$

Algorithm 10.4.2 may be invoked.

◇

Step 1 of Algorithm 10.4.1 may also be used to transform a nonunimodular para-Hermitian polynomial matrix with finite leading diagonal coefficient matrix that is not diagonally reduced unimodularly into a diagonally reduced polynomial matrix [2].

10.4.2 Algorithm (Factorisation of a special unimodular polynomial matrix.) We next concentrate on the factorisation $Z = P^*JP$ of the nonsingular unimodular para-Hermitian polynomial matrix

$$Z = \begin{bmatrix} 0 & Z_{12} \\ Z_{21} & Z_{22} \end{bmatrix}. \tag{10.44}$$

Note that Z_{12} is wide (that is, has at least as many columns as rows), because otherwise Z would be singular.

STEP 1. Find a unimodular polynomial matrix U such that $Z_{12}U = [z_{12} \;\; 0]$, with z_{12} square nonsingular. In fact, we show in what follows that z_{12} is unimodular and, hence, can be taken to be the unit matrix. Let

$$Z := \begin{bmatrix} I & 0 \\ 0 & U \end{bmatrix}^* Z \begin{bmatrix} I & 0 \\ 0 & U \end{bmatrix} = \begin{bmatrix} 0 & Z_{12}U \\ U^*Z_{21} & U^*Z_{22}U \end{bmatrix} = \left[\begin{array}{c|cc} 0 & z_{12} & 0 \\ \hline z_{21} & z_{22} & z_{23} \\ 0 & z_{32} & z_{33} \end{array}\right] \tag{10.45}$$

and

$$P := \begin{bmatrix} I & 0 \\ 0 & U^{-1} \end{bmatrix}. \tag{10.46}$$

It may easily be found that $\det Z = \det z_{12} \cdot \det z_{33} \cdot \det z_{21}$. Since by assumption Z is unimodular, also z_{12}, z_{22}, and z_{22} are unimodular. This proves the claim that z_{12} can be taken to be the unit matrix, and we now have reduced Z to the form

$$Z = \begin{bmatrix} 0 & I & 0 \\ I & z_{22} & z_{23} \\ 0 & z_{32} & z_{33} \end{bmatrix}. \tag{10.47}$$

STEP 2. Use the two unit matrices to remove the blocks z_{23} and z_{32} symmetrically and unimodularly and correspondingly let

$$Z = \begin{bmatrix} 0 & I & 0 \\ I & z_{22} & 0 \\ 0 & 0 & z_{33} \end{bmatrix}, \qquad P := \begin{bmatrix} I & 0 & z_{23} \\ 0 & I & 0 \\ 0 & 0 & I \end{bmatrix} P. \tag{10.48}$$

We have

$$\begin{bmatrix} 0 & I \\ I & z_{22} \end{bmatrix} = \begin{bmatrix} I & \frac{1}{2}z_{22} \\ 0 & I \end{bmatrix}^* \begin{bmatrix} 0 & I \\ I & 0 \end{bmatrix} \begin{bmatrix} I & \frac{1}{2}z_{22} \\ 0 & I \end{bmatrix} \tag{10.49}$$

$$= \begin{bmatrix} I & \frac{1}{2}z_{22} \\ 0 & I \end{bmatrix}^* \cdot \frac{1}{2} \begin{bmatrix} I & I \\ I & -I \end{bmatrix} \begin{bmatrix} I & 0 \\ 0 & -I \end{bmatrix} \begin{bmatrix} I & I \\ I & -I \end{bmatrix} \cdot \begin{bmatrix} I & \frac{1}{2}z_{22} \\ 0 & I \end{bmatrix} \tag{10.50}$$

$$= \tfrac{1}{2}\sqrt{2} \begin{bmatrix} I & \frac{1}{2}z_{22} + I \\ I & \frac{1}{2}z_{22} - I \end{bmatrix}^* \cdot \begin{bmatrix} I & 0 \\ 0 & -I \end{bmatrix} \cdot \tfrac{1}{2}\sqrt{2} \begin{bmatrix} I & \frac{1}{2}z_{22} + I \\ I & \frac{1}{2}z_{22} - I \end{bmatrix}. \tag{10.51}$$

The factorisation of the unimodular para-Hermitian polynomial matrix z_{33} follows (recursively) by Algorithm 10.4.1. ◇

10.4.3 Example (Factorisation of a unimodular matrix.) By way of example we consider the factorisation of the unimodular para-Hermitian polynomial matrix

$$Z(s) = \begin{bmatrix} \frac{-3+s^2}{4} & \frac{1-s^2}{4} \\ \frac{1-s^2}{4} & \frac{1+s^2}{4} \end{bmatrix}. \tag{10.52}$$

Algorithm 10.4.1, Step 1. The half virtual diagonal degrees of Z are $\delta_1 = \delta_2 = 1$, and, accordingly, its leading diagonal coefficient matrix is

$$Z_L = \begin{bmatrix} -\frac{1}{4} & \frac{1}{4} \\ \frac{1}{4} & -\frac{1}{4} \end{bmatrix}. \tag{10.53}$$

The leading coefficient matrix has the single null vector $e = \text{col}(1,\ 1)$. The highest degree active index set is $\mathcal{M} = \{1,\ 2\}$, and we may select the pivot as $k = 1$. Correspondingly, we write $Z = U_1^* Z_1 U_1$, where

$$Z_1(s) = \begin{bmatrix} 0 & \frac{1}{2} \\ \frac{1}{2} & \frac{1+s^2}{4} \end{bmatrix}, \quad U_1(s) = \begin{bmatrix} 1 & 0 \\ -1 & 1 \end{bmatrix}. \tag{10.54}$$

Since the degree of the diagonal entry is zero on the highest degree active set $\{1\}$ of the null vector of the leading diagonal coefficient matrix of Z_1, we terminate Step 1.

Step 2. Z_1 has no nonzero constant diagonal entries so Step 2 is skipped.

Step 3. We invoke Algorithm 10.4.2 to factor Z_1.

Algorithm 10.4.2, Step 1. Transforming the (1, 2) and (2, 1) entries of Z_1 to 1 we obtain $Z_1 = U_2^* Z_2 U_2$, with

$$Z_2(s) = \begin{bmatrix} 0 & 1 \\ 1 & 1+s^2 \end{bmatrix}, \quad U_2(s) = \begin{bmatrix} 1 & 0 \\ 0 & \frac{1}{2} \end{bmatrix}. \tag{10.55}$$

Step 2. Z_2, finally, may be factored as $Z_2 = U_3^* J U_3$, with

$$J = \begin{bmatrix} 1 & 0 \\ 0 & -1 \end{bmatrix}, \quad U_3(s) = \tfrac{1}{2}\sqrt{2} \begin{bmatrix} 1 & \frac{3+s^2}{2} \\ 1 & \frac{-1+s^2}{2} \end{bmatrix}. \tag{10.56}$$

Thus, we have the factorisation $Z = P^*JP$, with

$$P(s) = U_3(s)U_2(s)U_1(s) = \frac{1}{4\sqrt{2}}\begin{bmatrix} 1-s^2 & 3+s^2 \\ 5-s^2 & -1+s^2 \end{bmatrix}. \tag{10.57}$$

◇

10.5 Diagonalization

In this section we discuss an algorithm for the factorisation of Z that is based on diagonalization. The algorithm avoids calculation of the zeros of Z needed in the extraction and interpolation algorithms but involves more elaborate polynomial operations. On the other hand, it works whenever the conditions of Theorem 10.3.1 are satisfied, whether or not Z is diagonally reduced. The algorithm follows Jakubovič's proof of Theorem 10.3.1, except for the first step, where instead of the Smith form we allow for more general diagonal forms.

10.5.1 Algorithm (Diagonalization.) For a given Z satisfying conditions (i) and (ii) of Theorem 1 we calculate the desired P as well as the corresponding J in the following steps:

STEP 1. Find a diagonal form $D = \text{diag}\{d_l\}$ of Z and the corresponding unimodular matrices V and W such that

$$Z = VDW \tag{10.58}$$

and D is

(a) para-Hermitian ($D^* = D$) as well as

(b) nonnegative definite on Re $s = 0$.

In addition, if some D_l appears to be a zero polynomial (due to the singularity of given Z), replace it by 1 while zeroing the corresponding column of V at the same time. Such a way, D is always nonsingular while V is a generalized unimodular matrix[3] (singular if Z is so).

STEP 2. Perform scalar spectral factorisations of diagonal entries of D

$$d_l = f_l^* f_l \tag{10.59}$$

and form a (nonsingular) Hurwitz polynomial matrix

$$F = \text{diag}\{f_l\} \tag{10.60}$$

Clearly

$$D = F^*F \tag{10.61}$$

[3]A (possibly singular) polynomial matrix is called *generalized unimodular* if all its invariant polynomials are just real constants (possibly zeros).

STEP 3. Compute matrices

$$X = W^{-*}V \tag{10.62}$$

$$Y = F^{-*}XF \tag{10.63}$$

both of which appear to be polynomial and generalized unimodular. Moreover, Y is para-Hermitian ($Y^* = Y$).

STEP 4. Find a nonsingular (unimodular) polynomial matrix U along with a (possibly singular) real signature matrix J so that

$$Y = U^* J U \tag{10.64}$$

STEP 5. Form a desired Hurwitz spectral factor to be

$$P = UFW \tag{10.65}$$

The factorisation is not necessarily canonical even if one exist. ◇

10.5.2 Example (Factorisation by diagonalization). By way of example we consider the (non-canonical) factorisation of the polynomial matrix

$$Z(s) = \begin{bmatrix} 0 & 1-s \\ 1+s & 1-s^2 \end{bmatrix}, \tag{10.66}$$

which was also discussed in Example 10.3.5. This matrix may be diagonalized as

$$\underbrace{\begin{bmatrix} 0 & 1-s \\ 1+s & 1-s^2 \end{bmatrix}}_{\text{Z}} = \underbrace{\begin{bmatrix} -\frac{1}{2} & \frac{1-s}{2} \\ \frac{1}{2} & \frac{1+s}{2} \end{bmatrix}}_{\text{V}} \underbrace{\begin{bmatrix} 1-s^2 & 0 \\ 0 & 1 \end{bmatrix}}_{\text{D}} \underbrace{\begin{bmatrix} 1 & -s \\ 1+s & 2-s-s^2 \end{bmatrix}}_{\text{W}}. \tag{10.67}$$

Given V, D and W, the polynomial matrices F, X and Y easily follow as

$$F(s) = \begin{bmatrix} 1+s & 0 \\ 0 & 1 \end{bmatrix}, \tag{10.68}$$

$$X(s) = \begin{bmatrix} \frac{-3+s^2}{4} & \frac{(1-s)^2(1+s)}{4} \\ \frac{1+s}{4} & \frac{1+s^2}{4} \end{bmatrix}, \tag{10.69}$$

and

$$Y(s) = \begin{bmatrix} \frac{-3+s^2}{4} & \frac{1-s^2}{4} \\ \frac{1-s^2}{4} & \frac{1+s^2}{4} \end{bmatrix}. \tag{10.70}$$

Y is para-Hermitian as expected. According to Example 10.4.3 the factorisation of Y is $Y = U^* J U$, with

$$U(s) = \tfrac{1}{8}\sqrt{2} \begin{bmatrix} 1-s^2 & 3+s^2 \\ 5-s^2 & -1+s^2 \end{bmatrix} \tag{10.71}$$

and

$$J = \begin{bmatrix} 1 & 0 \\ 0 & -1 \end{bmatrix}. \tag{10.72}$$

It follows that the desired spectral factor is given by

$$P(s) = \tfrac{1}{2}\sqrt{2}\begin{bmatrix} 1+s & \frac{(1-s)(3+s)}{2} \\ 1+s & -\frac{(1+s)^2}{2} \end{bmatrix}. \tag{10.73}$$

This spectral factor is not the same as that given by Example 10.3.5. The two factors are related by

$$P_{\text{Example 10.3.5}} = U_0 P_{\text{This example}}, \tag{10.74}$$

with the J-unitary polynomial matrix

$$U_0(s) = \tfrac{1}{2}\begin{bmatrix} 2+s & -s \\ s & 2-s \end{bmatrix}. \tag{10.75}$$

10.6 Successive Factor Extraction

The algorithm presented in this section is an adaptation of a procedure developed by Callier [2] for standard spectral factorisation.

10.6.1 Algorithm (Successive factor extraction.) The method requires that $\det Z$ have no roots on the imaginary axis (and, *a fortiori*, be nonsingular) and that Z be diagonally reduced. Since any para-Hermitian matrix with finite leading diagonal coefficient matrix can be made diagonally reduced by a symmetric unimodular transformation (see Section 10.4) the second assumption causes little loss of generality.

STEP 0. The algorithm is initialized by determining the half virtual diagonal degrees $\delta_1, \delta_2, \cdots, \delta_n$ of Z as defined in Section 10.3. Furthermore, the *zeros* $\zeta_1, \zeta_2, \cdots, \zeta_m$ of Z, that is, the roots of $\det Z$, need to be calculated. The zeros also play an important role in the interpolation method for factorisation. Note that because Z is para-Hermitian, if ζ is a zero, so is $-\zeta$. Because Z has real coefficients, if ζ is a complex zero, so is its complex conjugate $\bar{\zeta}$.

STEPS 1 THROUGH m. The extraction algorithm consists of extracting symmetrically and successively elementary factors that correspond to the each of the zeros. The elementary factor is real only if the root is real. If the zero is complex, complex conjugate pairs may be combined to form a real factor of degree two ([2, 4]). In this paper, we limit ourselves to first-order, possibly complex factors.

If T_j is the elementary factor that is extracted at step j of the algorithm we have

$$Z_{m-j+1} = T_j^* Z_{m-j} T_j, \quad j = 1, 2, \cdots, m, \tag{10.76}$$

with m the number of zeros and $Z_m = Z$. The subscript on Z indicates the number of remaining extractions, and if A is any polynomial matrix with real or complex coefficients, A^* is the polynomial matrix defined by[4] $A^*(s) = (A(\bar{s}))^{\mathrm{H}}$.

[4]If A has coefficient matrices A_j, $j = 0, 1, \cdots, K$, then those of A^* are $(-1)^{-j}A_j^{\mathrm{H}}$, $j = 0, 1, \cdots, K$.

Each factor T_j corresponds to a zero ζ of Z. To simplify the presentation we consider the extraction

$$Z = T^* Z'' T, \tag{10.77}$$

without identifying the sequential number of the extraction. The elementary factor T corresponding to the zero ζ has the form

$$T(s) = \begin{bmatrix} I & -a_1 & 0 \\ 0 & s-\zeta & 0 \\ 0 & -a_2 & I \end{bmatrix}. \tag{10.78}$$

The two blocks I are unit matrices of generally different dimensions. The second column block is the kth column of T, with k to be determined. The entries a_1 and a_2 in this column are constant (generally complex valued) vectors that are obtained as follows. Because ζ is a root of $\det Z$, clearly $Z(\zeta)$ is singular, and there exists a vector e such that $Z(\zeta)e = 0$. The vector e is called the *null vector* of Z corresponding to the zero ζ. Because $Z(\zeta)e = T^*(\zeta)Z'(\zeta)T(\zeta)e = 0$, we may determine a_1 and a_2 by letting $T(\zeta)e = 0$. Writing out this identity component-by-component it is easily found that a_1 and a_2 follow from

$$a = e/e_k = \begin{bmatrix} a_1 \\ 1 \\ a_2 \end{bmatrix}, \tag{10.79}$$

where e_k is the kth component of e and the 1 is in the kth position of a. The elementary factor T as given by (10.78) may be viewed as a variant of the Hermite standard form [12] of a polynomial matrix of degree 1.

The following rule determines which column k is selected. Before doing an extraction define the *active index set* $\mathcal{A}$ as $\mathcal{A} = \{i : \; e_i \neq 0\}$, where e_i is the ith entry of the null vector e. The active index set contains the indices of all nonzero entries of the null vector e. Next, introduce the *highest degree active index set* $\mathcal{M}$ as $\mathcal{M} = \{i \in \mathcal{A} : \; \delta_i \geq \delta_j \text{ for all } j \in \mathcal{A}\}$. The highest degree active index set contains the indices of the diagonal elements of Z of highest degree within the active set $\mathcal{A}$. The column index k may be now chosen as any element of $\mathcal{M}$. If there are several such elements, for numerical reasons it is recommended to choose the element such that the magnitude $|e_k|$ of the kth entry of the null vector e is maximal.

For spectral factorisation, naturally each zero ζ that is extracted on the right is chosen to have negative real part.

We next discuss how to determine the "remaining factor" Z'' in (10.77). To this end, we first extract the factor T "on the right" and write $Z = Z'T$, with the square polynomial matrix Z' to be determined. Multiplying the equality $Z = Z'T$ out element-by-element it is easy to see that all entries of Z and Z' are equal except those in their kth columns. Denoting the kth column of Z' as z_k, it follows that

$$Z(s)a = z_k(s)(s-\zeta). \tag{10.80}$$

From this, z_k may easily be computed by dividing the left-hand side by $s - \zeta$.

The polynomial matrix Z'' now may be obtained by the left extraction $Z' = T^* Z''$, which we rewrite as the right extraction $(Z')^* = (Z'')^* T$. This extraction follows by the same procedure as before. Because Z'' is para-Hermitian, it is sufficient to compute the kth diagonal entry z_{kk} of Z''. This entry may be obtained by solving the equation

$$z_k^*(s)a = z_{kk}(s)(s - \zeta) \tag{10.81}$$

for z_{kk} by dividing the left-hand side by $s - \zeta$. The nondiagonal elements of the kth column of Z'' equal the corresponding entries of z_k, the nondiagonal elements of the kth row of Z'' follow by adjugation, while the remaining elements of Z'' equal the corresponding elements of Z. This defines Z''. As the last step in the extraction we modify the half virtual diagonal degree of the kth diagonal element as $\delta_k := \delta_k - 1$.

The extraction procedure is repeated until the supply of zeros is exhausted. The order in which factors are extracted is not important.

STEP $m + 1$. Eventually, Z is reduced to the form

$$Z = T_1^* T_2^* \cdots T_m^* Z_0 T_m \cdots T_2 T_1 = T^* Z_0 T, \tag{10.82}$$

with $T = T_m T_{m-1} \cdots T_1$. Generally, T and Z_0 are complex (that is, polynomial matrices with complex coefficients.)

If the factorisation is *canonical,* the half virtual diagonal degrees remain nonnegative during the process of extracting elementary factors. In this case the half virtual diagonal degrees eventually are reduced to zero, so that Z_0 is a *constant* Hermitian matrix. Moreover, T is column reduced with column degrees equal to the half virtual diagonal degrees of Z. T and Z_0 may both be made real by replacing T with $U^{-1}T$ and Z_0 with $U^{\mathrm{H}} Z_0 U$, with U a suitable constant matrix. A convenient choice is to take U equal to the leading column coefficient matrix [9] of T so that Z_0 becomes the leading diagonal coefficient matrix of Z. Alternatively, U may be chosen as the constant coefficient matrix of T, so that Z_0 becomes the constant coefficient matrix of Z. The final step of the factorisation is to factor the constant matrix Z_0.

If the factorisation is *non-canonical,* during the extraction process one or several of the half virtual diagonal degrees become *negative.* The result is that the "remaining factor" Z_0 is no longer a constant matrix, but polynomial unimodular. This unimodular para-Hermitian matrix may be J-factored as described in Section 10.4. The column degrees of the resulting spectral factor no longer equal the half real diagonal degrees of Z. ◇

10.6.2 Example (Factorisation by factor extraction). We consider the example studied in Example 10.3.5, where

$$Z(s) = \begin{bmatrix} \varepsilon & 1 - s \\ 1 + s & 1 - s^2 \end{bmatrix}, \tag{10.83}$$

with ε a real number such that $|\varepsilon| < 1$. Since $\det Z(s) = (\varepsilon - 1)(1 - s^2)$, the polynomial matrix Z has the roots ± 1. The null vector corresponding to the left-half plane root -1 is $e = \text{col}(2,\ -\varepsilon)$.

Case 1: $\varepsilon \neq 0$. If $\varepsilon \neq 0$, the active index set is $\mathcal{A} = \{1,\ 2\}$, and the highest degree active index set is $\mathcal{M} = \{2\}$. Since as a result we need to take $k = 2$ it follows that $a = \text{col}(-2/\varepsilon,\ 1)$, and we extract the factor

$$T(s) = \begin{bmatrix} 1 & \frac{2}{\varepsilon} \\ 0 & 1+s \end{bmatrix}. \tag{10.84}$$

By straightforward computation it follows from (10.80) that $z_2(s) = \text{col}(-1,\ -\frac{2}{\varepsilon} + 1 - s)$, while from (10.81) we obtain that $z_{22}(s) = 1$. Thus, after extracting the factor T we are left with

$$Z_0(s) = \begin{bmatrix} \varepsilon & -1 \\ -1 & 1 \end{bmatrix}. \tag{10.85}$$

This is a constant matrix and, hence, the factorisation of Z is canonical. Since

$$Z_0(s) = \begin{bmatrix} \varepsilon & -1 \\ -1 & 1 \end{bmatrix} = \begin{bmatrix} -1 & \sqrt{1-\varepsilon} \\ 1 & 0 \end{bmatrix} \begin{bmatrix} 1 & 0 \\ 0 & -1 \end{bmatrix} \begin{bmatrix} -1 & 1 \\ \sqrt{1-\varepsilon} & 0 \end{bmatrix}, \tag{10.86}$$

the desired spectral factor P is

$$P(s) = \begin{bmatrix} -1 & 1 \\ \sqrt{1-\varepsilon} & 0 \end{bmatrix} T(s) = \begin{bmatrix} -1 & -\frac{2}{\varepsilon} + 1 + s \\ \sqrt{1-\varepsilon} & \frac{2}{\varepsilon}\sqrt{1-\varepsilon} \end{bmatrix}. \tag{10.87}$$

This is the result shown in Example 10.3.5. The factorisation is canonical.

Case 2: $\varepsilon = 0$. If $\varepsilon = 0$, the active and highest degree index sets reduce to $\mathcal{A} = \mathcal{M} = \{1\}$, so that $k = 1$ and $a = \text{col}(1,\ 0)$. As a result, we extract the factor

$$T(s) = \begin{bmatrix} 1+s & 0 \\ 0 & 1 \end{bmatrix}. \tag{10.88}$$

By inspection, the remaining factor is seen to be given by

$$Z_0(s) = \begin{bmatrix} 0 & 1 \\ 1 & 1-s^2 \end{bmatrix}. \tag{10.89}$$

Z_0 is unimodular polynomial, so that the factorisation of Z is non-canonical. By Algorithm 10.4.2, Step 2, we have

$$Z_0(s) = \tfrac{1}{2}\sqrt{2} \begin{bmatrix} 1 & 1 \\ \frac{3-s^2}{2} & -\frac{1+s^2}{2} \end{bmatrix} \cdot \begin{bmatrix} 1 & 0 \\ 0 & -1 \end{bmatrix} \cdot \tfrac{1}{2}\sqrt{2} \begin{bmatrix} 1 & \frac{3-s^2}{2} \\ 1 & -\frac{1+s^2}{2} \end{bmatrix}. \tag{10.90}$$

Hence, the desired spectral factor P is

$$P(s) = \tfrac{1}{2}\sqrt{2} \begin{bmatrix} 1 & \frac{3-s^2}{2} \\ 1 & -\frac{1+s^2}{2} \end{bmatrix} \begin{bmatrix} 1+s & 0 \\ 0 & 1 \end{bmatrix} = \tfrac{1}{2}\sqrt{2} \begin{bmatrix} 1+s & \frac{3-s^2}{2} \\ 1+s & -\frac{1+s^2}{2} \end{bmatrix}, \tag{10.91}$$

which is the result shown in Example 10.3.5. ◇

10.7 Interpolation

Suppose that the para-Hermitian polynomial matrix Z is diagonally reduced and has no zeros on the imaginary axis (and, hence, necessarily is nonsingular). As explained in Section 10.6, by successively extracting elementary factors corresponding to each of the zeros of Z with negative real part we obtain a factorisation of Z of the form $Z = T^* Z_0 T$. If the factorisation is canonical, the matrix Z_0 is constant, and T is column reduced with column degrees equal to the half virtual diagonal degrees $\delta_1, \delta_2, \cdots, \delta_n$ of Z.

The factor T may be obtained in one calculation, without the successive procedure of the symmetric extraction algorithm, by exploiting the property that if ζ is any of the left-half plane zeros of Z, the corresponding null vector e of Z is also a null vector of T. Let $\zeta_1, \zeta_2, \cdots, \zeta_N$ be the left-half plane zeros of Z, and $e_1, e_2, \cdots, e_N$ the corresponding null vectors. Then,

$$T(\zeta_i)e_i = 0, \quad i = 1, 2, \cdots, N. \tag{10.92}$$

Together with the fact that the column degrees of T are known this permits the calculation of T as follows. The degree of T equals $M = \max_i \delta_i$. Define T in terms of its coefficient matrices as $T(s) = T_0 + T_1 s + T_2 s^2 + \cdots + T_M s^M$. Then from (10.92) we have

$$[T_0 \; T_1 \; \cdots \; T_M]E = 0, \tag{10.93}$$

where the constant matrix E is given by

$$E = \begin{bmatrix} e_1 & e_2 & \cdots & e_N \\ e_1\zeta_1 & e_2\zeta_2 & \cdots & e_N\zeta_N \\ e_1\zeta_1^2 & e_2\zeta_2^2 & \cdots & e_N\zeta_N^2 \\ \cdots & \cdots & \cdots & \cdots \\ e_1\zeta_1^M & e_2\zeta_2^M & \cdots & e_N\zeta_N^M \end{bmatrix}. \tag{10.94}$$

It follows from the fact that the degree of the kth column of T is δ_k that the columns with numbers $k + in$, $i = \delta_k + 1, \cdots, M$ of the matrix $[T_0 \; T_1 \; \cdots \; T_M]$ consist of zeros only. Hence, define the constant matrix $\hat{T}$ by removing from the matrix $[T_0 \; T_1 \; \cdots \; T_M]$ all the columns numbered $k + in$, $k = 1, 2, \cdots, n$, where $i = \delta_k + 1, \cdots, M$. Similarly, define the matrix $\hat{E}$ by removing from the matrix E all the rows numbered $k + in$, $k = 1, 2, \cdots, n$, with $i = \delta_k + 1, \cdots, M$. Then

$$\hat{T}\hat{E} = 0, \tag{10.95}$$

and the constant matrix $\hat{T}$ may be computed such that its rows form a real basis for the left null space of $\hat{E}$. Once $\hat{T}$ has been obtained, the coefficient matrix $[T_0 \; T_1 \; \cdots \; T_M]$, and, hence, the polynomial matrix T, follows by re-inserting the appropriate zero columns.

Complex arithmetic may be avoided as follows. If ζ is a complex left-half plane zero with null vector e, also its complex conjugate $\bar{\zeta}$ is a left-half plane zero, with null vector $\bar{e}$. We may then replace the two corresponding complex conjugate columns of E with two real columns, the first of which is the real part and the second the imaginary part of the complex valued column.

The algorithm yields a unique solution T (within multiplication on the left by a nonsingular constant matrix), provided $\hat{E}$ has full column rank. Under the assumption that the factorisation is canonical this condition is always satisfied if the zeros of Z are all distinct. If the zeros are nondistinct the condition may be violated.

It remains to complete the factorisation $Z = T^* Z_0 T$ by finding the matrix Z_0. A simple way of doing this is to left multiply the factor T by the inverse of its leading column coefficient matrix, so that the leading coefficient matrix of T is normalized to the unit matrix. Then Z_0 is the leading diagonal coefficient matrix of the diagonally reduced para-Hermitian matrix Z. The factorisation may be put into standard form by transforming Z_0 to its signature matrix.

It is not difficult to extend the algorithm so that it also applies when there are repeated zeros.

10.7.1 Algorithm (Interpolation.) Suppose that the para-Hermitian polynomial matrix Z has no zeros on the imaginary axis and is diagonally reduced with half virtual diagonal degrees $\delta_1, \delta_2, \cdots, \delta_n$.

STEP 1. Let $M = \max_i \delta_i$, and define the diagonal matrix

$$D(s) = \text{diag}(s^{M-\delta_1}, s^{M-\delta_2}, \cdots, s^{M-\delta_n}).$$

Make the highest coefficient matrix of Z nonsingular by considering $S = D^* Z D$. Next, make S monic (that is, with highest coefficient matrix equal to the unit matrix) by letting $S = S_{2M}^{-1} S$, with S_{2M} the highest coefficient matrix of S.

STEP 2. Form the block companion matrix

$$C = \begin{bmatrix} 0 & I & 0 & \cdots & \cdots & 0 \\ 0 & 0 & I & 0 & \cdots & 0 \\ \cdots & \cdots & \cdots & \cdots & \cdots & \cdots \\ 0 & \cdots & \cdots & \cdots & 0 & I \\ -S_0 & -S_1 & \cdots & \cdots & \cdots & -S_{2M-1} \end{bmatrix}, \tag{10.96}$$

with $S_0, S_1, \cdots, S_{2M-1}$ the coefficient matrices of S.

STEP 3. Calculate the Schur decomposition of C in the form

$$C = RUR^{\text{H}}, \tag{10.97}$$

with R unitary (that is, $R^{\text{H}} R = RR^{\text{H}} = I$) and U upper triangular with the eigenvalues of C along the diagonal, arranged in order of increasing real part. The nonzero eigenvalues of C are the zeros of the polynomial matrix Z.

STEP 4. Partition $R = [R_1 \;\; R_2]$, where the number of columns of R_1 equals the number of eigenvalues of S with negative real part, which, in turn, equals the number of zeros of Z with negative real part. The columns of R_1 span the maximal invariant subspace of C corresponding to the eigenvalues with negative real part.

Retain the first $(M+1)n$ rows of R_1 and denote the result as E. Next, remove those rows from E that correspond to the zeros at the origin that were introduced in Step 1. These are the rows numbered $i+kn$, $k=0, 1, \cdots, M-\delta_i-1$, with $i=1, 2, , \cdots, n$. The result is denoted $\hat{E}$. Next compute a real matrix $\hat{T}$ of maximal full row rank whose rows span the left null space of $\hat{E}$, that is,

$$\hat{T}\hat{E}=0. \tag{10.98}$$

STEP 5. Form the matrix T by inserting into $\hat{T}$ columns consisting of zeros at the column locations corresponding to the rows that have been deleted from E. The resulting polynomial matrix with matrix coefficients defined by $T=[T_0\ T_1\ \cdots\ T_M]$ has a number of spurious zeros at the origin, which were introduced in Step 1. These zeros need to be removed by interchanging the columns $i+kn$ and $i+(M-\delta_i+k)n$, for $k=0, 1, \cdots, M-\delta_i-1$, with $i=1, 2, \cdots, n$.

STEP 6. It remains to complete the factorisation $Z=T^*Z_0T$ by finding the constant matrix Z_0. This may be done by multiplying T on the left by the inverse of its leading column coefficient matrix, which is nonsingular if and only if the factorisation of Z is canonical. Z_0 then is the leading diagonal coefficient matrix Z_{L} of Z. The factorisation is finalized by factoring Z_{L}.

If the factorisation is non-canonical, the factor T as computed in Step 5 is well-defined, but its leading coefficient matrix is singular. As a result, Step 6 fails. The nearly non-canonical case, that is, when the factorisation is close to non-canonical, is discussed in Section 10.9.

10.7.2 Example (Factorisation by interpolation.) We consider the factorisation of the matrix

$$Z(s)=\begin{bmatrix}\varepsilon & 1-s\\ 1+s & 1-s^2\end{bmatrix}, \tag{10.99}$$

with $]\varepsilon]<1$, which was also considered in Example 10.6.2. Since $\delta_1=1$ and $\delta_2=0$, we let $D(s)=\mathrm{diag}(s,\ 1)$, so that, before and after making S monic as in Step 1, we have

$$S(s)=\begin{bmatrix}-\varepsilon s^2 & -s(1-s)\\ s(1+s) & 1-s^2\end{bmatrix}, \qquad S(s)=\begin{bmatrix}s^2+\frac{s}{1-\varepsilon} & \frac{1-s}{1-\varepsilon}\\ \frac{\varepsilon s}{1-\varepsilon} & s^2+\frac{\varepsilon-s}{1-\varepsilon}\end{bmatrix}. \tag{10.100}$$

As a result, the block companion matrix C is given by

$$C=\left[\begin{array}{cc|cc}0 & 0 & 1 & 0\\ 0 & 0 & 0 & 1\\ \hline 0 & -\frac{1}{1-\varepsilon} & -\frac{1}{1-\varepsilon} & \frac{1}{1-\varepsilon}\\ 0 & -\frac{\varepsilon}{1-\varepsilon} & -\frac{\varepsilon}{1-\varepsilon} & \frac{1}{1-\varepsilon}\end{array}\right]. \tag{10.101}$$

C has the eigenvalues ± 1 and a double eigenvalue 0. The matrix R_1 of Step 4 of the algorithm equals the eigenvector of C corresponding to the eigenvalue -1, and may be found to be given by

$$R_1 = \begin{bmatrix} 1 \\ \frac{\varepsilon}{2} \\ -1 \\ -\frac{\varepsilon}{2} \end{bmatrix}. \tag{10.102}$$

Retaining the first $(M+1)n = 4$ of R_1 rows yields $E = R_1$. The only row of E that needs to be deleted is the first, which results in $\hat{E}$ and corresponding $\hat{T}$ given by

$$\hat{E} = \begin{bmatrix} \frac{\varepsilon}{2} \\ -1 \\ -\frac{\varepsilon}{2} \end{bmatrix}, \quad \hat{T} = \begin{bmatrix} 1 & \frac{\varepsilon}{2} & 0 \\ 1 & 0 & 1 \end{bmatrix}. \tag{10.103}$$

Re-inserting the first column of zeros into $\hat{T}$ and interchanging the first and third columns of the result we obtain

$$[T_0 \;\; T_1] = \left[\begin{array}{cc|cc} \frac{\varepsilon}{2} & 1 & 0 & 0 \\ 0 & 1 & 0 & 1 \end{array}\right]. \tag{10.104}$$

This defines the polynomial matrix factor

$$T(s) = \begin{bmatrix} \frac{\varepsilon}{2} & 1 \\ 0 & 1+s \end{bmatrix}. \tag{10.105}$$

Premultiplication by the inverse of the leading column coefficient matrix results in

$$T(s) = \begin{bmatrix} 1 & \frac{2}{\varepsilon} \\ 0 & 1+s \end{bmatrix}. \tag{10.106}$$

Factoring the leading diagonal coefficient matrix of Z as

$$Z_0 = \begin{bmatrix} \varepsilon & -1 \\ -1 & 1 \end{bmatrix} = \begin{bmatrix} \sqrt{\varepsilon} & 0 \\ -\frac{1}{\sqrt{\varepsilon}} & \sqrt{\frac{1-\varepsilon}{\varepsilon}} \end{bmatrix} \begin{bmatrix} 1 & 0 \\ 0 & -1 \end{bmatrix} \begin{bmatrix} \sqrt{\varepsilon} & -\frac{1}{\sqrt{\varepsilon}} \\ 0 & \sqrt{\frac{1-\varepsilon}{\varepsilon}} \end{bmatrix} \tag{10.107}$$

we find the spectral factor

$$P(s) = \begin{bmatrix} \sqrt{\varepsilon} & -\frac{1}{\varepsilon} \\ 0 & \sqrt{\frac{1-\varepsilon}{\varepsilon}} \end{bmatrix} \begin{bmatrix} 1 & \frac{2}{\varepsilon} \\ 0 & 1+s \end{bmatrix} = \begin{bmatrix} \sqrt{\varepsilon} & \frac{1}{\sqrt{\varepsilon}}(1-s) \\ 0 & \sqrt{\frac{1-\varepsilon}{\varepsilon}}(1+s) \end{bmatrix}. \tag{10.108}$$

This result is not the same as that found in Example 10.6.2 (case 1). The two (canonical) spectral factors are related by

$$P_{\text{Example 10.6.2}} = U P_{\text{This example}}, \tag{10.109}$$

where U is the constant J-unitary matrix

$$U = \begin{bmatrix} -\frac{1}{\sqrt{\varepsilon}} & -\sqrt{\frac{1-\varepsilon}{\varepsilon}} \\ \sqrt{\frac{1-\varepsilon}{\varepsilon}} & \frac{1}{\sqrt{\varepsilon}} \end{bmatrix}. \tag{10.110}$$

◇

10.8 Riccati Equation

Factorisation by solution of an algebraic Riccati equation is most convenient if the para-Hermitian polynomial matrix Z is given in the "pre-factored" form

$$Z = Q^* W Q, \tag{10.111}$$

where W is a constant symmetric matrix, and Q a tall polynomial matrix that is column reduced, with column degrees $\delta_1, \delta_2, \cdots, \delta_n$. This situation is typical for the factorisations that arise in $\mathcal{H}_\infty$ optimization problems [4].

If Z is diagonally reduced nonsingular without zeros on the imaginary axis but not pre-factored, a pre-factorisation may easily be obtained by introducing an arbitrary[5] square column reduced strictly Hurwitz polynomial matrix N of the same dimensions as Z whose column degrees equal the half virtual diagonal degrees of Z, and solving the symmetric bilateral polynomial matrix equation

$$Z = N^* M + M^* N \tag{10.112}$$

for the polynomial matrix M, with column degrees again equal to the half virtual diagonal degrees of Z. Then the desired pre-factorisation is

$$Z = \underbrace{[N^* \; M^*]}_{Q^*} \underbrace{\begin{bmatrix} 0 & I \\ I & 0 \end{bmatrix}}_{W} \underbrace{\begin{bmatrix} N \\ M \end{bmatrix}}_{Q}. \tag{10.113}$$

The algorithm discussed in this section relies on the following well-known connection between algebraic Riccati equations and factorisations (compare [13]).

10.8.1 Theorem (Algebraic Riccati equation and factorisation.) Consider the linear time-invariant system $\dot{x} = Ax + Bu$, $y = Cx + Du$, with transfer matrix $H(s) = C(sI - A)^{-1}B + D$, and let R and W be given symmetric constant matrices. Suppose that the algebraic matrix Riccati equation

$$0 = A^{\mathrm{T}}X + XA + C^{\mathrm{T}}WC - (XB + C^{\mathrm{T}}WD)(D^{\mathrm{T}}WD + R)^{-1}(B^{\mathrm{T}}X + D^{\mathrm{T}}WC) \tag{10.114}$$

has a symmetric solution X. Then

$$R + H^* W H = V^* L V, \tag{10.115}$$

where the constant symmetric matrix L and the rational matrix function V are given by

$$L = R + D^{\mathrm{T}}WD, \qquad V(s) = I + F(sI - A)^{-1}B, \tag{10.116}$$

with $F = L^{-1}(B^{\mathrm{T}}X + D^{\mathrm{T}}WC)$. The zeros of the numerator of $\det V$ are the eigenvalues of the matrix $A - BF$. ◇

The factorisation algorithm may now be outlined as follows.

[5] It is convenient to take N diagonal. For numerical reasons it may be desirable to arrange that the coefficients of N are of the same order of magnitude as those of Z.

10.8.2 Algorithm (Factorisation of a pre-factored polynomial matrix by solution of a Riccati equation.) Suppose that the diagonally reduced nonsingular para-Hermitian polynomial matrix Z has no zeros on the imaginary axis, and is given in the pre-factored form

$$Z = Q^* W Q. \tag{10.117}$$

STEP 1. The first step in the algorithm is to convert the polynomial factorisation to a rational factorisation. To this end, introduce the diagonal polynomial matrix[6]

$$E(s) = \operatorname{diag}(s^{\delta_1},\ s^{\delta_2},\ \cdots,\ s^{\delta_n}), \tag{10.118}$$

and define the rational para-Hermitian matrix Π as

$$\Pi = (E^*)^{-1} Q^* W Q E^{-1} = H^* W H. \tag{10.119}$$

Because Q is column reduced, the rational matrix $H = QE^{-1}$ is proper with full rank at infinity.

STEP 2. Next represent H in the form

$$H(s) = D + C(sI - A)^{-1} B, \tag{10.120}$$

with A, B, C, and D constant matrices such that $\dot{x} = Ax + Bu$, $y = Cx + Du$ is a minimal realization of the system with transfer matrix H (see [12, Chapter 6.]). Also determine the polynomial matrix K such that $(sI - A)^{-1}B = K(s)E^{-1}(s)$ by solving the polynomial equation $(sI - A)K(s) = BE(s)$ for K.

STEP 3. Setting $R = 0$, application of Theorem 10.8.1 yields the rational factorisation $H^*WH = V^*LV$, which follows by solution of the algebraic Riccati equation

$$0 = A^{\mathrm{T}}X + XA + C^{\mathrm{T}}WC - (XB + C^{\mathrm{T}}WD)(D^{\mathrm{T}}WD)^{-1}(B^{\mathrm{T}}X + D^{\mathrm{T}}WC). \tag{10.121}$$

This rational factorisation reduces to the polynomial factorisation $Z = T^*LT$, with

$$T(s) = E(s) + FK(s), \tag{10.122}$$

where $L = D^{\mathrm{T}}WD$ and $F = L^{-1}(B^{\mathrm{T}}X + D^{\mathrm{T}}WC)$. If the solution of the Riccati equation (10.114) is chosen such that $A - BF$ has all its eigenvalues in the open left-half complex plane, T is strictly Hurwitz.

STEP 4. The factorisation is completed by the factorisation of the constant symmetric matrix L. ◇

Practically, the algebraic Riccati equation is solved by finding the Schur decomposition [14] of the associated Hamiltonian matrix

$$\mathcal{H} = \begin{bmatrix} A - BL^{-1}S^{\mathrm{T}} & -BL^{-1}B^{\mathrm{T}} \\ -C^{\mathrm{T}}WC - SL^{-1}S^{\mathrm{T}} & -(A - BL^{-1}S^{\mathrm{T}})^{\mathrm{T}} \end{bmatrix}, \tag{10.123}$$

[6] For numerical reasons it may be useful to modify E to $E(s) = \operatorname{diag}(\alpha_1 s^{\delta_1},\ \alpha_2 s^{\delta_2},\ \cdots,\ \alpha_n s^{\delta_n})$, with $\alpha_1,\ \alpha_2,\ \cdots,\ \alpha_n$ suitably chosen real constants.

with $S = C^{\mathrm{T}}WD$. The Robust-Control Toolbox [15] provides a routine. The solution of the Riccati equation then is $X = X_2X_1^{-1}$, where the square matrices X_1 and X_2 follow from the matrix

$$\begin{bmatrix} X_1 \\ X_2 \end{bmatrix} \tag{10.124}$$

whose columns form a basis for the maximal stable invariant subspace of the Hamiltonian.

If the Riccati equation has a solution, the factor P is column reduced with the correct column degrees, and the factorisation is canonical. The Riccati equation fails to have a solution, however, if the "top coefficient" X_1 is singular. In this case, the factorisation is noncanonical.

10.8.3 Example (Factorisation by solution of a Riccati equation.) Again we consider the factorisation of

$$Z(s) = \begin{bmatrix} \varepsilon & 1-s \\ 1+s & 1-s^2 \end{bmatrix}, \tag{10.125}$$

with $|\varepsilon| < 1$. Choosing

$$N(s) = \begin{bmatrix} 1 & 0 \\ 0 & 1+s \end{bmatrix} \tag{10.126}$$

it easily follows by solution of (10.112) that

$$M(s) = \begin{bmatrix} \frac{\varepsilon}{2} & 1-s \\ 0 & \frac{1+s}{2} \end{bmatrix}. \tag{10.127}$$

As a result, Z may be prefactored as $Z = Q^*WQ$, with

$$W = \begin{bmatrix} 0 & 0 & 1 & 0 \\ 0 & 0 & 0 & 1 \\ 1 & 0 & 0 & 0 \\ 0 & 1 & 0 & 0 \end{bmatrix}, \quad Q(s) = \begin{bmatrix} 1 & 0 \\ 0 & 1+s \\ \frac{\varepsilon}{2} & 1-s \\ 0 & \frac{1+s}{2} \end{bmatrix}. \tag{10.128}$$

Next, choosing the polynomial matrix E as $E(s) = \mathrm{diag}(1,\ s)$, we obtain

$$H(s) = Q(s)E^{-1}(s) = \begin{bmatrix} 1 & 0 \\ 0 & \frac{1+s}{s} \\ \frac{\varepsilon}{2} & \frac{1-s}{s} \\ 0 & \frac{1+s}{2s} \end{bmatrix}. \tag{10.129}$$

It may easily be established that $H(s) = D + C(sI - A)^{-1}B$, with

$$D = \begin{bmatrix} 1 & 0 \\ 0 & 1 \\ \frac{\varepsilon}{2} & -1 \\ 0 & \frac{1}{2} \end{bmatrix}, \quad C = \begin{bmatrix} 0 \\ 1 \\ 1 \\ \frac{1}{2} \end{bmatrix}, \quad A = 0, \quad B = [0 \ \ 1]. \tag{10.130}$$

After making the appropriate substitutions, the algebraic Riccati equation turns out to have the form $\varepsilon X^2 + 2(1+\varepsilon)X + 4 = 0$, which for $\varepsilon \neq 0$ has the two solutions $X = -2$ and $X = -2/\varepsilon$. The corresponding "gains" are

$$F = \begin{bmatrix} 0 \\ -1 \end{bmatrix}, \quad F = \begin{bmatrix} \frac{2}{\varepsilon} \\ 1 \end{bmatrix}. \tag{10.131}$$

We need the second, because it makes $A - BF = -1$ stable. It follows that T as given by (10.122) equals

$$T(s) = E(s) + FK(s) = \begin{bmatrix} 1 & 0 \\ 0 & s \end{bmatrix} + \begin{bmatrix} \frac{2}{\varepsilon} \\ 1 \end{bmatrix} \begin{bmatrix} 0 & 1 \end{bmatrix} = \begin{bmatrix} 1 & \frac{2}{\varepsilon} \\ 0 & 1+s \end{bmatrix}. \tag{10.132}$$

The remaining computations are identical to those in Example 10.6.2 (Case 1) and the result is identical. The factorisation is not the same as that obtained in Example 10.7.2. ◇

10.9 Nearly Non-Canonical Factorisations

In certain applications, in particular in $\mathcal{H}_\infty$-optimization, given a para-Hermitian matrix Z, the purpose of J-factorisation is to find stable rational matrices V of maximal and full column rank such that

$$V^* Z V \geq 0 \quad \text{on the imaginary axis.} \tag{10.133}$$

If Z has the spectral factorisation $Z = P^* J P$ with $J = \text{diag}(I_1, \ -I_2)$, all such V are given by

$$V = P^{-1} \begin{bmatrix} A \\ B \end{bmatrix}, \tag{10.134}$$

where A and B are stable rational matrices, with A square and nonsingular of the same dimensions as I_1, such that $A^*A \geq B^*B$ on the imaginary axis.

A slightly more general form of the factorisations $Z = P^*JP$ we consider is

$$Z = P^* K^{-1} P, \tag{10.135}$$

with K a constant diagonal matrix. In the canonical case, $K = J$. If the factorisation is "close to" non-canonical, for numerical reasons it may be advisable to use the alternate form (10.135) of the spectral factorisation, with some of the diagonal entries of K small. K may always be arranged in the form

$$K = \begin{bmatrix} K_1^{\mathrm{T}} K_1 & 0 \\ 0 & -K_2^{\mathrm{T}} K_2 \end{bmatrix}, \tag{10.136}$$

with K_1 and K_2 again diagonal. All stable rational V such that $V^*ZV \geq 0$ on the imaginary axis may now be expressed as

$$V = P^{-1} \begin{bmatrix} K_1 A \\ K_2 B \end{bmatrix}, \tag{10.137}$$

with A and B stable rational such that $A^*A \geq B^*B$ on the imaginary axis. This expression, in turn, is equivalent to

$$PV = \begin{bmatrix} K_1 A \\ K_2 B \end{bmatrix}. \tag{10.138}$$

In the case of a non-canonical factorisation, the algorithms based on interpolation (Section 10.5) and the solution of a Riccati equation (Section 10.8) are capable of producing a (singular) polynomial matrix P whose column degrees equal the half virtual diagonal degrees of Z and a (singular) constant diagonal matrix K, such that all stable V of maximal column rank that satisfy (10.138) have the property that $V^*ZV \geq 0$ on the imaginary axis. These are not all V with this property, but all that are needed for the solution of the $\mathcal{H}_\infty$ problem.

10.9.1 Example (Nearly non-canonical factorisation.) In Example 10.7.2 we found that for $\varepsilon \neq 0$ the polynomial matrix

$$Z(s) = \begin{bmatrix} \varepsilon & 1-s \\ 1+s & 1-s^2 \end{bmatrix}, \tag{10.139}$$

with ε real such that $|\varepsilon| < 1$, has a canonical factorisation with spectral factor and signature matrix

$$P(s) = \begin{bmatrix} \sqrt{\varepsilon} & \frac{1}{\sqrt{\varepsilon}}(1-s) \\ 0 & \sqrt{\frac{1-\varepsilon}{\varepsilon}}(1+s) \end{bmatrix}, \qquad J = \begin{bmatrix} 1 & 0 \\ 0 & -1 \end{bmatrix}. \tag{10.140}$$

Inspection shows that as ε approaches 0, the coefficients of the spectral factor approach ∞. It is also easily seen that the factorisation can be rewritten as $Z = P^*K^{-1}P$, with

$$P(s) = \begin{bmatrix} \varepsilon & 1-s \\ 0 & 1+s \end{bmatrix}, \qquad K = \begin{bmatrix} \varepsilon & 0 \\ 0 & -\frac{\varepsilon}{1-\varepsilon} \end{bmatrix}. \tag{10.141}$$

As $\varepsilon \to 0$, the matrix K approaches the zero matrix, and (10.138) reduces to

$$\begin{bmatrix} 0 & 1-s \\ 0 & 1+s \end{bmatrix} V(s) = \begin{bmatrix} 0 \\ 0 \end{bmatrix}. \tag{10.142}$$

Inspection shows that this equation has the general stable rational solution

$$V(s) = \begin{bmatrix} v(s) \\ 0 \end{bmatrix}, \tag{10.143}$$

with v any stable scalar rational function. It follows that if $\varepsilon = 0$ then $V^*ZV = 0$. ◇

We now discuss how the interpolation algorithm of Section 10.7 may be used to obtain the alternate form

$$Z = P^*K^{-1}P. \tag{10.144}$$

10.9.1 Algorithm (Modification of the interpolation algorithm 10.7.1.)

STEPS 1–5. Steps 1–5 of Algorithm 10.7.1 may be followed without modification to construct the polynomial matrix T.

STEP 6. If the factorisation is close to non-canonical, Step 6 may become perilous because the leading column coefficient matrix T_L of T is close to singular. As long as T_L is nonsingular we have $Z = T^*(T_L^{-1})^T Z_L T_L^{-1} T$, where Z_L is the leading diagonal coefficient matrix of Z. Rewriting this as $Z = T^*(T_L Z_L^{-1} T_L^T)^{-1} T$, and introducing the Schur decomposition

$$T_L Z_L^{-1} T_L^T = Q K Q^T, \tag{10.145}$$

with Q orthogonal and K diagonal, we obtain the desired factorisation

$$Z = P^* K^{-1} P, \tag{10.146}$$

with $P = Q^T T$. K is well-defined even if the factorisation is non-canonical. ◇

Algorithm 10.8.2 based on the solution of a Riccati equation may also be modified to handle nearly non-canonical factorisations.

10.9.2 Algorithm (Modification of the Riccati equation algorithm 10.8.2.)

STEPS 1–2. Steps 1–2 of Algorithm 10.8.2 may be followed without modification.

STEP 3. Step 3 is initiated by computing a real matrix

$$\begin{bmatrix} X_1 \\ X_2 \end{bmatrix} \tag{10.147}$$

with X_1 and X_2 square, whose columns form a basis for the maximal invariant stable subspace of the Hamiltonian matrix (10.123). When the factorisation is nearly non-canonical, the "top coefficient" X_1 is close to singular. In this case, rather than computing $F = G X_1^{-1}$, with $G = B^T X_2 + S^T X_1$, we let $G X_1^{-1} = \hat{X}_1^{-1} \hat{G}$, where $\hat{G}$ and $\hat{X}_1$ are computed from $\hat{X}_1 G = \hat{G} X_1$, or, equivalently

$$[\hat{X}_1 \quad -\hat{G}] \begin{bmatrix} G \\ X_1 \end{bmatrix} = 0. \tag{10.148}$$

This amounts to computing a (real) basis for the left null space of

$$\begin{bmatrix} G \\ X_1 \end{bmatrix}, \tag{10.149}$$

which is well-defined even if X_1 is singular. Accordingly, we replace T as given by (10.122) with

$$T(s) = \hat{X}_1 E(s) + \hat{G} K(s). \tag{10.150}$$

STEP 4. Step 4 is identical to Step 6 of Algorithm 10.9.1. ◇

10.10 Conclusions

Of the four algorithms for the J-spectral factorisation of para-Hermitian polynomial matrices reviewed in this chapter that based on diagonalization (Section 10.5) is the most universal. It can be applied to matrices that are singular and have zeros on the imaginary axis, does not require diagonal reducedness, and can be used both for canonical and for non-canonical factorisations. Its disadvantages are that it involves rather elaborate polynomial operations, and that in the case of canonical factorisations no control seems available over the column degrees of the spectral factor.

The algorithm based on successive factor extraction (Section 10.6) requires the matrix to be diagonally reduced and to have no zeros on the imaginary axis (and, hence, to be nonsingular). The method can handle non-canonical factorisations, and produces a column reduced spectral factor for canonical factorisations. On the negative side, it requires explicit computation of the zeros of the polynomial matrix.

The interpolation method (Section 10.7) also requires the matrix to be diagonally reduced and to have no zeros on the imaginary axis. It appears to be quite efficient. Although it cannot handle non-canonical factorisations the algorithm can deal with nearly non-canonical factorisations.

The factorisation algorithm based on the solution of an algebraic Riccati equation (Section 10.8) likewise requires diagonal reducedness and does not allow zeros on the imaginary axis. It can deal with nearly non-canonical factorisations but does not apply to non-canonical factorisations. The algorithm appears quite efficient but requires prefactorisation and conversion to a state space representation as an intermediate step.

Several of the algorithms have been implemented in MATLAB ([16, 17]). It is obvious that much more work needs to be done on the numerical efficiency and robustness of the algorithms.

Bibliography

[1] V. Kučera, *Discrete Linear Control: The Polynomial Approach.* Chichester: Wiley, 1980.

[2] F. M. Callier, "On polynomial matrix spectral factorization by symmetric extraction," *IEEE Transactions on Automatic Control*, vol. 30, pp. 453–464, May 1985.

[3] J. Ježek and V. Kučera, "Efficient algorithm for matrix spectral factorization," *Automatica*, vol. 21, no. 6, pp. 663–669, 1985.

[4] H. Kwakernaak, "The polynomial approach to $\mathcal{H}_\infty$-optimal regulation," in *$\mathcal{H}_\infty$-Control Theory* (E. Mosca and L. Pandolfi, eds.), Lecture Notes in Mathematics, Berlin: Springer-Verlag, 1991.

[5] M. Šebek and H. Kwakernaak, "J-spectral factorization," in *Proceedings of the 30th IEEE Conference on Decision and Control*, (Brighton, England), pp. 1278–1283, Dec. 1991.

[6] H. Kwakernaak and M. Šebek, "Polynomial J-spectral factorization," to appear in the *IEEE Transactions on Automatic Control*, 1993.

[7] M. Šebek and H. Kwakernaak, "Further numerical methods for J-spectral factorization," in *Preprints of the 2nd IFAC Workshop on System Structure and Control*, (Prague, Czechoslovakia), Sept. 1992.

[8] V. A. Jakubovič, "Factorization of symmetric matrix polynomials," *Dokl. Acad. Nauk. SSSR*, vol. 194, no. 3, pp. 1261–1264, 1970.

[9] F. M. Callier and C. Desoer, *Multivariable Feedback Systems.* New York: Springer-Verlag, 1982.

[10] I. Gohberg, P. Lancaster, and I. Rodman, *Matrix Polynomials.* New York: Academic Press, 1982.

[11] G. Golub and C. van Loan, *Matrix Computations.* Baltimore, Maryland: The John Hopkins University Press, 1983.

[12] T. Kailath, *Linear Systems.* Englewood Cliffs: Prentice Hall, 1980.

[13] P. Lancaster and L. Rodman, "Solutions of the continuous and discrete time algebraic Riccati equations: A review," in *The Riccati Equation* (S. Bittanti, A. Laub, and J. Willems, eds.), Berlin: Springer-Verlag, 1991.

[14] A. Laub, "Invariant subspace methods for numerical solutions of Riccati equations," in *The Riccati Equation* (S. Bittanti, A. Laub, and J. Willems, eds.), Berlin: Springer-Verlag, 1991.

[15] R. Y. Chiang and M. Safonov, *Robust-Control Toolbox User's Guide.* South Stick, Mass.: The MathWorks, 1988.

[16] H. Kwakernaak, "MATLAB macros for polynomial $\mathcal{H}_\infty$-optimal regulation," Memorandum No. 881, Department of Applied Mathematics, University of Twente, The Netherlands, Sept. 1990.

[17] M. Šebek, "An algorithm for spectral factorization of polynomial matrices with any signature," Memorandum No. 912, Department of Applied Mathematics, University of Twente, The Netherlands, Dec. 1990.

Index

Heterick Memorial Library
3 5111 00399 0359